CMOS CURRENT AMPLIFIERS

THE KLUWER INTERNATIONAL SERIES IN ENGINEERING AND COMPUTER SCIENCE

ANALOG CIRCUITS AND SIGNAL PROCESSING

Consulting Editor*: Mohammed Ismail. *Ohio State University

Related Titles:

DATA CONVERTERS FOR WIRELESS STANDARDS
C. Shi and M. Ismail
ISBN: 0-7923-7623-4

AUTOMATIC CALIBRATION OF MODULATED FREQUENCY SYNTHESIZERS
D. McMahill
ISBN: 0-7923-7589-0

MODEL ENGINEERING IN MIXED-SIGNAL CIRCUIT DESIGN
S. Huss
ISBN: 0-7923-7598-X

CONTINUOUS-TIME SIGMA-DELTA MODULATION FOR A/D CONVERSION IN RADIO RECEIVERS
L. Breems, J.H. Huijsing
ISBN: 0-7923-7492-4

DIRECT DIGITAL SYNTHESIZERS: THEORY, DESIGN AND APPLICATIONS
J. Vankka, K. Halonen
ISBN: 0-7923 7366-9

SYSTEMATIC DESIGN FOR OPTIMISATION OF PIPELINED ADCs
J. Goes, J.C. Vital, J. Franca
ISBN: 0-7923-7291-3

OPERATIONAL AMPLIFIERS: Theory and Design
J. Huijsing
ISBN: 0-7923-7284-0

HIGH-PERFORMANCE HARMONIC OSCILLATORS AND BANDGAP REFERENCES
A. van Staveren, C.J.M. Verhoeven, A.H.M. van Roermund
ISBN: 0-7923-7283-2

HIGH SPEED A/D CONVERTERS: Understanding Data Converters Through SPICE
A. Moscovici
ISBN: 0-7923-7276-X

ANALOG TEST SIGNAL GENERATION USING PERIODIC ΣΔ-ENCODED DATA STREAMS
B. Dufort, G.W. Roberts
ISBN: 0-7923-7211-5

HIGH-ACCURACY CMOS SMART TEMPERATURE SENSORS
A. Bakker, J. Huijsing
ISBN: 0-7923-7217-4

DESIGN, SIMULATION AND APPLICATIONS OF INDUCTORS AND TRANSFORMERS FOR Si RF ICs
A.M. Niknejad, R.G. Meyer
ISBN: 0-7923-7986-1

SWITCHED-CURRENT SIGNAL PROCESSING AND A/D CONVERSION CIRCUITS: DESIGN AND IMPLEMENTATION
B.E. Jonsson
ISBN: 0-7923-7871-7

RESEARCH PERSPECTIVES ON DYNAMIC TRANSLINEAR AND LOG-DOMAIN CIRCUITS
W.A. Serdijn, J. Mulder
ISBN: 0-7923-7811-3

CMOS DATA CONVERTERS FOR COMMUNICATIONS
M. Gustavsson, J. Wikner, N. Tan
ISBN: 0-7923-7780-X

DESIGN AND ANALYSIS OF INTEGRATOR-BASED LOG -DOMAIN FILTER CIRCUITS
G.W. Roberts, V. W. Leung
ISBN: 0-7923-8699-X

VISION CHIPS
A. Moini
ISBN: 0-7923-8664-7

CMOS CURRENT AMPLIFIERS

Speed versus Nonlinearity

by

Kimmo Koli
Nokia Research Center, Finland

and

Kari Halonen
Helsinki University of Technology, Finland

KLUWER ACADEMIC PUBLISHERS
BOSTON / DORDRECHT / LONDON

A C.I.P. Catalogue record for this book is available from the Library of Congress.

ISBN 1-4020-7045-4

Published by Kluwer Academic Publishers,
P.O. Box 17, 3300 AA Dordrecht, The Netherlands.

Sold and distributed in North, Central and South America
by Kluwer Academic Publishers,
101 Philip Drive, Norwell, MA 02061, U.S.A.

In all other countries, sold and distributed
by Kluwer Academic Publishers,
P.O. Box 322, 3300 AH Dordrecht, The Netherlands.

Printed on acid-free paper

Printed in the Netherlands.

Contents

Chapter 1

Introduction to current-mode circuit techniques

1.1 Development of integration technologies

The development of modern integration technologies is normally driven by the needs of digital CMOS circuit design. As the sizes of integrated devices decrease, so maximum voltage ratings also rapidly decrease. Although decreased supply voltages do not restrict the design of digital circuits, it is harder to design high performance analogue integrated circuits using new processes.

In digital integration technologies, there are fewer integrated devices available for circuit design. In a worst case situation, this means that only transistors are available for analogue circuit design. There may occasionally bee capacitances and resistors but their values may be small and there are significant parasitic components present. Thus, if we want to utilise the fastest integration technologies available, we are usually restricted to active components in the design of integrated analogue circuits.

Since the introduction of integrated circuits, the operational amplifier has served as the basic building block in analogue circuit design. Since then, new integrated analogue circuit applications have emerged and the performance requirements for analogue circuits have changed. Voltage-mode operational amplifier circuits have limited bandwidth at high closed-loop gains due to the constant gain-bandwidth product. Furthermore, the limited slew-rate of the operational amplifier affects the large-signal, high-frequency operation.

When wide bandwidth, low power consumption and low voltage operation are needed simultaneously, the voltage-mode operational amplifier easily becomes too complex and has characteristics that are not needed, for example DC-accuracy. On the other hand, circuit techniques used in radio frequency applications are usually too simple to reach the required accuracy. Therefore, there is a growing need for new, low voltage analogue circuit techniques.

1.2 Motivation for current-mode circuit design

One procedure for finding alternative, preferably simpler, circuit realisations is to use current signals rather than voltage signals for signal processing [1,2]. MOS-transistors in particular are more suitable for processing currents rather than voltages because the output signal is current both in common-source and common-gate amplifier configurations and common-drain amplifier configuration is almost useless at low supply voltages because of the bulk-effect present in typical CMOS-processes. Moreover, MOS current-mirrors are more accurate and less sensitive to process variation than bipolar current-mirrors because with the latter the base currents limit the accuracy. Therefore, at the very least, MOS-transistor circuits should be simplified by using current signals in preference to voltage signals. For this reason, integrated current-mode system realisations are closer to the transistor level than the conventional voltage-mode realisations and therefore simpler circuits and systems should result.

When signals are widely distributed as voltages, the parasitic capacitances are charged and discharged with the full voltage swing, which limits the speed and increases the power consumption of voltage-mode circuits. Current-mode circuits cannot avoid nodes with high voltage swing either but these are usually local nodes with less parasitic capacitances. Therefore, it is possible to reach higher speed and lower dynamic power consumption with current-mode circuit techniques. Current-mode interconnection circuits in particular show promising performance [3].

When the signal is conveyed as a current, the voltages in MOS-transistor circuits are proportional to the square-root of the signal, if saturation region operation is assumed for the devices. Similarly, in bipolar transistor circuits the voltages are proportional to the logarithm of the signal. Therefore, a compression of voltage signal swing and a reduction of supply voltage are possible. This feature is utilised for example in log-domain filters [4], switched current filters [5], and in non-linear current-mode circuits in general. Unfortunately, as a consequence of the device mismatches this non-linear operation may generate an excessive amount of distortion for applications with high linearity requirements. Thus, in certain current-mode circuits, linearisation techniques are utilised to reduce the nonlinearity of the transistor transconductance, in which case the voltage signal swing is not reduced.

However, new solutions invariably entail new problems. The compression of the voltage signal swing, for example, increases sensitivity to mismatches. Similarly, a large amount of reported current-mode circuits require advanced complementary bipolar integration processes utilising vertical npn- and pnp-transistors with a high f_T, circuits which need excessively high supply voltages in order to be useful in most battery operated applications. Furthermore, some current-mode techniques such as the current-feedback are very old (compare to cathode-feedback in electron tube amplifiers) and are used as enhanced voltage-mode signal processing building blocks rather than as true current-mode signal processing building blocks. At radio frequencies, current-mode circuit techniques are limited to on-chip signal processing in integrated circuits as off-chip impedance levels are fixed, typically 50 Ω. However, the aggressive scaling of integration technologies ensures that current-mode circuit techniques will remain useful in the future, while some longer on-chip sub-system interconnections may need RF design techniques.

1.3 Evolution of current-mode building blocks

The current-conveyor, published in 1968 [6], represented the first building block intended for current signal processing. In 1970 appeared the enhanced version of the current-conveyor: the second-generation current-conveyor CCII [7]. Neither of these building blocks became popular as a consequence of the introduction of the integrated operational amplifier at the time. As the voltage-mode operational amplifier concept had already been introduced in the forties, it is no wonder that the current-conveyor did not become a success overnight. Additionally, integrated current-conveyors were difficult to realise due to the lack of high performance pnp-devices in the integration technologies of the nineteen seventies.

In the nineteen eighties, fast vertical pnp-devices were introduced in bipolar integration technologies. During that time, research societies started to notice that the voltage-mode operational amplifier is not necessarily the best solution to all analogue circuit design problems. New research findings regarding current-mode signal processing using current-conveyors were presented. Furthermore, a commercial product became available: the current-feedback operational amplifier [1, 8]. The high slew rate and wide bandwidth of this amplifier resulted in its popularity in video amplifier applications.

Most reported current-conveyors and current-feedback operational amplifiers rely on the complementary bipolar process technology. In order to realise current-mode circuits with the less expensive CMOS-technology, different circuit topologies and operation principles are required. In 1988 the principle of a MOS current copier was presented [9], which enabled sampled data analogue circuits using only MOS-transistors. In 1989 the switched-current (SI) principle was presented [5]. The SI-circuits represent an alternative to the switched-capacitor (SC) circuits that do not need linear capacitors.

SI-circuits can therefore be realised with a standard digital CMOS-process. Several improvements to this circuit technique have been presented, for example the second generation SI-integrator [10] and the S^2I-technique [11] to reduce current memory errors with a two-step sampling method.

Following the introduction of sampled-data signal processing using current-mirrors continuous-time filter realisations based on current-mirrors were also reported [12–14]. Furthermore, various proposals for a CMOS current-mode operational amplifier have been published, either with a differential input and single-ended output [15] or with a single-ended input and differential output [16].

1.4 Adjoint principle

As a wide range of voltage-mode analogue circuits already exist, a straight forward method of converting these voltage-mode circuits to current-mode circuits would be very useful. In such a method a circuit using voltage amplifiers and passive components is converted into one that contains current amplifiers and passive components. An ideal voltage amplifier has infinite input impedance and zero output impedance, while an ideal current amplifier has zero input impedance and infinite output impedance. Consequently, direct replacement of a voltage amplifier with a current amplifier will lead to different circuit behaviour.

A voltage-mode circuit can be converted into a current-mode circuit by constructing an interreciprocal network by using the adjoint principle [1, 17]. According to this principle, a network N is replaced with an adjoint network N_a, the voltage excitation is interchanged to a current response, and the voltage response is interchanged to a current excitation, as demonstrated in Figure 1.1. Thus, the resulting transfer functions of these two networks N and N_a are identical:

$$H_v(s) = \frac{v_{out}}{v_{in}} = \frac{i_{out}}{i_{in}} = H_i(s). \tag{1.1}$$

The networks N and N_a are thus said to be inter-reciprocal to one another. When the networks N and N_a are identical, for example in the case of passive networks, the networks are said to be reciprocal.

Since all passive networks are reciprocal, all passive circuit elements have themselves as their adjoint elements i.e., passive elements are inter-reciprocal. In order to maintain identical transfer functions for both the original network N and the adjoint network N_a the impedance levels in the corresponding nodes of both networks should be identical. Therefore, the signal flow is reversed in the adjoint network and a voltage source is converted to a current sensing element as they both behave as short circuits. Similarly, a voltage sensing element is converted to a current source. A list of circuit

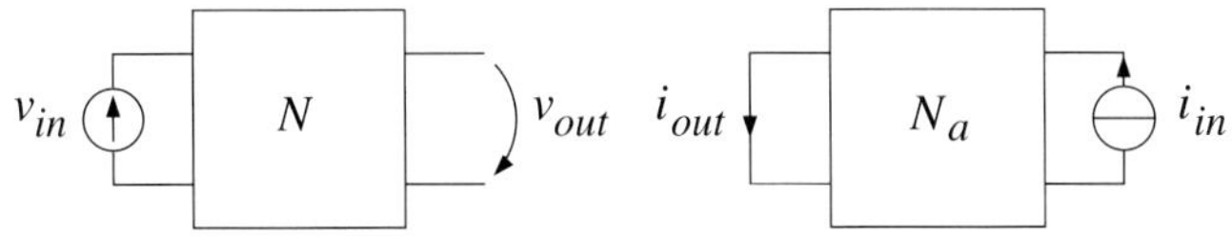

Figure 1.1 Interreciprocal networks N and N_a.

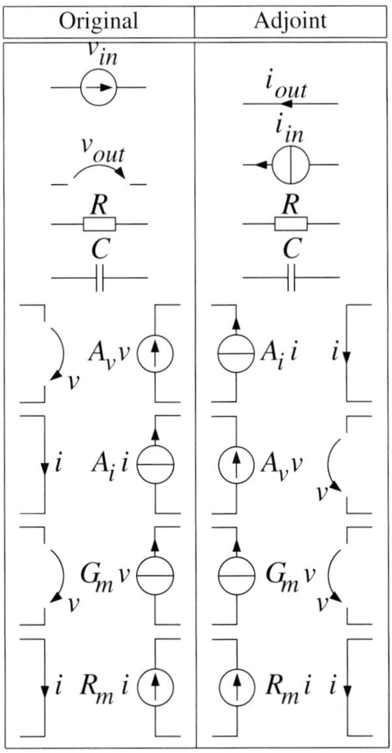

Table 1.1 Some circuit elements with their corresponding adjoint elements.

elements and their adjoint elements are presented in the Table 1.1.

In addition, controlled sources can be converted with the same principles: the signal flow is reversed and the impedance level is kept the same. In this way, a voltage amplifier is converted to a current amplifier and a current amplifier is converted to a voltage amplifier, respectively. However, since transresistance and transconductance amplifiers are inter-reciprocal, networks containing only transresistance or transconductance amplifiers and passive elements differ only in signal direction and type.

The adjoint principle can also be applied to transistor level circuits. In this case, a bipolar transistor in a common-emitter amplifier configuration is inter-reciprocal to itself and the common-collector amplifier configuration has the common-base configuration as its adjoint. Converting a voltage-mode bipolar transistor circuit to a current-mode MOS-transistor circuit could be beneficial as it minimises the use of source-follower stages which have poor low-voltage performance due to the bulk-effect. Bipolar transistor circuits are conventionally constructed of common-emitter and common-collector amplifier stages and the resulting MOS-transistor adjoint circuit is constructed of common-source and common-gate amplifier stages.

1.5 Scope of this book

Because the development of new integration technologies is driven by the needs of digital integrated circuit design, deep-submicron CMOS technologies will be the main integration technology in the near future. Furthermore, because the high density of integration leads to low supply voltages, digital circuits will no longer benefit from the use of bipolar transistors. Consequently, the significance of BiCMOS integration technology is diminishing. Thus, in this work, predominantly standard low-cost n-well CMOS-processes are used to realise the developed current-mode circuits and different circuit topologies. Certain circuits have been realised with BiCMOS-processes with only npn-type vertical bipolar transistors, as a simple bipolar add-on to an analogue CMOS-process may yet remain a commercially feasible option long after the more specialised complementary BiCMOS-processes become too expensive to develop.

In this book we concentrate on the design of linear current-mode amplifiers and on different linear and non-linear signal processing applications utilising these amplifiers. Such applications are comparable to signal processing techniques based on voltage-mode operational amplifiers. Therefore, a circuit designer familiar with circuit design with voltage-mode operational amplifiers would be able to see the analogy between current- and voltage-mode amplifiers and would be able to choose between both design methodologies in integrated circuit design.

This book does not cover the entire field of current-mode circuit design, which is extensive indeed. One such current-mode design technique is the translinear circuit principle, originally developed by Barrie Gilbert for synthesis of non-linear signal processing functions by bipolar integrated circuits [18] and extended to MOS integrated circuits that use transistors operating in weak inversion or in saturation [19]. However, in Chapter 7 logarithmic amplifiers are discussed as an example of non-linear signal processing with current amplifiers. Although logarithmic amplifiers do not belong to the category of translinear circuits, certain translinear circuit techniques can be used in conjunction with such amplifiers, for example in post-processing of the logarithmic output signal. Neither the log-domain filters [4] nor other dynamic translinear circuits is covered in this book because such circuit techniques are best suited to bipolar integrated circuits or micropower MOS integrated circuits which are beyond the scope of this book.

Since 1968, a confusing amount of different current-mode building blocks have been proposed. Consequently, the most significant current-mode building blocks are reviewed and compared in Chapters 3 and 4. As many of these current amplifiers operate without any global feedback, the linearity of the amplifier becomes an important design parameter. Consequently, the distortion mechanisms of different current amplifier topologies are discussed in detail. Current-mode amplifier macromodels are also discussed in Chapters 3 and 4 as they are an important tool in evaluating amplifier performance in different applications.

The input voltage-to-current and output current-to-voltage conversions play an important role in all current-mode signal processing systems and are discussed in Chapter 5 together with various differential to single-ended and single-ended to differential conversion techniques. Several design examples, with experimental results, are included in order to demonstrate the performance of the enhancement techniques discussed. Similarly, the noise in current-amplifiers is discussed as the noise behaviour of the system is strongly linked to the input interface.

As an application example, active continuous-time filter realisations using current-amplifiers are discussed in Chapter 6. In addition, first conventional active filter techniques based on voltage-mode amplifiers are reviewed. Subsequently, different active filter techniques based on current-conveyors are reviewed. Furthermore, three different methods for implementing active continuous-time filters using high-gain current-conveyors are presented. Finally, two high-gain current-conveyors based continuous-time filter chips are designed and fabricated and the filter design procedure and the resulting filter performance is discussed in detail.

In Chapter 7 two different logarithmic amplifier operating principles are discussed. Two logarithmic amplifiers are designed and fabricated using current amplifiers and other current-mode circuitry. These two design examples serve to demonstrate that efficient logarithmic amplifier realisations can be designed with both operating principles using current-mode design techniques.

References

[1] C. Toumazou, F. J. Lidgey, D. G. Haigh (ed.), *Analogue IC design: the current-mode approach*, London, Peter Peregnirus Ltd, 1990, 646 p.

[2] A. Sedra, G. Roberts, F. Gohn, "The current-conveyor: history, progress and new results," *IEE Proceedings G*, vol.137, pp. 78-87, April 1990.

[3] S. Long, J. Zhang, "Low power GaAs current-mode 1.2 Gb/s interchip interconnections," *IEEE J. Solid-State Circuits*, vol. 32, pp.890-897, June 1997.

[4] D. Frey, "Log-domain filtering: an approach to current-mode filtering," *IEE Proceedings G*, vol. 140, pp. 406-416, Dec. 1993.

[5] J. Hughes, N. Bird, I. Macbeth, "Switched currents - a new technique for analog sampled-data signal processing," in *Proc. IEEE Int. Symposium on Circuits and Systems* (ISCAS'89), Portland, USA, 1989, pp. 1584-1587.

[6] K. Smith, A. Sedra, "The current-conveyor-a new circuit building block," *IEEE Proc.,* vol. 56, pp. 1368-69, 1968.

[7] A. Sedra, K. Smith, "A second-generation current-conveyor and its applications," *IEEE Trans.,* vol. CT-17, pp. 132-134, 1970.

[8] Comlineàr Corporation, A new approach to op-amp design, Application Note 300-1, March 1985.

[9] S. Daubert, D. Vallancourt, Y. Tsividis, "Current copier cells," *Electronics Letters*, vol. 24, pp. 1560-1561, Dec. 1988.

[10] J. Hughes, I. Macbeth, D. Pattullo, "Second generation switched-current signal processing," in *Proc. IEEE Int. Symposium on Circuits and Systems* (ISCAS'90), New Orleans, USA, 1990, pp. 2805-2808.

[11] J. Hughes, K. Moulding, "S^2I: A two-step approach to switched-currents," in *Proc. IEEE Int. Symposium on Circuits and Systems* (ISCAS'93), Chicago, USA, 1993, pp. 1235-1238.

[12] J. Ramírez-Angulo, M. Róbinson, E. Sánchez-Sinencio, "Current-mode continuos-time filters: two design approaches," *IEEE Trans. Circuits and Systems-II*, CAS-39, June 1992, pp.337-341.

[13] S. Lee, R. Zele, D. Allstot, G. Liang, "CMOS continuous-time current-mode filters for high-frequency applications," *IEEE J. Solid-State Circuits*, vol. 28, pp.323-329, March 1993.

[14] S. Smith, E. Sánchez-Sinencio, "Low voltage integrators for high-frequency CMOS filters using current mode techniques," *IEEE Trans. Circuits and Systems-II*, vol. CAS-43, pp.39-48, Jan. 1996.

[15] E. Bruun, "Bandwidth optimization of a low power, high speed CMOS current op amp," *Int. J. Analog Integrated Circuit Design and Signal Processing*, vol. 7, pp. 11-19, Jan. 1995.

[16] R. Zele, S. Lee, D. Allstot, "A high gain current-mode operational amplifier," in *Proc. IEEE Int. Symposium on Circuits and Systems* (ISCAS'92), San Diego, USA, IEEE, pp. 2852-2855.

[17] G. Roberts, "All current-mode frequency selective circuits," *Electronics Letters*, vol. 25, pp. 759-761, June 1989.

[18] B. Gilbert, "Translinear circuits: a proposed classification," *Electronics Letters*, vol. 11, pp. 14-16, 1975.

[19] E. Seevink, R. Wiegerink, "Generalized Translinear Circuit Principle," *IEEE J. Solid State Circuits*, vol. 26, pp. 1098-1102, 1991.

Chapter 2

Basic current amplifiers

In this chapter, different basic current amplifier topologies are discussed. With these amplifiers, more complicated current-mode amplifiers are constructed in the CMOS integration technology. There are two types of basic CMOS current amplifiers: inverting and noninverting current amplifiers. The inverting current amplifiers are realised with different current-mirror topologies whereas the noninverting current amplifiers are either realised with two cascaded current-mirrors or with a one-transistor amplifier stage in a common-gate configuration. Furthermore, it is well know that, with a common-gate amplifier stage, only unity gain noninverting current amplifiers, i.e. current buffers, can be realised. Therefore, the basic current amplifiers involved are different realisations of current mirrors and current buffers.

Basic amplifier building blocks such as current-mirrors have already been discussed in much of the literature on analogue integrated circuit design [1–3]. As these textbooks assume that the basic amplifiers are used to construct high gain voltage amplifiers, an essential nonideality for current amplifiers is omitted: distortion. When current signals are assumed rather than the traditional voltage signals several assumptions that have been made in the literature are not valid or at least have a different significance. Therefore, in this chapter is evaluated which mechanisms in practice dominate the dynamic range of these current amplifier building blocks. As signal frequencies increase, different mechanisms begin to dominate the operation of the amplifier and consequently the amplifier nonidealities are analysed at both high and low frequencies.

One such basic amplifier building block is however not included in this chapter: the differential amplifier stage, also known as the source-coupled pair. The reason for this omission is that this building block has very limited application in current-mode signal processing, which typically involves gain boosting of a local feedback loop. As this amplifier stage is now used as a traditional voltage amplifier, the same assumptions as those already mentioned in the literature [1–3] also apply here. In certain cases, as with the circuit topology depicted in Figure 2.17b, the topology may only resemble a source coupled pair, but in practise its operation is completely different.

2.1 Current-mirror

Inverting current amplifiers with moderate and easily controlled gain can be realised with current-mirrors. A simple realisation of a MOS current-mirror [1] is presented in Figure 2.1. If one assumes that the transistors operate in the saturation region ($v_{DS} > v_{GS} - V_T$), so that the drain current equation is $i_D = \frac{\mu_o C_{ox} W}{2L} (v_{GS} - V_T)^2 (1 + \lambda v_{DS})$, where the parameters are:

μ_o	surface mobility of the channel,
$C_{ox} = \frac{\varepsilon_{ox}}{t_{ox}}$	gate oxide capacitance density,
W	effective channel width,
L	effective channel length,
V_T	threshold voltage,
λ	channel length modulation parameter.

Then, if the effect of source and load impedances is first neglected, the current-mirror large signal equation shows inverting current gain proportional to the aspect ratios of transistors M_1 and M_2, with certain additional dependencies

$$A_i = \frac{i_{OUT}}{i_{IN}} = -\frac{W_2 L_1}{L_2 W_1} \left(\frac{v_{GS} - V_{T2}}{v_{GS} - V_{T1}} \right)^2 \frac{1 + \lambda v_{DS2}}{1 + \lambda v_{DS1}} \frac{\mu_{o2} C_{ox2}}{\mu_{o1} C_{ox1}}. \tag{2.1}$$

In the current-mirror, the process variation of the channel width W, channel length L, mobility μ_o, and oxide thickness t_{ox} produce linear gain error comparable to the variation of the resistor ratio errors in the closed loop operational amplifier circuits.

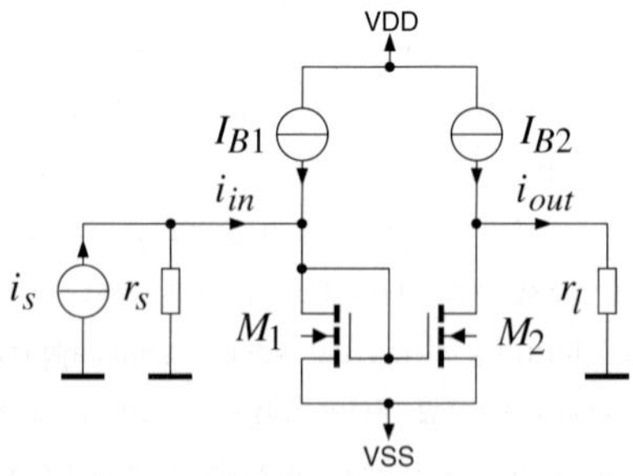

Figure 2.1 Inverting current amplifier with a simple MOS current-mirror.

Other parameters in the current gain equation likewise have an effect on the circuit nonlinearity and are discussed in detail in the following sections.

2.1.1 Nonidealities due to the channel length modulation

In addition to the random process variation, finite input and output impedances have a significant effect on the gain accuracy. The small-signal input impedance of the inverting current amplifier depends on the transconductance of the input transistor M_1

$$r_{in} \approx \frac{1}{g_{m1}} = \frac{1}{\sqrt{2\mu_{o1}C_{ox1}\frac{W_1}{L_1}I_{B1}}}, \tag{2.2}$$

and the small-signal output impedance depends on the drain-source conductance of the output transistor M_2, accordingly

$$r_{out} \approx \frac{1}{g_{ds2}} = \frac{1}{\lambda I_{B2}}. \tag{2.3}$$

In modern sub-micron CMOS processes the g_m/g_{ds} ratio is less than 100 and consequently a significant gain error results when more of these current amplifier stages are cascaded. This gain error is usually reduced by increasing the output impedance using different cascode mirror topologies rather than the simple two transistor current-mirror.

Three different cascode current-mirror topologies are presented in Figure 2.2. The simplest uses only one more transistor at the output that can be biased quite freely (Figure 2.2a). In this case, however, the drain-source voltages $v_{DS1} = v_{GS}$ and v_{DS2} are different and gain error may result. Moreover, these drain voltages are signal dependent and thus the channel length modulation λ, which continuously increases as progressively shorter channel length devices are introduced, can additionally produce significant amount of distortion. Therefore, topological improvements must be made in the current-mirror in order to maintain the drain-source voltages v_{DS1} and v_{DS2} as equal as possible.

This is achieved by using other cascode current-mirror topologies rather than the simple current-mirror or the cascode current-mirror of Figure 2.2a. In the circuit of Figure 2.2b, an additional current-mirror constructed of transistors M_3 and M_4 is added on top of the original current mirror in order to force the drain voltage v_{DS2} of the transistor M_2 equal to $v_{DS1} = v_{GS}$ [1]. Unfortunately, this reduces significantly the input and output voltage ranges and thus the minimum supply voltage of the circuit is quite high. It is therefore better to add the cascode transistors, M_3 and M_4, directly to the drains of both mirror transistors, M_1 and M_2, as presented in Figure 2.2c and bias the cascode transistors with an additional voltage so that the drain voltages v_{DS1} and v_{DS2} are slightly above the saturation voltages of the mirror transistors M_1 and M_2, in

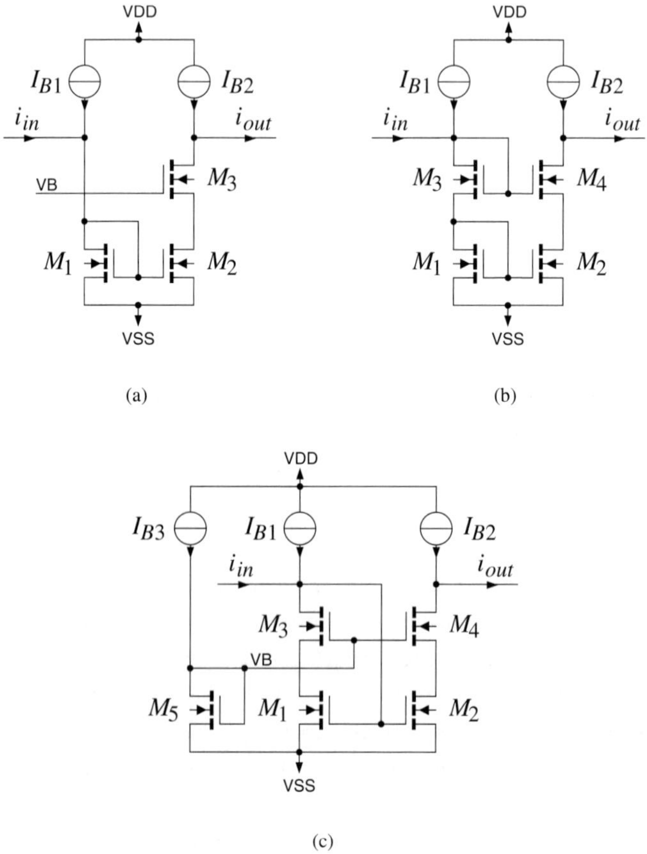

Figure 2.2 Different cascode current-mirror topologies. (a) A simple three transistor cascode current-mirror. (b) A simple four transistor cascode mirror [1]. An enhanced current-mirror capable of operating with low supply voltages [4, 5].

all signal conditions [4,5].

If $\beta = \mu_o C_{ox}\frac{W}{L}$, λ and V_T of all of the transistors in the three current mirror circuits are assumed equal, so that the drain-source voltages of transistors M_1 and M_2 are the only unmatched parameters, the output current i_{OUT} equals

$$i_{OUT} = -\frac{1+\lambda v_{DS2}}{1+\lambda v_{DS1}} i_{IN} \approx -i_{IN} - \lambda\left(v_{DS2} - v_{DS1}\right) i_{IN}. \tag{2.4}$$

Although this equation is an approximation, the distortion of these three cascode current mirrors can be compared with adequate precision. Furthermore, the drain-source voltages v_{DS1} and v_{DS2} of the main mirror transistors M_1 and M_2 can be expressed as a function of the small signal input current i_{in} with additional constants that depend on the mirror topology

$$v_{DS1} \approx V_{DS} + \frac{i_{IN}}{g_m}, \tag{2.5}$$

$$v_{DS2} \approx V_{DS} + \Delta V_{DS} + a\frac{i_{IN}}{g_m}. \tag{2.6}$$

For the current-mirror of Figure 2.2a, the constant a is approximately -1 because the drain-source voltage v_{DS2} drops as the output current increases. With this topology the static difference of the drain-source DC-voltages, ΔV_{DS}, can be relative high depending on the cascode transistor bias voltage. For current mirrors of Figures 2.2b and c, the constant a is approximately 1 and ΔV_{DS} is only a few millivolts.

Given these assumptions, the output current can be now expressed as

$$i_{OUT} = -i_{in}\left(1 + \lambda \Delta V_{DS}\right) - i_{in}^2 \frac{\lambda}{g_m}\left(1-a\right). \tag{2.7}$$

This equation demonstrates that the drain-source DC-voltage mismatch term b produces only gain error as the term a has a significant effect on the current-mirror distortion. If the input signal is rewritten as $i_{in} = \hat{\imath}\sin(\omega t)$, where $\hat{\imath}$ is the current signal amplitude and ω the signal frequency, the second order harmonic distortion of the mirror can be solved:

$$HD2\left(\lambda\right) = \frac{\lambda \hat{\imath}}{2g_m}\left(1-a\right) = m\frac{g_{ds}}{g_m}\frac{1-a}{2}, \tag{2.8}$$

where $m = \frac{\hat{\imath}}{I_B}$ is the modulation index. If we assume a moderate modulation index $m = 0.2$ and a modern submicron CMOS process, where the $\frac{g_m}{g_{ds}}$ ratio may be as low as 40, the distortion of the three transistor cascode current-mirror of Figure 2.2a is 0.5 %, while in both four transistor cascode current-mirrors the distortion almost vanishes. Therefore, the distortion arising from channel length modulation can be neglected if the appropriate mirror topology is selected and the main mirror transistors M_1 and M_2 are designed so that they have significantly longer channel lengths than the minimum channel length. One must be careful in bias circuit design because it is easy to construct a bias circuit where the same distortion mechanism as in the mirror of Figure 2.2a occurs.

In practice, the λ-model used to approximate the drain-source conductance does not work well with submicron channel lengths as the drain-source conductance is no longer proportional to the drain current near the saturation voltage. The same is also true in the case of drain-source voltages in the range of two volts or more. Therefore, the λ of the two mirror transistors can be assumed to be equal only if the static difference of the drain-source DC-voltages ΔV_{DS} is minimal. However, the final conclusion that the effect of the channel length modulation is minimised by tracking the drain-source voltages of the main mirror transistors is still true as, in this case, the parameter λ can also be assumed equal for both transistors. Therefore, the use of proper cascode topologies is even more important than the calculations may show as the parameter λ itself can generate distortion in certain current-mirror topologies.

2.1.2 Nonidealities due the V_T mismatch

If $\beta = \mu_o C_{ox} \frac{W}{L}$, λ and v_{DS} of both transistors are assumed to be equal, so that the transistor threshold voltage mismatch $\Delta V_T = V_{T2} - V_{T1}$ is the only nonideality, then the output current i_{OUT} equals

$$i_{OUT} = -i_{IN} - \frac{\beta}{2}\Delta V_T^2 - \Delta V_T \sqrt{2\beta\left(i_{IN} + I_B\right)}. \tag{2.9}$$

From this equation, it can be seen that the V_T mismatch produces a static offset current

$$I_{OFF}(\Delta V_T) = \frac{\beta}{2}\Delta V_T^2 + \Delta V_T \sqrt{2\beta I_B}, \tag{2.10}$$

but as the square root term is signal dependent, the V_T mismatch also produces gain error and distortion. This term can be approximated with a Taylor series

$$\sqrt{i_{IN} + I_B} = \sqrt{I_B}\left(1 + \frac{i_{IN}}{2I_B} - \frac{i_{IN}^2}{8I_B^2} + \frac{i_{IN}^3}{16I_B^3} - \frac{5i_{IN}^4}{128I_B^4} + \frac{7i_{IN}^5}{256I_B^5} + \ldots\right). \tag{2.11}$$

If we assume that the input signal is

$$i_{IN} = \hat{i}\sin(\omega t) = mI_B \sin(\omega t), \tag{2.12}$$

where $\hat{i}$ is the current signal amplitude, ω the signal frequency, and $m = \frac{\hat{i}}{I_B}$ the modulation index, different harmonic components can be collected

$$\begin{aligned}\sqrt{i_{IN}+I_B} \;\approx\; & \sqrt{I_B}\Big(1+\frac{m}{2}\sin(\omega t) \\ & +\frac{m^2}{16}\cos(2\omega t)-\frac{m^3}{64}\sin(3\omega t) \\ & -\frac{5m^4}{1024}\cos(4\omega t)+\frac{7m^5}{4096}\sin(5\omega t)+\ldots\Big).\end{aligned} \tag{2.13}$$

By using this approximation, we can solve the current gain as a function of the threshold voltage mismatch from Equation (2.9):

$$A_i(\Delta V_T) \approx -1-\Delta V_T\sqrt{\frac{\beta}{2I_B}} = -1-\frac{\Delta V_T}{V_{GS}-V_T}. \tag{2.14}$$

This shows that the gain error depends on the saturation voltage $V_{DSAT} = V_{GS} - V_T$. For example, a threshold voltage mismatch of 5 mV produces 0.5 % gain error with a saturation voltage of 500 mV.

The second order harmonic distortion due to the threshold voltage is

$$HD2(\Delta V_T) \approx \frac{m}{8}\frac{\Delta V_T}{V_{GS}-V_T}, \tag{2.15}$$

and the third order harmonic distortion is accordingly

$$HD3(\Delta V_T) \approx \frac{m^2}{32}\frac{\Delta V_T}{V_{GS}-V_T}. \tag{2.16}$$

The above equations show that it is relatively easy to design a current-mirror that has harmonic distortion lower than -60 dB at low frequencies. At higher frequencies, other distortion mechanisms start to dominate the nonlinearity of the current mirror.

To test the validity of the equations, the current mirror distortion was simulated as a function of the threshold voltage mismatch. Both a simple current-mirror and the cascode current-mirror of Figure 2.2c are compared with the harmonic distortion equations (2.15) and (2.16) in Figure 2.3. In the simulations, Level 2 Spice-models of a 1.2μm CMOS-process are used; the aspect ratio of the mirror transistors M_1 and M_2 is $100\mu/5\mu$ while the aspect ratio of the cascode transistors M_3 and M_4 is $100\mu/2\mu$. The saturation voltage of the mirror transistors is approximately 300 mV and the modulation index is 0.2. In order to avoid high frequency nonidealities in the simulations, the input signal frequency is only one Hz.

The simulations show that, in the cascode current-mirror, the threshold voltage mismatch is almost the only distortion mechanism present at low frequencies. For the simple current-mirror, the g_m/g_{ds} ratio is approximately 167 and thus the second

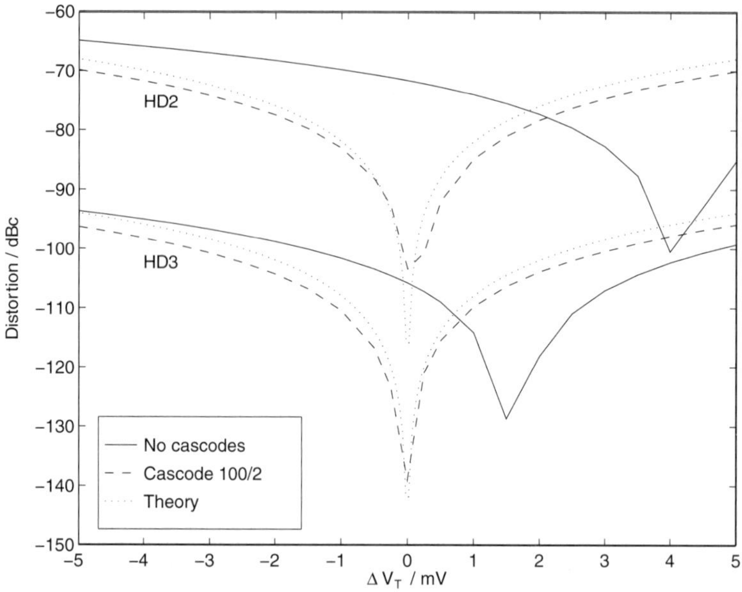

Figure 2.3 Simulated low frequency distortion of a simple and a cascode current mirror as a function of the threshold voltage mismatch. The mirror transistor V_{DSAT} is approximately 300 mV and the cascode mirror uses the topology of Figure 2.2c. The input signal frequency is one Hz and the modulation index is 0.2.

order distortion due to the channel length modulation is approximately 64 dB, which is in the same range as the threshold voltage mismatch distortion. The two distortion mechanisms either cancel out or sum up, depending on the sign of the threshold mismatch. Besides, the threshold mismatch distortion alone predicts quite well the level of distortion also in the simple current-mirror case.

2.1.3 High frequency nonidealities

Linear effects

The small-signal equivalent circuit of the simple MOS current-mirror of Figure 2.1 is presented in Figure 2.4. In this schematic, the capacitance C_{in} consists of the gate capacitances of the mirror transistors, M_1 and M_2, and all other parasitic capacitances in the current-mirror input node, excluding the gate-drain capacitance C_{gd2} i.e. the Miller-capacitance. In order to minimise the inaccuracies arising from the threshold voltage mismatch, the mirror transistors usually have relatively large gate areas and therefore the gate-source capacitances are the dominant capacitances and consequently

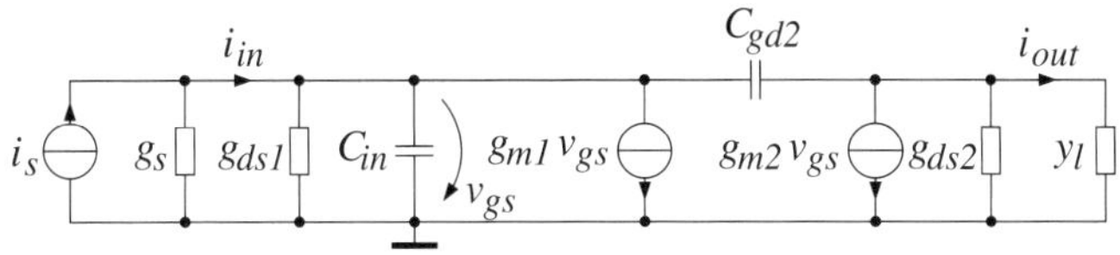

Figure 2.4 The small-signal equivalent circuit of the simple current mirror.

the input capacitance can be approximated as

$$C_{in} \approx C_{gs1} + C_{gs2} = \frac{2}{3} C_{ox} W_1 L_1 (A_i + 1), \tag{2.17}$$

where $A_i \approx -\frac{W_2 L_1}{L_2 W_1}$ and $L_1 = L_2$.

When the current-mirror is driving another current-mirror, the load conductance g_l in the output admittance $y_l = g_l + sC_l$ is comparable to g_{m1} and similarly the load capacitance C_l is comparable to the mirror input capacitance C_{in}. Furthermore, all capacitances at the output other than C_{gd2} are included in the load capacitance C_l and thus the minimal effect of the drain diffusion capacitances of the output mirror transistor and the output current source is omitted. Hence, we can assume the gate-drain capacitance C_{gd2} to be far smaller than the capacitances C_{in} and C_l. As the drain-source conductances g_{ds1} and g_{ds2} are also much smaller than the transconductances g_{m1} and g_{m2} and the load conductance g_l, the current-transfer function can be approximated as

$$\frac{i_{out}}{i_{in}} \approx \frac{g_{m2}}{g_{m1}} \frac{\left(1 + s\frac{C_l}{g_l}\right)\left(s\frac{C_{gd2}}{g_{m2}} - 1\right)}{1 + s\left(\frac{C_{in}}{g_{m1}} + \frac{g_{m1}C_l + g_{m2}C_{gd2}}{g_{m1}g_l}\right) + s^2\frac{C_{in}C_l}{g_{m1}g_l}}. \tag{2.18}$$

With current gains close to one, the pole and the zero due to the load capacitance cancel each other resulting in a first order transfer function with a pole at

$$p_0 = \frac{g_{m1}}{C_{in}}, \tag{2.19}$$

and a zero at

$$z_0 = -\frac{g_{m2}}{C_{gd2}}. \tag{2.20}$$

The right half-plane zero does not contribute a great deal to the amplitude response near the corner frequency. Furthermore, the relatively insignificant Miller effect can be reduced even further if cascode current-mirror topologies are used. Therefore, the corner frequency can be expressed as a function of device dimensions and bias current

$$\omega_0 \approx \frac{3}{A_i + 1}\sqrt{\frac{\mu_0 I_{B1}}{2C_{ox}W_1L_1^3}}, \tag{2.21}$$

if the Miller effect is neglected. The equation shows that there is a similar trade-off between gain and bandwidth as with operational amplifiers. In this case, however, the frequencies are much higher. As the corner frequency is strongly dependent on the channel length, there is a strong trade-off between bandwidth and gain accuracy as well as bandwidth and distortion. With high current gains, the Miller effect in the simple current-mirror reduces the bandwidth more than is indicated in Equation 2.21, as the capacitance C_{gd2} is proportional to W_2 and A_i, whereas the transconductance g_{m2} is only proportional to the square root of W_2 and A_i.

Certain cascode current-mirror topologies, such as the low voltage cascode current-mirror in Figure 2.2c, may reveal a degree of peaking of the current gain near the corner frequency, as the input mirror transistor M_1 and the input cascode transistor M_3 form a feedback loop with a second order transfer function. If the cascode transistors have the same aspect ratio as the mirror transistors, the non-dominant pole frequency is still twice the dominant pole frequency or even more. Moreover, the cascode transistors should have larger aspect ratios than the mirror transistors in order to maintain good biasing conditions for larger signal variations. Furthermore, the threshold voltage mismatch of the cascode transistors has very little effect on the mirror accuracy. Consequently, minimum channel lengths can be used for the cascode transistors, as is not the case with mirror transistors. As a result, the non-dominant pole is shifted even further and thus, in properly designed cascode current-mirrors, the cascode transistors have a negligible effect on the frequency response.

Nonlinearity in limited frequency ranges

High frequency distortion can be easily and intuitively derived if only a limited frequency range is considered. Then, instead of deriving complex equations for each distortion component valid for a very wide frequency range, a few simple asymptotic equations can be derived.

First, frequencies significantly below the mirror corner frequency are considered. Although the frequency response is flat below the corner frequency, the parasitic capacitances start to generate a significant amount of distortion even at relatively low frequencies. As the signal frequency increases, the input parasitic capacitances gradually start to convert the non-linear input voltage v_{IN} into a non-linear current i_{C1} [6,7]. Consequently, the large signal output current equation for high frequencies can be approximated for the simple current-mirror as [8]

$$i_{OUT} = -i_{IN} + i_{C1} = -i_{IN} + C_{IN}\frac{dv_{IN}}{dt}. \tag{2.22}$$

For the cascode current mirror of Figure 2.5b, the parasitic capacitances C_{D1} and C_{D2}

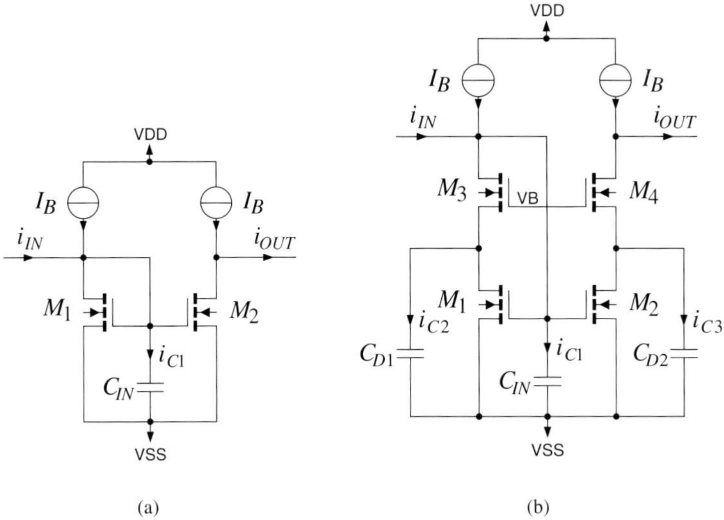

Figure 2.5 High frequency effects (a) in a simple current-mirror and (b) in a low voltage cascode current-mirror.

generate distortion in the same way as the input capacitance C_{IN}. However, the currents i_{C2} and i_{C3} cancel each other out because the parasitic capacitances at the mirror drain nodes, in addition to the cascode transistor transconductances, are well matched. Therefore, the same output current equation is also valid for the low voltage cascode current-mirror.

At signal frequencies significantly lower than the mirror pole frequency $\omega_0 = \frac{g_m}{C_{IN}}$, it can be assumed that the current flowing through the gate capacitance is merely a fraction of the input current, so that it does not affect the mirror input voltage v_{IN}. As the gate-source capacitance is relatively linear in the saturation region, it can be assumed that the derivative of the mirror input voltage $\frac{dv_{IN}}{dt}$ is the only non-linear element of the large signal current equation. Therefore, the nonlinear output current can be approximated

$$i_{OUT} = -i_{IN} + C_{IN}\sqrt{2\beta}\frac{d}{dt}\sqrt{i_{IN}+I_B}. \tag{2.23}$$

If the term $\sqrt{i_{IN}+I_B}$ is approximated with a Taylor-series of Equation (2.11) and after substituting $i_{IN} = \hat{i}\sin(\omega t)$, as in Equation (2.13), the equation can be derived and used to calculate harmonic distortion arising from the input capacitance. The second order harmonic distortion is

$$HD2(C_{IN})\Big|_{\omega<\omega_0} \approx \frac{1}{4}m\frac{\omega}{\omega_0}, \tag{2.24}$$

where $m = \frac{\hat{i}}{I_B}$. Accordingly, the third order distortion is

$$HD3(C_{IN})\Big|_{\omega<\omega_0} \approx \frac{3}{32}m^2\frac{\omega}{\omega_0}. \tag{2.25}$$

At signal frequencies significantly above the mirror pole frequency ω_0, the input capacitance C_{IN} dominates the input impedance and thus the input voltage can be expressed as

$$\begin{aligned} v_{IN} &= V_T + \sqrt{\frac{2I_B}{\beta}} + \frac{1}{C_{IN}}\int i_{IN}dt \\ &= V_T + \sqrt{\frac{2I_B}{\beta}} - \frac{mI_B\cos(\omega t)}{\omega C_{IN}}. \end{aligned} \tag{2.26}$$

By substituting this into the output current equation $i_{OUT} = \frac{\beta}{2}(v_{IN} - V_T)^2$ and collecting different harmonic components, one arrives at

$$\begin{aligned} i_{OUT} &= -\frac{\omega_0 m I_B}{\omega}\cos(\omega t) + \frac{\omega_0^2 m^2 I_B}{8\omega^2}\cos(2\omega t) \\ &\quad + \frac{\omega_0^2 m^2 I_B}{8\omega^2} + I_B. \end{aligned} \tag{2.27}$$

The equation shows that there is ideally exclusively second order distortion present, which equals

$$HD2(C_{IN})\Big|_{\omega>\omega_0} = \frac{1}{8}m\frac{\omega_0}{\omega}. \tag{2.28}$$

Mirror as a feedback amplifier

The current-mirror input transistor M_1 can be considered as a feedback amplifier and thus the dynamic distortion equations derived in Appendix A.3 can be applied to depict the high frequency distortion also in this case. Then, as in Appendix A.3, the feedback amplifier can be divided into two parts, a linear integrator stage and a nonlinear output stage. The capacitance C_{IN} integrates the input current forming the mirror input voltage. The mirror input voltage is then amplified by the transistor transconductance into a nonlinear drain current. The diode-connected input transistor M_1 forms the feedback alone so that, for the distortion equations, the feedback factor f is always one regardless of current mirroring ratio. The current-mirror mirroring ratio affects only the total input capacitance C_{IN} thus moving the mirror pole frequency ω_0.

Letting $f = 1$, $b_1 = \sqrt{2\beta_2 I_B}$, $b_2 = \frac{1}{2}\beta_2$ and $b_3 = 0$, the second-order distortion equation (A.38) as a function of frequency can be simplified to

$$HD2 = \frac{1}{4} m \frac{\omega \omega_0}{\sqrt{\omega^2 + \omega_0^2} \sqrt{4\omega^2 + \omega_0^2}}, \tag{2.29}$$

and, respectively, the third order harmonic distortion equation (A.39) is simplified to

$$HD3 = \frac{3}{32} m^2 \frac{\omega \omega_0^3}{\left(\omega^2 + \omega_0^2\right) \sqrt{4\omega^2 + \omega_0^2} \sqrt{9\omega^2 + \omega_0^2}}. \tag{2.30}$$

By assuming $\omega < \omega_0$, Equation (2.29) simplifies to Equation (2.24) and similarly by assuming $\omega > \omega_0$ the result equals Equation (2.28). Similarly, Equation (2.30) simplifies to Equation (2.27) when $\omega < \omega_0$. Therefore the simple equations agree with the more complex feedback equations.

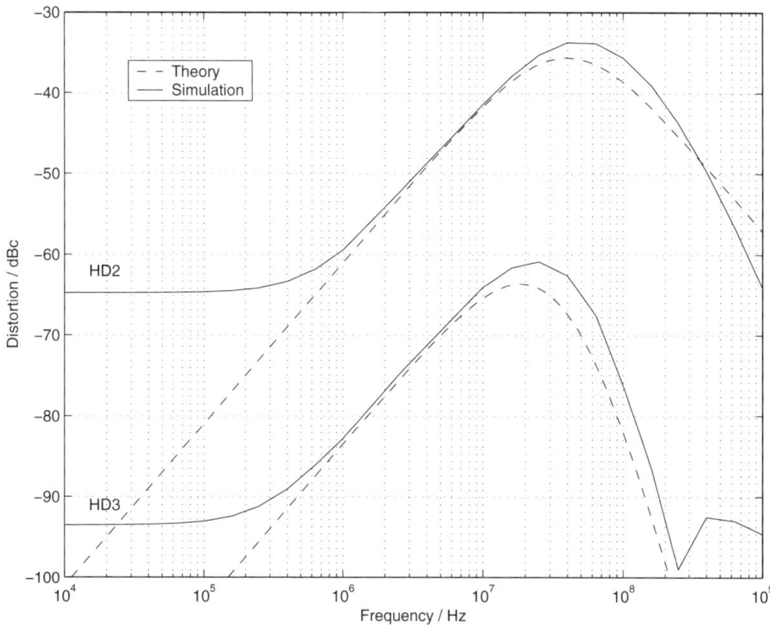

Figure 2.6 Simulated and theoretical second and third order harmonic distortion of a a simple NMOS current-mirror with a input signal modulation index of 0.2.

To test the validity of the distortion equations, the calculated distortion was plotted against simulated distortion in Figure 2.6. In the simulations the Level 2 Spice-models of a 1.2μm CMOS-process were used and the aspect ratio of the mirror transistors M_1 and M_2 was $100\mu/5\mu$. A threshold voltage mismatch of 5 mV was assumed and the saturation voltage of the mirror transistors was approximately 300 mV. In the calculations, the $\omega_0 \approx 2\pi 56$ MHz was extracted from the operating point simulation data. The simulated distortion of the simple mirror agrees quite well with the theoretical distortion although above the pole frequency ω_0 also other distortion generation mechanisms such as nonlinear drain and source diffusion capacitances begin to affect the simulated distortion. At low frequencies, of course, simulated distortion does not agree with the calculated distortion since DC-nonidealities such as the threshold mismatch was not included in Equations (2.29) and (2.30).

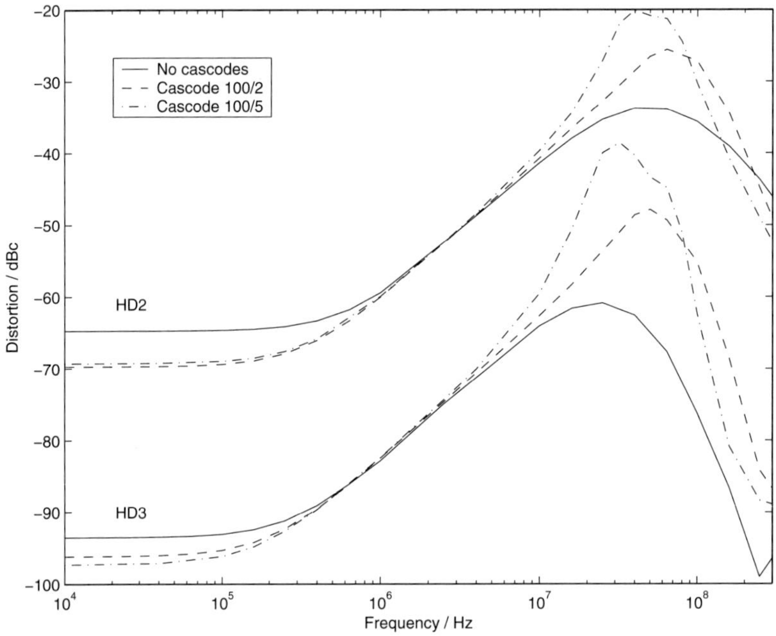

Figure 2.7 Simulated second and third order harmonic distortion of a cascode current-mirror of Figure 2.2c with a input signal modulation index of 0.2. The cascode mirror is simulated with two different cascode sizes $100\mu/5\mu$ and $100\mu/2\mu$.

In addition, the high frequency distortion of the cascode current-mirror topology of Figure 2.2c was compared with the simple current-mirror without cascode transistors (Figure 2.7). The aspect ratio of the cascode transistors M_3 and M_4 was either $100\mu/5\mu$ or $100\mu/2\mu$. With the cascode current-mirror, there is a significant increase in the simulated distortion near the corner frequency. Although the distortion currents i_{C2} and i_{C3} caused by the parasitic capacitances C_{D1} and C_{D2} cancel each other well below the pole frequency ω_0, this assumption is no longer valid near ω_0 as the input transistors form a feedback loop and the output transistors do not. As the currents i_{C2} and i_{C3} depend on the frequency in the same way as i_{C1}, the distortion peak can be shifted up by minimising the cascode transistor channel length. Furthermore, the peaking of

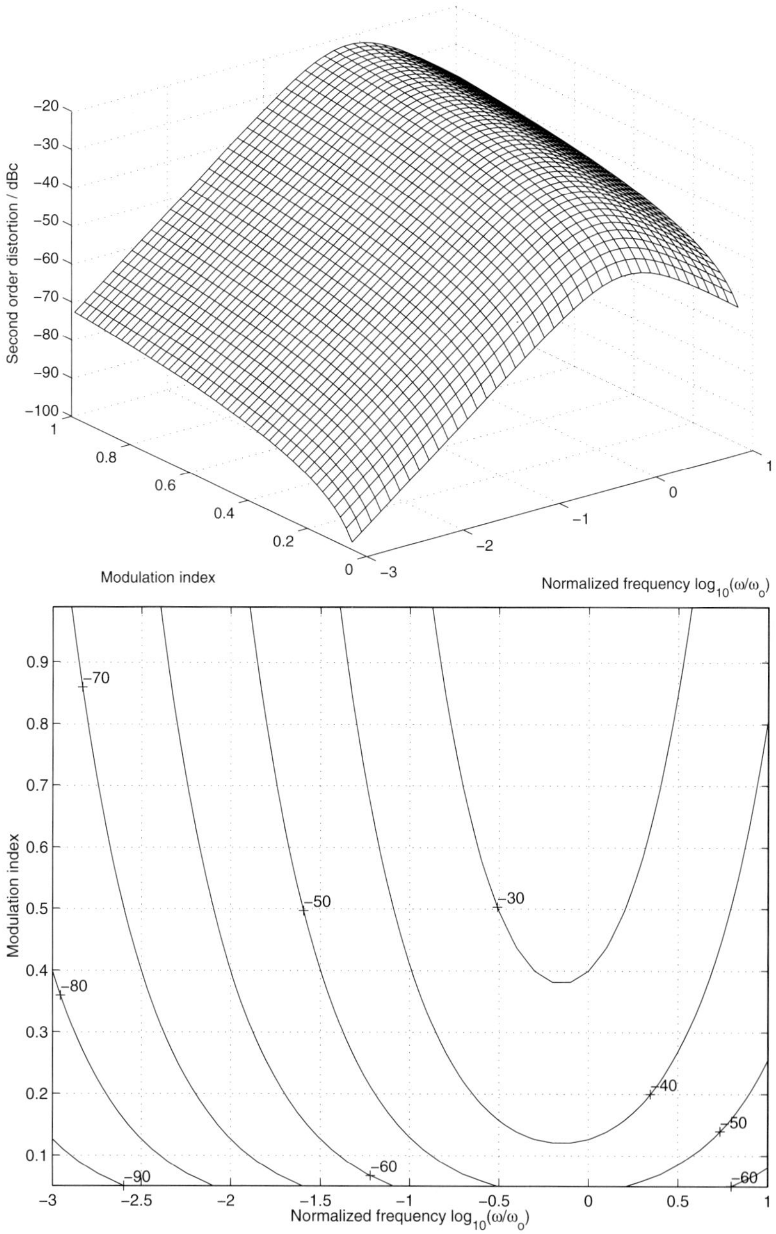

Figure 2.8 A three dimensional plot and its contour plot of the theoretical second order harmonic distortion of Equation (2.29) as a function of the modulation index m and the normalised frequency ω/ω_0.

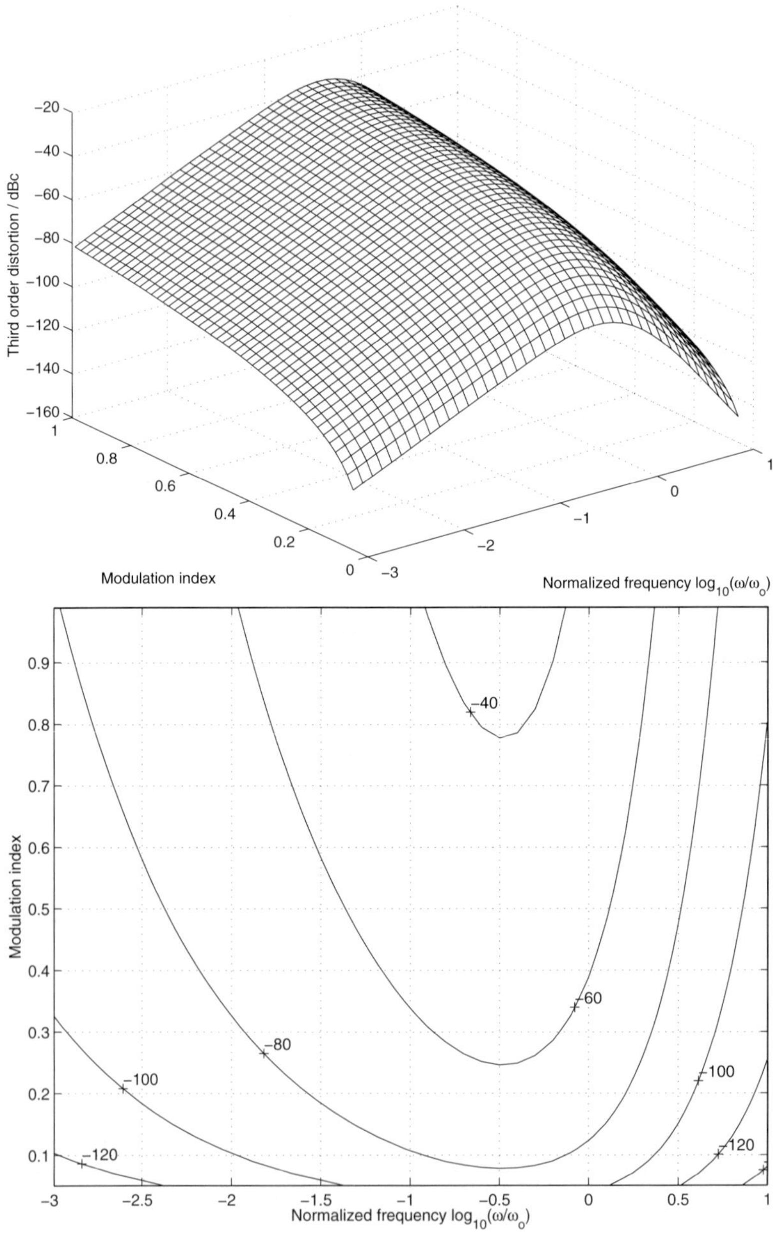

Figure 2.9 A three dimensional plot and its contour plot of the theoretical third order harmonic distortion of Equation (2.30) as a function of the modulation index m and the normalised frequency ω/ω_0.

the current gain near ω_0 increases the signal amplitude and consequently distortion, but the peaking decreases as the cascode transistor channel lengths are decreased. A significant increase in the main mirror transistor area will shift ω_0 down so that these two distortion peaks do not overlap anymore resulting in a lower maximum distortion.

The distortion Equations (2.29) and (2.30) can be presented as function of the normalised frequency $\frac{\omega}{\omega_0}$ and the modulation index m in a three dimensional graphical form. A contour plot representation of the three dimensional graph is presented in Figures 2.8 and 2.9. The second order distortion plot shows that a distortion level below -40 dBc is maintained with a modulation index of 0.1 near ω_0. On the other hand, at frequencies 30 times below ω_0, the distortion level below -40 dBc is maintained regardless of the modulation index. The third order distortion, however, is below -40 dBc almost up to a modulation index of 0.8 with the whole frequency range.

When the modulation index m approaches 1, the mirror transistors finally fall into the weak inversion region at negative input current peaks. Moreover, in the low voltage current-mirror, either the input cascode transistor or both mirror transistors may fall into the triode region if the cascode transistors are not properly biased or their aspect ratio is too small. For these reasons, these three dimensional plots are no longer valid as the modulation index reaches 1. However, these three dimensional plots depict the minimum distortion that is always present in a MOS current-mirror operating in strong inversion. As the cascode transistors add their own distortion peaks in the current mirrors, this minimum distortion is not easy to reach near ω_0 and if even lower distortion levels are required, a better circuit topology or a transistor with a better linearity must be found.

2.1.4 Distortion reduction methods

As the high frequency distortion depends only on the modulation index m and the normalised frequency ω/ω_0, we can reduce the high frequency distortion only by increasing the bandwidth or decreasing the maximum modulation index, which is realised only by increasing the bias current. However, there are at least three different possibilities to further reduce the high frequency distortion. The first method is to reduce the nonlinearity of transistor transconductance by deploying special circuit techniques. The second method is to reduce the non-linear current, either by lowering the input impedance or by reducing the input capacitance. The third method is to cancel the non-linear current by using a replica current.

Transconductance linearisation

The transconductance can be linearized by source-degeneration or by biasing the main mirror transistors in the linear region rather than the saturation region. The source degeneration method involves adding a resistance in series to the transistor source. This method is simple to realise and it has an additional advantage: sensitivity to the threshold voltage mismatch is reduced. The output impedance of the current-mirror is increased by source degeneration but the input impedance is similarly increased. Current-mirror topologies optimised for a low voltage operation, using triode region transistors as source degeneration resistors, have been published [9,10]. Unfortunately, these topologies are relatively complicated and thus their high frequency performance is not comparable to simple current mirrors.

As biasing the mirror transistors in the linear region results is a drastic lowering of the mirror output impedance, regulated cascode structures described in more detail in Section 2.2 are sometimes used in current-mirrors, as reported in [11–13]. Moreover, these topologies are relatively complicated and thus have low distortion only at low frequencies. Furthermore, in all regulated cascode current-mirrors, the dominant low frequency distortion mechanism remains the distortion due to the threshold voltage mismatch as in all other current-mirror topologies. Thus, the full potential of the regulated cascode technique is not utilised in current-mirror applications.

Another, albeit relatively theoretical, way to reduce the nonlinearity of the transconductance is to increase the bias current and thus also the transistor gate voltages, so that the i_D versus v_{GS} characteristic becomes linear as a result of mobility degradation. Nevertheless, this does not lead to a reliable or to a power efficient realisation. However, if we have to process very large currents the high frequency distortion may not be as bad as the calculations earlier have shown.

Nonlinear current reduction

The input impedance can be lowered and the input capacitance decreased simultaneously by isolating the mirror transistor gates and the input node with an external amplifier stage. The additional amplifier stage adds more poles to the input feedback path. Stability problems will occur and all successful compensation methods will reduce the loop gain at high frequencies and increases the input node capacitance. Unfortunately, this additional gain is only needed near the corner frequency and thus the non-linear current is reduced only slightly or not at all. Therefore, reducing the nonlinear current by decreasing the input capacitance or input impedance is probably possible only in BiCMOS technology, when very large MOS current-mirror transistors are buffered by a wide band bipolar amplifier.

Nonlinear current cancellation

There remains the non-linear current cancelling method. At high frequencies the current flowing through the gate-source capacitance in MOS-transistor circuits is a simi-

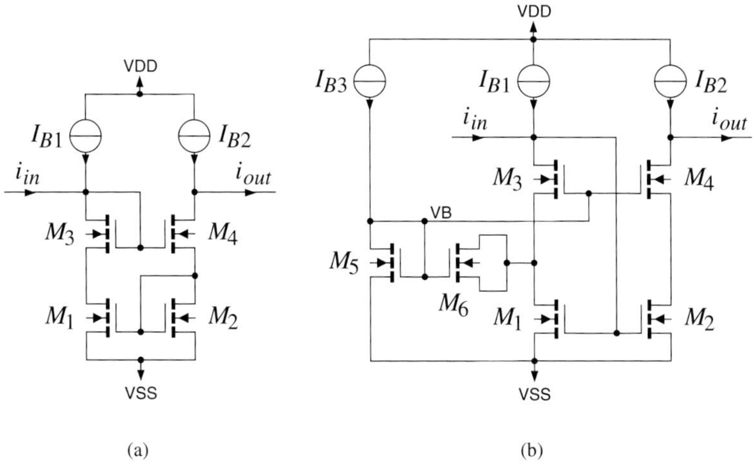

Figure 2.10 Cascode current-mirror topologies which cancel out the high frequency distortion due to the input capacitance. (a) Enhanced Wilson current-mirror. (b) Distortion compensation with an additional capacitance [8].

lar nonideality as the base current in bipolar circuits. This nonideality is decreased in bipolar integrated circuits, for example by the Wilson current-mirror. So the MOS version of the Wilson current mirror and the enhanced Wilson current mirror [2] in Figure 2.10a have several times lower high frequency distortion than the mirror topologies of Figure 2.2. This replica current method can also be utilised in the low voltage cascode current mirror by adding a MOS-capacitor to the drain of the input mirror transistor M_1 as shown in Figure 2.10b [8]. Unfortunately, such non-linear current cancellation is difficult to realise in practice while the input capacitance depends on the interconnection capacitances and on the parasitic capacitances of the signal source or driving amplifier stage.

In Figure 2.11 the two current-mirror topologies using the non-linear current cancellation are compared to the normal low-voltage cascode current mirror using cascode transistors with the aspect ratio of $100\mu/2\mu$. The device sizes of both mirror topologies are adjusted for minimum distortion level at a frequency of 2.5 MHz. The aspect ratios of the main mirror transistors M_1 and M_2 are $100\mu/5\mu$ in all mirror topologies.

The optimum distortion performance for the enhanced Wilson-mirror topology of Figure 2.10a was found with an aspect ratio of $100\mu/3.2\mu$ for the cascode transistors M_3 and M_4. Larger aspect ratio is needed to compensate for the reduction of the cascode transistor transconductance due to the bulk effect. The additional MOS-capacitor M_6 in the low voltage current-mirror has its optimal value at an aspect ratio of $190\mu/5\mu$ with $100\mu/2\mu$ cascode transistors. If cascode transistors have the same dimension as

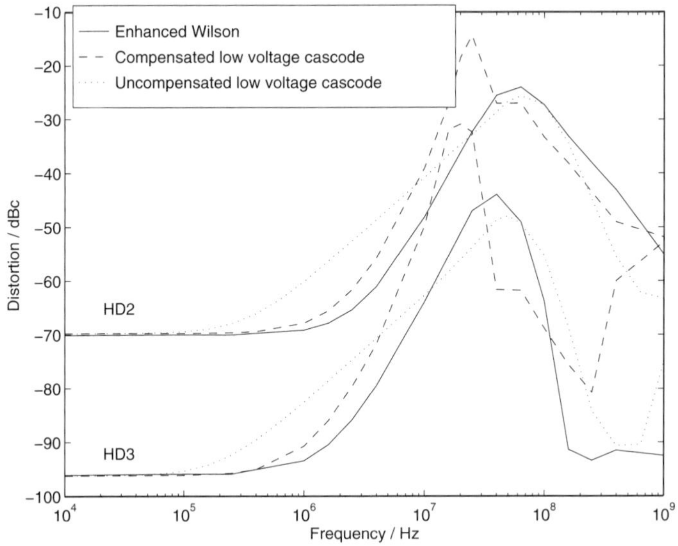

Figure 2.11 Simulated distortion (modulation index $m = 0.2$) of an enhanced Wilson current-mirror and a distortion compensated low voltage cascode current mirror compared against the earlier presented uncompensated low voltage cascode current mirror.

the mirror transistors ($100\mu/5\mu$), also the MOS-capacitor should preferably have the same dimensions, but because of the bulk effect, the minimum distortion is reached with an aspect ratio of $140\mu/5\mu$.

The comparison of the simulated distortions shows that the enhanced Wilson current-mirror has significantly better distortion performance below ω_0. At higher frequencies, this topology has similar distortion performance to the uncompensated low voltage current-mirror. The distortion compensated low voltage current-mirror does not reach as low distortion levels as the enhanced Wilson current-mirror, even at lower frequencies, and just below ω_0 it has a very strong distortion peak. Therefore, this current-mirror topology merely shapes the distortion rather than reducing the total amount of distortion.

The optimal non-linear current cancellation depends on the bulk effect and thus the optimal device aspect ratios depend on the cascode bias voltage and on the integration process used. Furthermore, the input capacitance strongly depends on the stray capacitances of other circuits connected to the current-mirror. Therefore, these non-linear current cancellation methods cannot be used in general purpose applications where the same current-mirror cell is used for many different purposes. But, if we want to minimise the distortion arising from the threshold voltage mismatch by using very large current-mirror transistors, it would be useful to decrease the distortion arising from the input capacitance by means of such non-linear current cancellation methods.

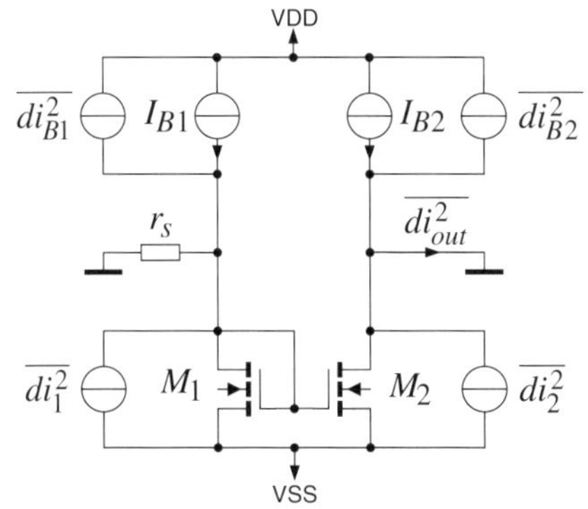

Figure 2.12 Noise in a simple MOS current-mirror.

2.1.5 Noise and dynamic range

The equivalent circuit for current-mirror noise behaviour is presented in Figure 2.12. As current-mirrors have limited current gain, it is not practical to reduce all noise to the input and thus the output current noise $\overline{di_{out}^2}$ is preferred in the calculations. The output current noise is

$$\overline{di_{out}^2} = A_i^2 \left(\overline{di_1^2} + \overline{di_{B1}^2} + \frac{4kT}{r_s} df \right) + \overline{di_2^2} + \overline{di_{B2}^2}, \tag{2.31}$$

where $A_i \approx -\frac{W_2 L_1}{L_2 W_1}$. As the current gain is moderate, the output noise sources cannot be neglected. Furthermore, the signal source impedance r_s is usually very high and thus its contribution to the noise is usually negligible.

The noise of a MOS-transistor, including both the thermal and the $1/f$ noise [2], is

$$\overline{di_d^2} = \frac{8kT}{3} g_m df + \frac{K_f}{C_{ox} W L} g_m^2 \frac{df}{f}, \tag{2.32}$$

where the coefficient K_f is the flicker noise coefficient. The current sources are typically realised with low transconductance PMOS transistors and therefore the noise of the current sources, $\overline{di_{B1}^2}$ and $\overline{di_{B2}^2}$, have little effect on the output noise. Furthermore, the $1/f$ noise normally makes only a minor contribution to the total wide bandwidth noise and therefore we approximate the output current noise as

$$\overline{di_{out}^2} \approx (A_i + 1) \frac{8kT}{3} g_{m1} df. \tag{2.33}$$

As the transconductance g_{m1} is

$$g_{m1} = \sqrt{2 \mu_0 C_{ox} \frac{W_1}{L_1} I_{B1}} = \frac{2 I_{B1}}{V_{GS1} - V_T}, \tag{2.34}$$

lowering the noise level means using small aspect ratios and high gate-source voltages. Furthermore, the dynamic range is maximised by using as high bias currents as possible. This, unfortunately, increases circuit area or supply voltage (or both) yet this applies regardless for all electronic circuits. Fortunately, at low frequencies, increasing the mirror gate area decreases both the $1/f$-noise and the distortion due to V_T mismatch.

If we assume one pole transfer function for the current mirror as in equation (2.19), its noise bandwidth is

$$\frac{\pi}{2} f_0 \approx \frac{g_{m1}}{4C_{IN}}, \tag{2.35}$$

and thus the total integrated output noise is

$$\begin{aligned} \overline{i_{out}} &= \sqrt{\int_0^{\infty} \overline{di_{out}^2}\, |H(f)|^2\, df} \\ &\approx g_{m1}\sqrt{(A_i+1)\frac{2kT}{3C_{IN}}}. \end{aligned} \tag{2.36}$$

If signal frequencies are assumed to be at least two decades lower than the mirror corner frequency, so that the maximum signal amplitude can be assumed equal to the bias current, I_{B1}. Furthermore, if the total input capacitance C_{IN} and the mirror input transconductance g_{m1} are expressed as a function of device dimensions and bias current with Equations (2.17) and (2.34), the dynamic range of the simple current-mirror is expressed as

$$DR \approx L_1\sqrt{\frac{I_{B1}}{2\mu_0 kT}}. \tag{2.37}$$

However, if the bandwidth of interest is limited so that the noise bandwidth does not increase with transconductance, the dynamic range increases with bias current more rapidly than the equation shows.

The cascode transistors also contribute to the current mirror output noise but with a different mechanism. The small-signal equivalent circuit of Figure 2.13 is used to calculate how the noise of the output cascode transistor $\overline{di_4^2}$ compares with the noise of the output mirror transistor $\overline{di_2^2}$. As a consequence, other noise sources are neglected in these calculations. The total output noise generated in the output cascode and mirror transistor equals

$$\overline{di_{out}^2} = \overline{di_2^2} + \overline{di_4^2}\left(\frac{g_{ds2}+sC_s}{g_{m4}+g_{ds4}+g_{ds2}+sC_s}\right)^2, \tag{2.38}$$

where $C_s = C_{gs4} + C_{s4} + C_{d2}$ represents the total capacitance at the cascode transistor

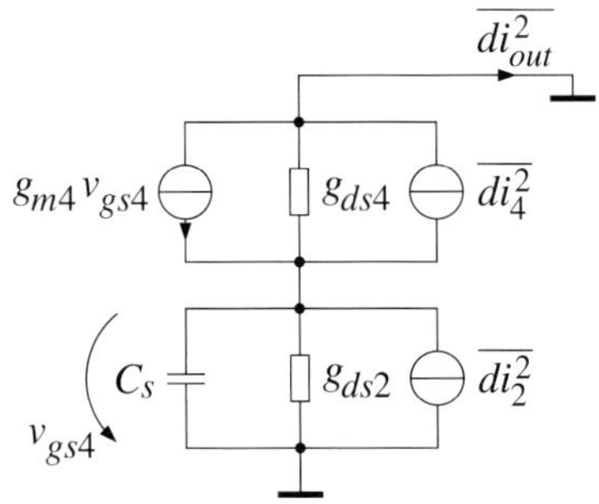

Figure 2.13 The small-signal equivalent circuit of the noise generated at the output of a cascode current-mirror.

source node. At low frequencies, the noise of the cascode transistor is attenuated by approximately $(g_{ds2}/g_{m4})^2$ and thus this noise source can be neglected. At high frequencies, the cascode transistor noise increases as a result of the parasitic capacitances at the cascode transistor source. However, then this noise is usually attenuated elsewhere in the signal path and thus the noise and dynamic range calculated for the simple current-mirror do not change significantly for the cascode current-mirror topologies.

2.1.6 Other mirror topologies

Accurate current-mirror topologies for large signal amplitudes

Occasionally, for example in the push-pull current-conveyors discussed in the next chapter, the current-mirrors must mirror very large current peaks compared to the bias current. Because of the large current variation, the channel length modulation has a considerable effect on the mirror accuracy and therefore cascode mirror topologies sensitive to channel length modulation should be avoided. Similarly, with large currents, the basic cascode current-mirror may require too much voltage to operate with low supply voltages.

The low-voltage cascode current-mirror earlier discussed has unfortunately one drawback: the input transistors are difficult to keep in saturation with large current signal amplitudes because the mirror gate voltage v_{GS} varies with input current and the cascode transistor M_3 may fall into the triode region at negative current signal peaks if the cascode bias voltage V_B is too high. Similarly, the mirror transistors M_1 and M_2 can fall into the triode region at positive current signal peaks if the cascode bias voltage V_B is too low.

The maximum current swing can be extended by using modified transistor-biasing schemes [14] such as the cascode current-mirror presented in Figure 2.14a. In this current-mirror, the cascode transistors are biased from the mirror input voltage with an

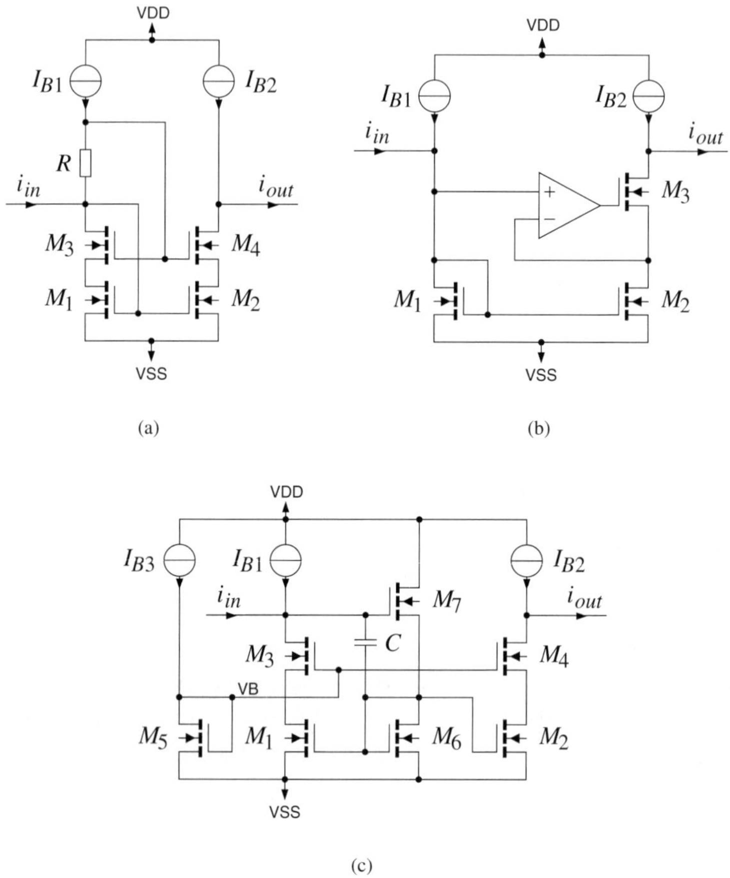

Figure 2.14 Other cascode current-mirror topologies insensitive to channel length modulation. (a) A low-voltage cascode current-mirror with self-adjusting cascode bias [14]. (b) Cascode current-mirror with large current handling capability [16]. (c) A cascode current-mirror with a source-follower input level shifter [15].

additional level shifter realised by a resistor. The nonlinearity of the resistor does not degrade the distortion performance of this current-mirror. Therefore, the resistor can be realised with a triode region MOS-transistor or a n-well resistor. However, the parasitic capacitances involved in such integrated resistors add to the input capacitance and thus reduce the bandwidth of the mirror. Unfortunately, quite large resistance values are needed and thus the parasitic capacitance can similarly be large. Moreover, these parasitic capacitances are voltage dependent and therefore high frequency distortion may increase.

An alternative method to ensure saturation region operation with large currents is to use a level shifter to increase the mirror input voltage. The mirror topology presented in Figure 2.14c, with a Wilson-mirror like input structure does just that [15], where the level shifter is realised with a source-follower with a diode load (transistors M_6 and M_7). This amplifier stage additionally isolates the large gate-source capacitances of transistors M_1, M_2 and M_6 from the input node. This would lead to reduced input capacitance and hence to lowered high frequency distortion unless the resulting feedback loop does not cause problems of stability. To avoid any peaking in the current transfer function, the feedback loop must be stabilised with an additional capacitor in parallel with the gate-source capacitance of the source-follower transistor M_7. This increases the input capacitance again and if the transistors M_1, M_2 and M_6 have equal dimensions, the input capacitance is increased 50% compared to the simple current-mirror. Therefore the frequency response and distortion of this Wilson-input mirror resembles that of the normal low-voltage cascode current-mirror, which has a 50% larger input capacitance.

In the case of very large currents, we can no longer use a cascode transistor in the drain of the input mirror transistor M_1. The only way to render the mirror insensitive to channel length modulation is to force the drain voltage of the output mirror transistor M_2 to follow the input voltage. This action can be realised with an additional voltage amplifier as depicted in Figure 2.14b [16]. Typically, the input capacitance of a differential voltage amplifier is quite large and this combines with the mirror input capacitance and thus the high frequency distortion is again increased.

Resistively compensated mirror

As in most cases, the current-mirror can be assumed to be a first-order low-pass system and thus it is an underdamped system. For this reason, its settling behaviour is not optimal. By using the low voltage cascode current-mirror of Figure 2.2c, a second-order low-pass system will result and the settling behaviour of the current-mirror can be tuned by the aspect ratio of the cascode transistors.

However, there is also a different method to enhance the settling behaviour, one which also extends the current-mirror bandwidth. This method inserts a resistor R between the current mirror input node and the gate of the mirror input transistor M_1 as presented in Figure 2.15 [17]. In this case, the gate-source capacitance C_{gs1} of the mirror input transistor M_1 is isolated from the rest of the total input capacitance $C'_{in} = C_{in} - C_{gs1}$ and a following second-order current transfer function results:

$$A_i(s) = \frac{g_{m2}}{g_{m1}} \frac{1 + sRC'_{in}}{1 + s\frac{C_{gs1}+C'_{in}}{g_{m1}} + s^2 \frac{RC_{gs1}C'_{in}}{g_{m1}}}. \tag{2.39}$$

When $R = \frac{1}{g_{m1}}$ the zero cancels one of the poles and a first-order low-pass function results:

$$A_i(s) = \frac{g_{m2}}{sC'_{in} + g_{m1}}. \tag{2.40}$$

Then the resulting pole frequency is at

$$\omega_0 = \frac{g_{m1}}{C'_{in}}, \tag{2.41}$$

which is approximately twice the pole frequency of the uncompensated current-mirror (i.e. $R = 0$) with a unity gain current-mirror. Unfortunately, the increase of the bandwidth decreases with current gain. This resistive compensation method also works with cascode current-mirrors, where the transfer function is a third-order low pass function with one zero. However, as stated earlier, in most cases the poles caused by the cascode transistors can be shifted to higher frequencies so that the cascode current-mirror can be considered a first-order low-pass system.

The resistor R is usually implemented with a NMOS transistor operating in triode region as other integrated resistors have large variations that do not correlate with the transconductance of the mirror NMOS transistors. Nevertheless, the NMOS resistor requires additional circuitry that tunes the resistance value with a control voltage and

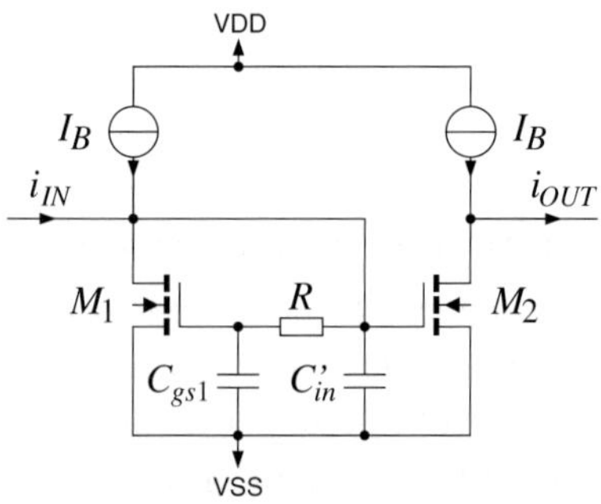

Figure 2.15 Resistive compensation technique enhancing current mirror bandwidth and settling time [17].

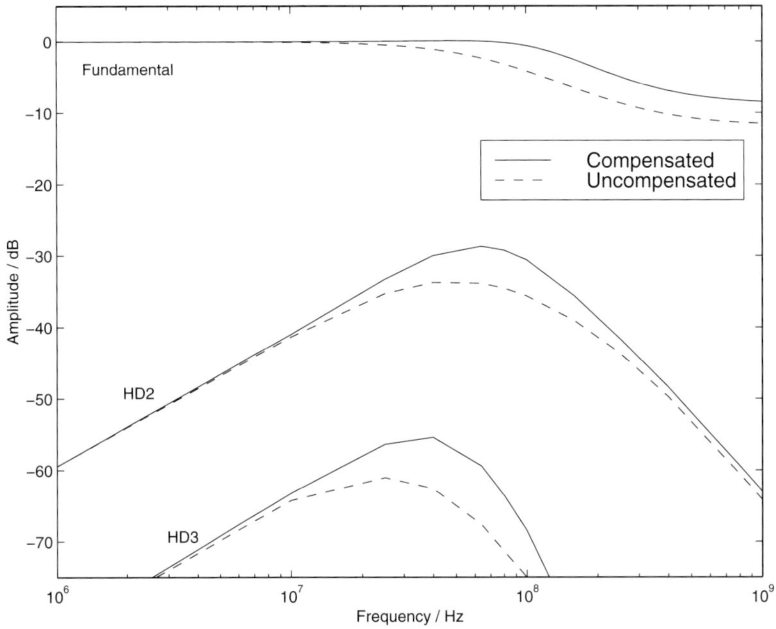

Figure 2.16 The frequency response and harmonic distortion of the resistively compensated current-mirror compared to the uncompensated mirror.

tracks temperature and process variations [18].

The tuning circuit must be relatively accurate because an unmatched pole-zero doublet degrades the settling behaviour of the current-mirror in the way documented earlier in the case of operational amplifiers [19]. However, the poles and zeros in an operational amplifier are normally almost signal independent because the currents inside an operational amplifier vary significantly only when the amplifier is slewing or when it must supply exceptionally large currents to the load. Unfortunately, this is not the case with current-mirrors with resistive compensation: the transconductances of the mirror transistors are signal dependent and it increases with the input current. The resistance of the NMOS compensation resistor similarly increases with the input current. Therefore, the lowest pole and the zero move away from each other with large input signals, and settling times are significantly increased.

As described earlier, the settling times can also be optimised by adjusting the aspect ratios of the cascode transistors in the low voltage cascode current-mirror. In this case, both the dominant and nondominant poles track each other with large current signals. Thus, in many cases, the resistive compensation method does not significantly improve the high frequency performance of a current-mirror, at least with most cascode current-mirror topologies and with large signal amplitudes.

The simulated frequency response and the harmonic distortion of the compensated current-mirror compared to the uncompensated mirror are presented in Figure 2.16. The current-mirror topology used is the simple current-mirror, i.e. no cascode transistors were used. Furthermore, the same aspect ratios (100/5) are used in mirror transistors as well as the same modulation index (0.2) and bias currents (100μA) as in earlier examples. The compensation resistor is not a MOS-resistor but an ideal linear resistor instead. The resistance is set to the inverse of the input transistor transconductance, which is approximately $1.98\text{k}\Omega$.

The results of the simulation reveal that, while the bandwidth of the current-mirror is doubled, the harmonic distortion peak location remains almost constant. Consequently, the distortion increases significantly compared to an uncompensated current-mirror with the same corner frequency as the compensated one. Moreover, the maximum distortion is a few decibels higher than in the compensated one. Therefore this compensation technique is not very suitable for low distortion applications.

2.2 Current buffer

In the case of current-mirrors, we can realise inverting amplifiers with arbitrary current gain. Noninverting current amplifiers with arbitrary gain are, however, more difficult to implement. If we want arbitrary current gain, we can cascade two current-mirrors to get positive current gain. But if unity gain is sufficient, we can realise noninverting current buffers with common-gate MOS-transistor amplifier stages. As the voltage buffers have a significant role in voltage-mode circuits, the current buffers have an equally significant role in current-mode circuits.

Two simple CMOS implementations of a noninverting unity gain current amplifier are presented in Figure 2.17. The amplifier in Figure 2.17a is a simple NMOS common-gate amplifier stage. In order to keep the transistor M_1 in saturation the output DC-voltage level must be at least

$$V_{OUT}(min) = V_{DSAT1} + V_{IN} = V_{B1} - V_{T1}. \tag{2.42}$$

In certain applications, it is impossible to maintain these bias conditions and therefore we need a voltage level shifter either at the input or at the output. A typical realisation is to use a diode-connected transistor at the input as a voltage level shifter as presented in Figure 2.17b. This current buffer resembles a simple current-mirror, except that the mirror transistor sources are connected to a high impedance node (current-source) rather than a low impedance node (negative supply voltage).

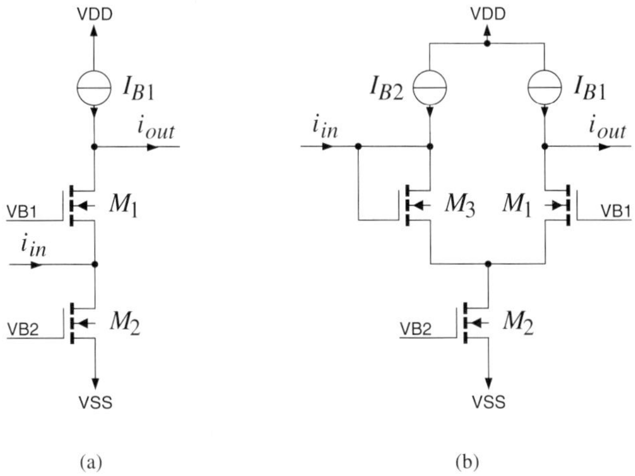

Figure 2.17 Simple CMOS noninverting unity gain current amplifiers. (a) A simple common-gate amplifier stage. (b) A common-gate amplifier stage with an input level shifter.

2.2.1 Linear nonidealities

The NMOS common-gate amplifier stage of Figure 2.17a produces an input impedance

$$Z_{in} = \frac{g_o + g_{ds1}}{g_o\left(g_{m1} + g_{ds1} + g_{ds2}\right) + g_{ds1}g_{ds2}}, \tag{2.43}$$

where g_o is the sum of the output conductance of current-source load I_{B1} and the load conductance. Usually, the current buffer drives a low impedance load because $g_m \gg g_{ds}$ the input impedance reduces to

$$Z_{in} \approx \frac{1}{g_{m1}}. \tag{2.44}$$

and thus the DC current gain of the amplifier stage is approximately

$$A_i \approx 1 - \frac{g_{ds2}}{g_{m1}}.$$

In order to increase the DC-accuracy the current source transistor M_2 should be a long channel device.

In many applications, however, such an amplifier stage is used to drive a high impedance load, typically a MOS-transistor gate, in which case $g_o \ll g_{ds1}$ and the input impedance is approximately

$$Z_{in} \approx \frac{1}{g_{ds2} + g_{m1}\frac{g_o}{g_{ds1}}}, \tag{2.45}$$

and the amplifier stage provides a DC transimpedance gain approximately

$$R_m \approx \frac{1}{g_o + g_{ds2}\frac{g_{ds1}}{g_{m1}}}. \tag{2.46}$$

This increased input impedance does not affect the high frequency behaviour as the parasitic capacitances start to reduce the high output impedance at relative low frequencies and thus the current buffer has a one-pole transfer function with a pole frequency of $\omega_0 = \frac{g_{m1}}{C_{in}}$ regardless of the load impedance.

The operation of the level-shifted current buffer of Figure 2.17b is similar to the simple common-gate amplifier stage, where the input impedance is higher because of the added series resistance of $1/(g_{m3} + g_{ds3})$. When comparing this current-buffer to the simple current-mirror with comparable device sizes the dominant poles ω_0 of both amplifiers are at the same frequency. While the input impedance of the level shifted current-buffer is typically half of the current-mirror input impedance, the total input capacitance C_{IN} is similarly halved.

2.2.2 Nonlinearity

At low frequencies the drain conductance of the current source transistor M_2 converts the non-linear input voltage to a non-linear current and thus the low frequency second order harmonic distortion is

$$HD2 = \frac{1}{8} m \frac{g_{ds2}}{g_{m1}}, \tag{2.47}$$

and the third order harmonic distortion is consequently

$$HD3 = \frac{1}{32} m^2 \frac{g_{ds2}}{g_{m1}}, \tag{2.48}$$

where m is the modulation index. At high frequencies the input capacitance C_{IN} converts the non-linear input voltage to a non-linear current as in the current-mirror and therefore the distortion equations (2.24), (2.25), and (2.28) also predict the high frequency distortion of the current-buffer. Regardless, as this amplifier is not sensitive to device mismatch, the main buffer transistor M_1 can be a minimum channel length device and thus the distortion due to the input capacitance is not as big a problem as with the current-mirror. However, as the linear gate-source capacitance is now significantly smaller, the non-linear drain and source diode capacitances make a larger contribution to the high frequency distortion than in the current-mirror.

If we compare a simple current buffer (i.e. a common-gate NMOS amplifier stage) to a simple voltage buffer (i.e. a common-drain NMOS amplifier stage), and we assume that no floating well devices are used, the operation of the voltage buffer is much more sensitive to the bulk effect. With floating well devices or with more complex

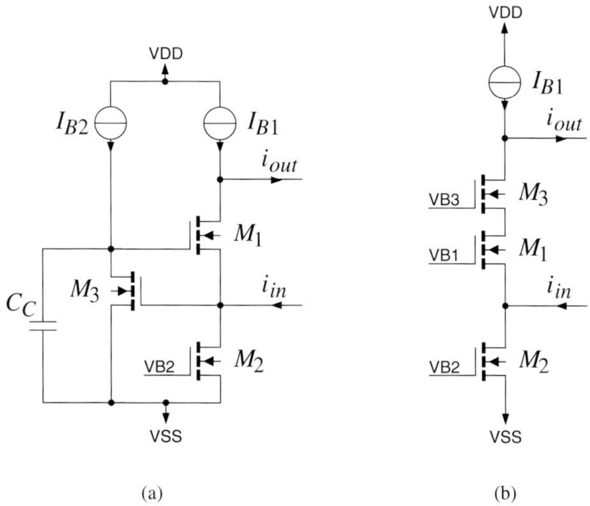

Figure 2.18 Alternative current buffer topologies. (a) Decreased input impedance by a regulated cascode circuit [11]. (b) Decreased sensitivity to output impedance by an additional cascode transistor.

circuit topologies, the nonlinearity caused by the bulk effect can be minimised but this leads to reduced amplifier bandwidth.

2.2.3 Noise

At low frequencies the noise in current buffers is mainly arising from the current-source transistor M_2 and the current-source load as the noise calculations on page 32 for the cascode current-mirror have earlier shown. The cascode transistor source node is typically an internal node which has significantly less parasitic capacitance than the current buffer input node and therefore the high frequency noise arising from the main buffer transistor M_1 is higher. If the current buffer is used as a voltage-to-current converter the impedance level at the input is quite low and the noise caused by the main buffer transistor increases.

2.2.4 Alternative topologies

The performance the current buffer can be enhanced by a modification to the circuit topology. The input impedance can be reduced by adding an amplifier stage to keep the input voltage variation as low as possible. This is easily realised by the regulated cascode circuit of Figure 2.18a [11]. In this circuit the input voltage variation is amplified by the transistor M_3 and fed to the gate of the transistor M_1 and thus the transconductance of the transistor M_1 is virtually multiplied by the $\frac{g_m}{g_{ds}}$ ratio of the transistor

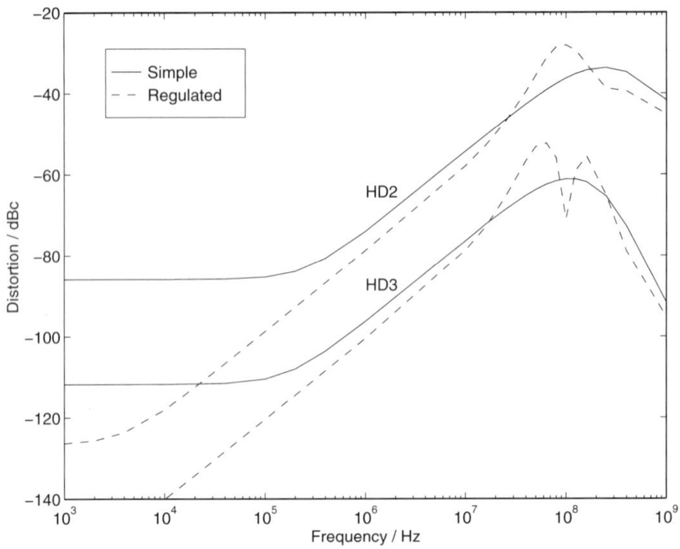

Figure 2.19 Simulated distortion of a simple current buffer of Figure 2.17a and a regulated current buffer of Figure 2.18a. The modulation index is 0.2 in both cases.

M_3. This lowers the low frequency input impedance, which is however dependent on the load impedance, albeit with reduced sensitivity. This dependency is effectively removed by adding a cascode transistor to the drain of the transistor M_1 as presented in Figure 2.18b. Unfortunately, the addition of this cascode transistor increases high frequency distortion as there are now two nodes where parasitic capacitances convert non-linear voltages to non-linear currents. On the other hand, the regulated cascode circuit effectively reduces distortion as it reduces the input voltage variation more than it adds to the input capacitance. As the regulated cascode feedback loop gain decreases at higher frequencies, the improvement in distortion performance gradually decreases. Furthermore, the feedback loop may need additional compensation which is realised with the capacitor C_C.

In Figure 2.19, the simulated distortion of a simple current buffer of Figure 2.17a is compared to the regulated current buffer of Figure 2.18a. In both circuits, the aspect ratio for the main buffer transistors M_1 is $100\mu/2\mu$ and for the current source transistor M_2 is $100\mu/5\mu$, the bias current I_{B1} is 100μA, and the modulation index is 0.2. In the regulated cascode buffer, the transistor M_3 is five times smaller than the main buffer transistor so that the aspect ratio is $20\mu/2\mu$ and accordingly the bias current I_{B2} is 20μA. These simulations reveal that the regulated cascode circuit reduces the distortion efficiently at low frequencies and, as the signal frequency increases, it still has 4-5 dB less distortion than the simple current buffer. Near the corner frequency, there is slight peaking in the frequency response of the regulated current buffer and hence the distortion peak at the same frequency range.

The noise performance of these two current buffer topologies does not differ greatly from the simple current buffer topology. As long as the $\frac{g_m}{g_{gs}}$ ratio of the transistor M_3 is high in the regulated cascode circuit the main buffer transistor M_1 has almost no effect on the total output noise [11]. Moreover, the additional cascode transistor in the current buffer of Figure 2.18b does not contribute to the total noise.

References

[1] P. Allen, D. Holberg, *CMOS Analog Circuits Design*, New York, Holt, Rinehart and Winston, Inc, 1987, 701 p.

[2] P. Gray, R. Meyer, *Analysis and Design of Analog Integrated Circuits*, second edition, New York, John Wiley & Sons, 1984, 771 p.

[3] K. Laker, W. Sansen, *Design of Analog Integrated Circuits and Systems*, New York, McGraw-Hill, Inc, 1994,898 p.

[4] J. Babanezhad, R. Gregorian, "A programmable gain/loss circuit," *IEEE J. Solid State Circuits*, vol. 22, pp. 1082-1089, Dec. 1987.

[5] P. Crawley, G. Roberts, "High-swing MOS current mirror with arbitrarily high output resistance," *Electronics Letters*, vol. 28, pp. 361-362, Feb. 1992.

[6] D. Haigh, C. Losala, A. Parker, D. Webster, "Systemic approach for the development and design of analogue communication circuits." In: H. Dedieu (ed), *EC-CTD'93 - Selected Topics in Circuits and Systems*, Elsevier Science Publishers B.V., 1993, pp. 3-74 (Part I).

[7] D. Webster, D. Haigh, G. Passiopoulos, A. Parker, "Distortion in short channel FET circuits," In: G. Machado (ed), *Low-power HF microelectronic: a unified approach*, IEE, 1996, pp.929-958 (Chapter 24).

[8] R. Balmford, W. Redman-White, "A new CMOS current mirror circuit with reduced distortion for high bandwidth applications," *European Conference on Circuits Design and Theory* (ECCTD'93), Davos, Switzerland, 1993, pp. 923-927.

[9] V. Prodanov, M. Green, "CMOS current mirrors with reduced input and output voltage requirements" *Electronics Letters*, vol. 32, pp. 104-105, Jan. 1996.

[10] J. Mulder, A. van der Woerd, W. Serdin, A. van Roermund, "High-swing cascode MOS current mirror," *Electronics Letters*, vol. 32, pp. 1251-1252, July 1996.

[11] E. Säckinger, W. Guggenbuhl, "A high-swing, high-impedance MOS cascode circuit," *IEEE J. Solid State Circuits*, vol. 25, pp. 289-298, Feb. 1990.

[12] T. Serrano, B. Linares-Barranco, "The active-input regulated-cascode current mirror," *IEEE Trans. Circuits and Systems-I*, vol. CAS-41, pp. 464-467, June 1994

[13] A. Zeki, H. Kuntman, "Accurate and high output impedance current mirror suitable for CMOS output stages," *Electronics Letters*, vol. 33, pp. 1042-1043, June 1997.

[14] E. Bruun, P. Shah, "Dynamic range of low-voltage cascode current mirrors," in *Proc. IEEE Int. Symposium on Circuits and Systems* (ISCAS'95), Seattle, USA, 1995, pp. 1328-1331.

[15] M. Steyaert, "An improved CMOS current mirror using a bulk-effect independent cascode biasing structure," *European Conference on Circuits Design and Theory* (ECCTD'89), 1989, pp. 656-660.

[16] G. Palmisano, G. Palumbo, S. Pennisi, "Harmonic distortion on class AB CMOS current output stages," *IEEE Trans. Circuits and Systems-II*, vol. CAS-45, pp. 243-250, Feb. 1998.

[17] T. Voo, C. Toumazou, "High-speed current mirror resistive compensation technique," *Electronics Letters*, vol. 31, pp. 248-250, Feb. 1995.

[18] T. Voo, C. Toumazou, "Precision temperature stabilized tunable CMOS current-mirror for filter applications," *Electronics Letters*," vol. 32, pp. 105-106, Jan. 1996.

[19] B. Kamath, R. Meyer, P. Gray, "Relationship between frequency response and settling time of operational amplifiers," *IEEE J. Solid-State Circuits*, vol. SC-9, pp. 347-352, Dec. 1974.

Chapter 3

Open-loop current amplifiers

In this chapter, the different open-loop current-mode amplifiers are discussed. By 'open-loop amplifiers', amplifiers with low and fixed current gain are referred to. This gain is set without any external feedback circuitry between the input and the output. Although no global feedback is used there nevertheless may be local feedback loops inside the amplifier itself. Despite the low and fixed current gain of these amplifiers it is possible to realise different types of amplification functions with a wide gain range. This gain is set directly by the transistor aspect ratios of the amplifier or alternatively by controlling the impedance levels at the input and output of the amplifier and thus realising either voltage-to-current or current-to-voltage conversions. In addition, most general purpose signal processing applications can be realized with these amplifiers. The amplifiers discussed in this chapter additionally have well set input DC voltages unlike the basic current amplifiers discussed in the previous chapter. Furthermore, because of the open-loop operation, amplifier nonlinearity has a greater effect on the signal processing performance and therefore different distortion mechanisms are discussed in detail.

3.1 First generation current-conveyor CCI

The concept of the current-conveyor was first presented in 1968 [1] and further developed to a second-generation current-conveyor in 1970 [2]. The current-conveyor is intended as a general building block as with the operational amplifier. Because of the operational amplifier concept has been current since the late 1940's, it is difficult to get any other similar concept widely accepted. However, operational amplifiers do not perform well in applications where a current output signal is needed and consequently there is an application field for current-conveyor circuits. Since current-conveyors operate without any global feedback, a different high frequency behaviour compared to operational amplifier circuits results.

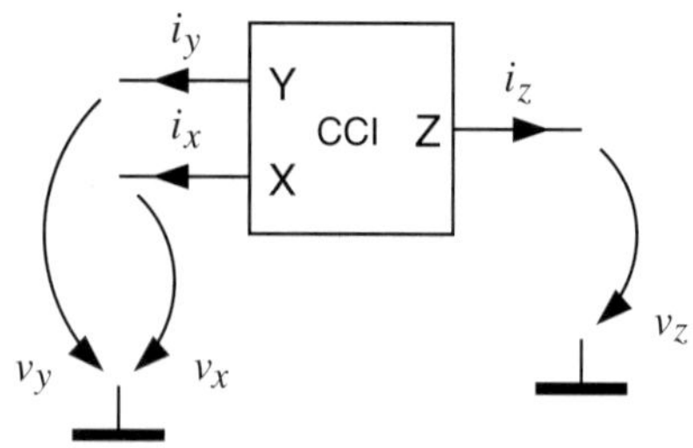

Figure 3.1 The first generation current-conveyor symbol and its signal definitions.

Current-conveyors are three-port networks with terminals X, Y and Z as represented in Figure 3.1. The network of the first generation current-conveyor CCI has been formulated in a matrix form as follows

$$\begin{bmatrix} i_y \\ v_x \\ i_z \end{bmatrix} = \begin{bmatrix} 0 & 1 & 0 \\ 1 & 0 & 0 \\ 0 & 1 & 0 \end{bmatrix} \begin{bmatrix} v_y \\ i_x \\ v_z \end{bmatrix}. \tag{3.1}$$

In other words, the first generation current conveyor CCI forces both the currents and the voltages in ports X and Y to be equal and a replica of the currents is mirrored (or conveyed) to the output port Z.

Figure 3.2a presents a simple MOS implementation of the first generation current-conveyor CCI[1]. In this circuit, the NMOS transistors M_1 and M_2 form a current mirror that forces the drain currents of the PMOS transistors M_3 and M_4 to be equal and hence the voltages at the terminals X and Y are forced to be identical.

3.1.1 Linear nonidealities

The two current-mirrors in the CCI form a feedback loop so that the currents and voltages in the terminals X and Y follow each other quite accurately, even without cascodes. If the terminal Y is grounded, the low-frequency impedance at the terminal X is expressed as:

$$Z_X(0)\Bigg|_{Z_Y=0} \approx \frac{1}{g_{m4}} - \frac{g_{m1}}{g_{m2}g_{m3}} + \frac{g_{ds1}+g_{ds3}}{g_{m3}g_{m4}} + \frac{g_{ds2}+g_{ds4}}{g_{m2}g_{m4}} \tag{3.2}$$

At first glance, it may be noticed in this equation that the impedance may become negative with certain mismatch conditions. However, as the transistors M_1, M_2, M_3 and M_4 form a loop that tries to balance itself the impedance in the X-terminal remains pos-

[1]The first current-conveyors were implemented using bipolar transistors but, because the goal in this paper is to compare a variety of circuit topologies, the corresponding integration technology is assumed for all discussed circuits.

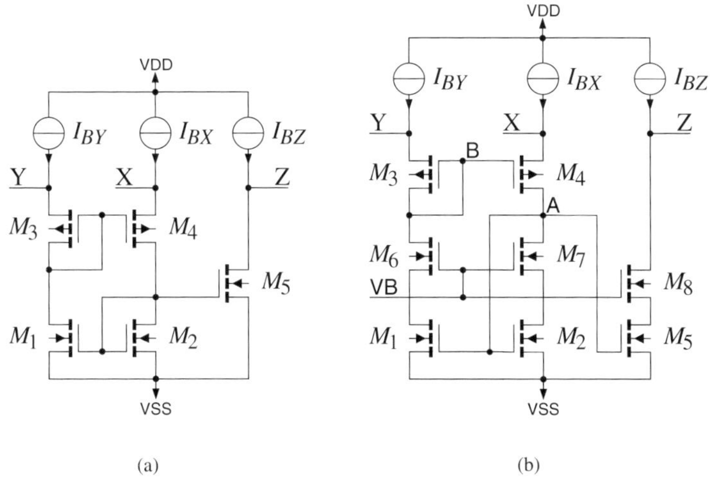

Figure 3.2 (a) A simple class-A implementation of the current conveyor CCI. (b) A more accurate class-A CCI with a cascode NMOS mirror.

itive even in the case of considerable transistor mismatch. Therefore, this circuit can readily provide an impedance in the range of ten ohms, which can be further reduced if cascode current-mirrors are used.

For the current-conveyor of Figure 3.2b, the errors because of the NMOS cascode current-mirror are minimal compared to the errors because of the transistors M_3 and M_4. Therefore, the drain currents of transistors M_3 and M_4 can be assumed equal but, as the drain-source voltages of such transistors are different, there is a systematic offset voltage between X and Y terminals, which is approximated as

$$V_{OFF} \approx \lambda V_{DSAT}\left(V_{GS3} - V_{DS4}\right). \tag{3.3}$$

These results reveal that, in order to maintain adequate DC-accuracy, the input transistors M_3 and M_4 should be long channel devices with a low transconductance.

Similarly, the calculations can be simplified by assuming the NMOS mirror transconductances equal $g_{m1} = g_{m2} = g_{mn}$ and by neglecting the NMOS drain-source conductances when calculating the input impedance as a function of frequency:

$$Z_X(s)\Big|_{Z_Y=0} \approx \frac{g_{mn}\left(g_{m3} - g_{m4} + g_{ds3}\right) + g_{m3}g_{ds4} + s g_{m3}C_A}{g_{mn}g_{m3}g_{m4} + s g_{m3}g_{m4}C_A + s^2 g_{m3}C_A C_X}. \tag{3.4}$$

The capacitances C_A and C_X in the equation are the total capacitances in nodes A and X, respectively. Nevertheless, the capacitance C_B at the node B is neglected, the explanation following a few paragraphs later.

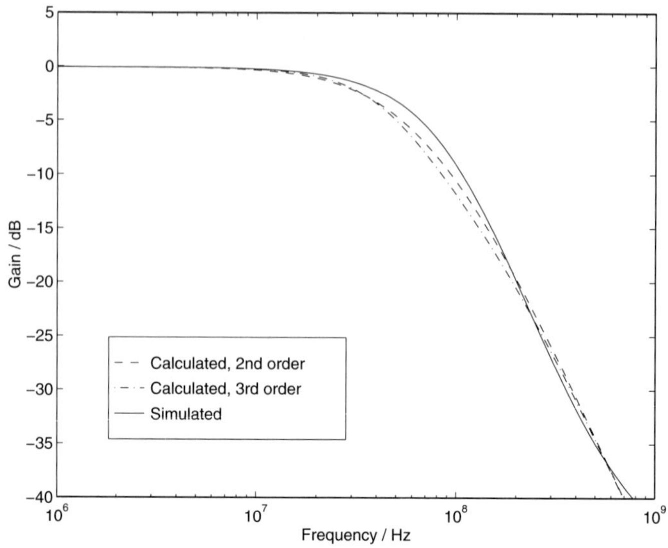

Figure 3.3 The simulated and calculated current transfer functions of the CCI of Figure 3.2b.

The current transfer function from X to Z can be calculated with identical assumptions to Equation (3.4)

$$A_i(s) \approx \frac{1}{1 + s\frac{C_A}{g_{mn}} + s^2 \frac{C_A C_X}{g_{mn} g_{m4}}}, \tag{3.5}$$

and thus the resulting poles of the transfer function are

$$p_{1,2} = -\frac{g_{m4}}{2C_X} \pm \frac{1}{2}\sqrt{\frac{g_{m4}}{C_X}\left(\frac{g_{m4}}{C_X} - 4\frac{g_{mn}}{C_A}\right)}. \tag{3.6}$$

Therefore, the current-conveyor bandwidth depends greatly on the input transistor transconductance g_{m4} and the input capacitance C_X, whereas the NMOS transconductance g_{mn} and the capacitance C_A affect primarily the step response of the conveyor.

If the capacitance at the node B is included, the current transfer function becomes a third order function with a left half-plane zero

$$A_i(s) \approx \frac{1 + s\frac{C_B}{g_{m3}}}{1 + a_1 s + a_2 s^2 + a_3 s^3}, \tag{3.7}$$

where the coefficients of the denominator are

$$a_1 \approx \frac{C_A}{g_{mn}} + \frac{C_B}{g_{m3}} + C_X \frac{g_{m3} - g_{m4}}{g_{m3} g_{m4}}, \tag{3.8}$$

$$a_2 \approx \frac{C_B C_X}{g_{m3} g_{m4}} + \frac{C_A C_X}{g_{mn} g_{m4}} + \frac{C_A C_B}{g_{mn} g_{m3}}, \tag{3.9}$$

$$a_3 \approx \frac{C_A C_B C_X}{g_{mn} g_{m3} g_{m4}}. \tag{3.10}$$

This third order current gain equation (3.7) is plotted against the second order gain equation (3.5) and the simulated current gain in Figure 3.3. Since the second and third order transfer functions are almost identical, the left half-plane zero cancels efficiently the pole due to the capacitance at node B. Unfortunately, both these equations differ from the simulated current gain. There are several reasons for this variation. The calculated current-gain equations neglect the NMOS cascode transistor high-frequency behaviour. In addition, all parasitic capacitances in node B are assumed to be grounded yet the gate-source capacitance C_{gs4} represents quite a large capacitance between the input node X and the internal node B. The impedance level at the input is sensitive to device mismatch and as a consequence this impedance level is relatively unpredictable near the corner frequency.

3.1.2 Nonlinearity

At low frequencies, the four-transistor loop attenuates effectively all nonlinearities and only the threshold mismatch in the NMOS current-mirror and the channel length modulation of the output transistor M_5 produce distortion. The distortion deriving from the channel length modulation is significant only if the conveyor Z-output is driving a high impedance load, an effect that can be avoided by using a cascode NMOS current-mirror instead, which additionally increases the output impedance of the Z-terminal and decreases the input impedances of the X- and Y-terminals at the same time.

At high frequencies, the signal path from the X-input to the Z-output increasingly resembles a cascade of a current buffer (transistor M_4) and a current-mirror (transistors M_2 and M_5) and consequently the high frequency distortion is roughly doubled from what the distortion of a single current buffer or a current mirror would be.

3.1.3 Applications of the CCI

Because of this low impedance at the input terminal this circuit can be used as an accurate current amplifier. In addition, the DC-voltage level at the current input X can be easily set to a desired value by the voltage at the Y-terminal and input voltage-to-current conversion is easier than in the case of a current-mirror.

The first generation current-conveyor can be used as a negative impedance converter (NIC) [5], as if the Y-terminal is terminated with a grounded resistance R, the impedance at the terminal X equals

$$Z_X\Big|_{Z_Y=R} \approx -R. \tag{3.11}$$

With this negative impedance, for example, the input impedance of terminal X can be nulled with a small resistance R value [5] or the Q-value of passive inductances can be enhanced [6]. Unfortunately, the large input impedance variation arising from device mismatch makes it necessary to trim this small resistance value for each circuit. However, if the simulated negative resistance value is maintained high enough, it can additionally be used to construct amplifiers, filters and oscillators.

Although not often realised, a popular application of the current-conveyor is that involving current- and voltage-reference circuits. The core of a typical CMOS PTAT- or Bandgap-reference is a first generation current-conveyor CCI [4]. Since this application generally has no specific speed requirements, both the variation of input impedance and the input offset voltage are easily minimised by using large input transistors. More detailed examples of these circuits can be found in Chapters 6 and 7.

One recent application for a CCI involves linearisation of MOS transistor transconductance in transconductance-C filters [7]. Because the transconductance of a MOS transistor in triode region strongly depends on its drain-source voltage, a CCI can be used to force the drain voltage to a fixed potential and mirror the drain current to the high-impedance Z-output.

3.1.4 Push-pull CCI topologies

The current handling capability of the first generation current-conveyor can be extended by constructing a push-pull current-conveyor, as presented in Figure 3.4a [8]. The positive half of an AC-signal goes primarily through the lower conveyor constructed of transistors $M_1 ... M_5$, whereas in the case of the negative half of the signal goes primarily through the upper conveyor, constructed of transistors $M_6 ... M_{10}$.

The quiescent current of the push-pull CMOS CCI is not easy to derive as it depends on several different mechanisms. However, an approximation for the quiescent current of this push-pull conveyor is presented in [8]:

$$I_Q \approx \frac{I'_Q + I''_Q}{2}, \tag{3.12}$$

where

$$I'_Q = \frac{\beta_8 + \beta_9}{2\left(\sqrt{\beta_8} + \sqrt{\beta_9}\right)^2} \left(V_{DD} - V_{T8} - |V_{T9}|\right)^2, \tag{3.13}$$

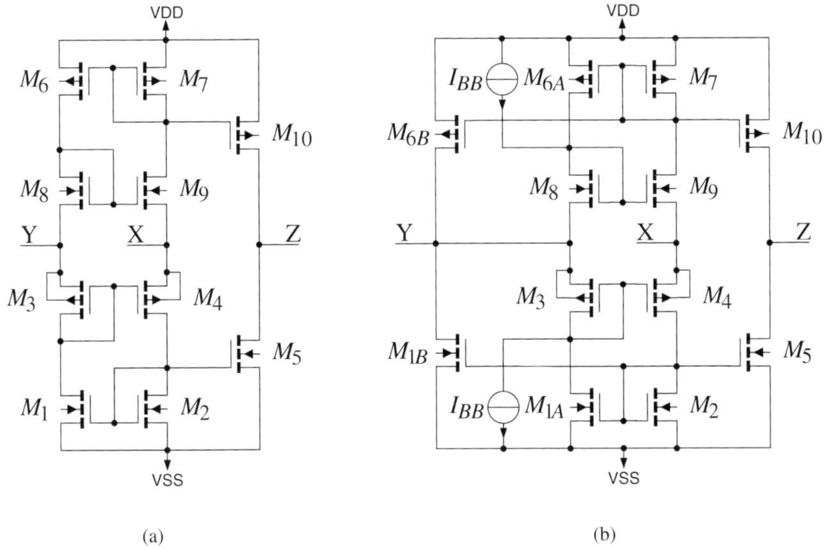

Figure 3.4 Push-pull CCI topologies. (a) Uncontrolled bias current [8]. (b) Controlled bias current [9].

$$I''_Q = \frac{\beta_1 + \beta_3}{2\left(\sqrt{\beta_1} + \sqrt{\beta_3}\right)^2} \left(|V_{SS}| - V_{T1} - |V_{T3}|\right)^2, \tag{3.14}$$

are the quiescent currents of the upper and the lower half circuits. These equations show that the quiescent current depends strongly on the supply voltages. Yet in these calculations transistors are assumed ideally matched and as a result the current mirroring errors in the four stacked current-mirrors add significant variation to the quiescent current [9]. This large quiescent current variation adds on to the already large input impedance variation. This quiescent current variation may even be so large that few millivolts of threshold voltage mismatch in the current mirrors may entirely turn off the quiescent current.

The quiescent current can be controlled by using an arrangement shown in Figure 3.4b [9], where the transistors M_1 and M_6 are split into two unequally sized transistors. The larger transistors, M_{1A} and M_{6A}, have their channel widths scaled by a factor λ_{FB} and the smaller transistor, M_{1A} and M_{6A}, have their channel widths scaled by a factor $1 - \lambda_{FB}$, so that the total channel width of these split transistors remains identical to the unsplit one. In this case, the X-terminal current is mirrored by a ratio λ_{FB} to the Y-terminal through the transistors M_3 and M_8. Additionally, a constant bias current I_{BB} is injected into these Y-input transistors. The smaller transistors M_{1B} and M_{6B} are needed only to maintain the conveyor relation $I_X = I_Y$ and can be omitted if Y-terminal

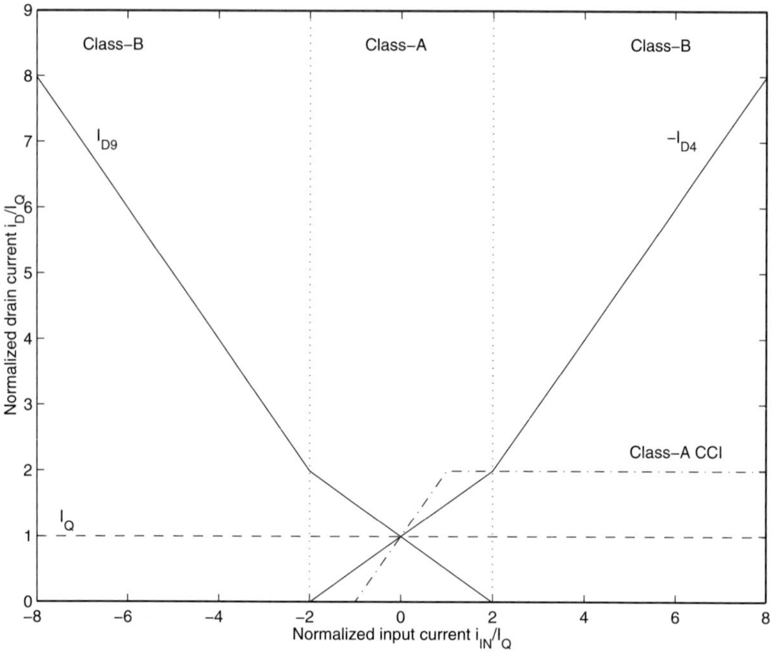

Figure 3.5 The division of the X-input current between transistors M_4 and M_9 in the push-pull conveyor of Figure 3.4a compared to the class-A CCI.

is grounded. Then the quiescent current settles approximately to

$$I_Q \approx \frac{I_{BB}}{1-\lambda_{FB}}, \tag{3.15}$$

and the quiescent current variation is effectively reduced. Similarly, the input impedance of the lower half-circuit can be expressed as

$$Z_X''(s)\bigg|_{Z_Y=0} \approx \frac{1-\lambda_{FB}}{g_{mp}} \frac{1+s\frac{C_A}{g_{mn}(1-\lambda_{FB})}}{1+s\frac{C_A}{g_{mn}}+s\frac{C_X(1-\lambda_{FB})}{g_{mp}}+s^2\frac{C_A C_X}{g_{mn}g_{mp}}}, \tag{3.16}$$

if all transistors are assumed ideally matched (apart from the deliberately scaled transistors), so that $g_{m1} = g_{mn}$, $g_{m2} = \lambda_{FB} g_{mn}$ and $g_{m3} = g_{m4} = g_{mp}$. The total X-input impedance is the parallel connection of the two half-circuit input impedances and is roughly half of the impedance of one half-circuit. Although this quiescent current controlling scheme increases the X-input impedance, this impedance level is easier to control. This technique can additionally be used to control the input impedance of the simple class-A CCI implementations of Figure 3.2.

The large signal behaviour as a function of the input signal current is depicted graphically in Figure 3.5. As a consequence of the symmetrical nature of the push-pull connection the distortion of a push-pull amplifier is significantly different that of one of the half-circuits composing the push-pull amplifier. As in Appendix B.1, the even order distortion components are cancelled in push-pull amplifiers as in differential amplifiers if the nonlinearities of the upper and lower half-circuit correlate. Additionally, as the push-pull conveyor stays in class-A region with two times larger signal currents than the half-circuits operating alone, there is a 6 dB improvement on the second order distortion and a 12 dB improvement on the third order distortion.

Unfortunately, the dominant distortion mechanism at low frequencies is the totally uncorrelated threshold voltage mismatch. Moreover, as NMOS transistors are approximately three times faster than PMOS-transistors, the two current amplifiers have different bandwidths and as a consequence of the high frequency distortion arising from current-mirror input capacitances is also different. Therefore, no dramatic improvement on the second order distortion at high frequencies is expected in the case of push-pull current-conveyors.

However, the errors deriving from the channel length modulation are quite systematic and as a result push-pull conveyors are not as sensitive to channel length modulation as simple class-A conveyors, particularly where nonlinearity is concerned. For this reason, it is possible to use simpler cascode structures such as that presented in Figure 2.2a. Furthermore, the push-pull conveyors should be able to amplify accurately significantly larger currents than the quiescent current and therefore more accurate current-mirror topologies cannot be used as explained in Chapter 2.1.6.

When the signal amplitude exceeds the quiescent current one of the half-circuits turns off quite abruptly as seen in Figure 3.5. Therefore, cross-over distortion will occur both at positive and negative currents near the quiescent current. This cross-over distortion is difficult to analyse and simulate but even without any detailed analysis can be assumed that this distortion will be greatly significant at high frequencies.

When the signal amplitude is significantly larger than the quiescent current of the conveyor, the distortion of the amplifier approaches the distortion of a push-pull class-B amplifier discussed in Appendix B.2. Although the power series coefficients describing the nonlinearities of the half-circuits in the push-pull conveyor cannot be exactly calculated in the case of signal currents larger than the quiescent current they nevertheless reveal a clue to the distortion performance of the push-pull connection. Thus, according to Equation (B.16), the second order distortion is because of the gain mismatch between the two half-circuits. Similarly, the equation (B.17) shows that the third order distortion depends on the average of the second order nonlinearities attenuated by approximately nine dB.

The class-B distortion has different behaviour in respect to the signal amplitude.

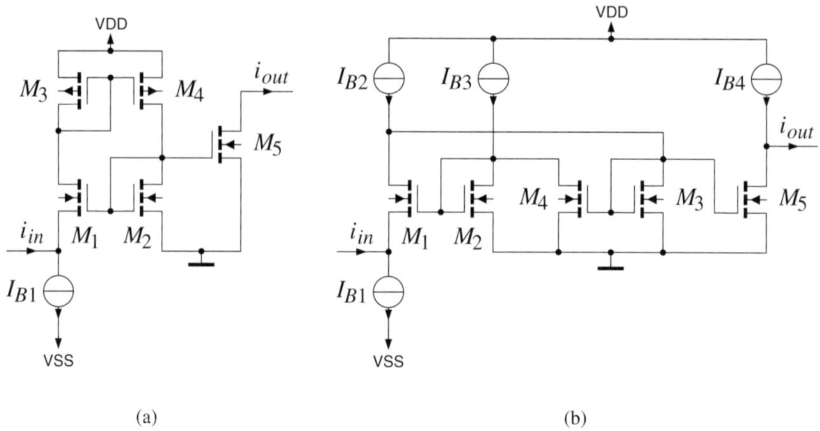

Figure 3.6 Low voltage CCI topologies. (a) Alternative input arrangement [10]. (b) A folded CCI [11].

While in the class-A region the second order distortion is proportional to the signal amplitude and the third order distortion is proportional to the square of the signal amplitude, in the class-B region the second order distortion remains constant with respect to the amplitude, and the third order distortion is proportional to the signal amplitude. As the signal amplitude is increased, other nonidealities such as mobility degradation begin to affect the distortion and some of the transistors fall off the saturation region and as a result the distortion begins to rise with amplitude more rapidly once more.

3.1.5 Low voltage CCI topologies

The input voltage range at the terminals X and Y of the first generation current conveyor is limited. The push-pull conveyor has particularly limited input range when realised with a standard n-well CMOS-process. Consequently, because of the bulk effect, the gate-source voltage of the input NMOS-transistors M_8 and M_9 is high. When the conveyor is used as an inverting current amplifier, so that the terminal Y is grounded the limited input voltage range is not such a critical parameter. However, the bulk effect causes the optimal input voltage to shift towards the negative supply rail and the minimum supply voltage remains relative high even in the current amplifier application.

One solution to the problem of extending the conveyor operation range to lower supply voltages is to alter the input structure. If X- and Y-terminals of the simple CCI of Figure 3.2 (i.e. the sources of the input PMOS-transistors M_3 and M_4) are connected to the positive supply voltage and the input signal is injected into the source of the NMOS-transistor M_1, a current amplifier as presented in Figure 3.6a results [10].

However, this circuit is no longer a general purpose current conveyor but a current-amplifier with low input impedance. Because the input voltage is fixed to the ground, several amplifier stages such as this cannot be cascaded. Since the input impedance of the circuit remains low and the input voltage level is tied to a well known and fixed voltage, this circuit is ideal as an input stage of a current-mode signal processing system.

In order to feed the bias current to the circuit, a negative supply voltage is needed, which will prevent this circuit from being used in certain applications. Alternatively, the bias current can be user programmed by a single external resistor inserted between the current input and the negative supply voltage. As the bias current and the input signal can share the same input pin, fewer pins are required for the chip package.

Lower supply voltage can alternatively be achieved by using an NMOS current-mirror rather than a PMOS mirror, i.e. folding the circuit as depicted in Figure 3.6b [11]. Since all transistors in the signal path are now NMOS transistors, the high frequency behaviour is improved. This circuit can additionally be used as a general purpose current-conveyor if a current source is inserted into the source of the input transistor M_2 and then cascading these amplifier stages is possible, although the minimum supply voltage is raised by a few hundreds of millivolts.

3.2 Second generation current-conveyor CCII

In many applications, only one of the virtual grounds in terminals X and Y of the first generation current-conveyor is used and the unused terminal must be grounded or otherwise connected to a suitable potential. This grounding must be done carefully since a poorly grounded input terminal may cause an unwanted negative impedance at the other input terminal. Moreover, for many applications a high impedance input terminal is preferable. For these reasons, the second generation current-conveyor was developed. It has one high and one low impedance input rather than the two low impedance inputs of the CCI [2].

The matrix representation of the second-generation current-conveyor CCII is

$$\begin{bmatrix} i_y \\ v_x \\ i_z \end{bmatrix} = \begin{bmatrix} 0 & 0 & 0 \\ 1 & 0 & 0 \\ 0 & \pm 1 & 0 \end{bmatrix} \begin{bmatrix} v_y \\ i_x \\ v_z \end{bmatrix}. \tag{3.17}$$

This current-conveyor differs from the first generation conveyor in that the terminal Y is a high impedance port, i.e. there is no current flowing into Y. While the Y-terminal of the second generation current-conveyor is a voltage input and the Z-terminal is a current output, the X-terminal can be used both as a voltage output or as a current

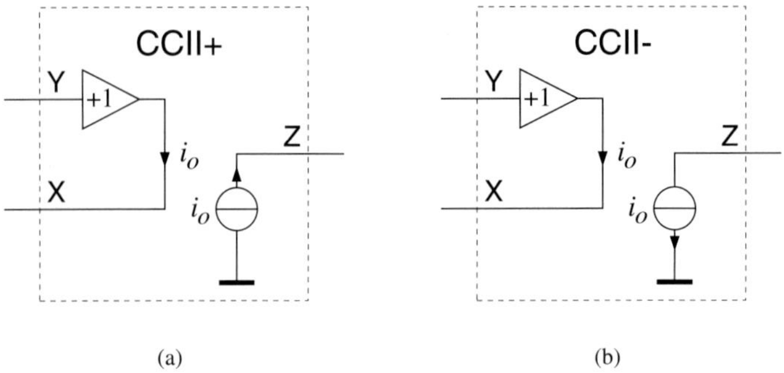

Figure 3.7 The principle of the second generation current-conveyors. (a) The positive conveyor CCII+, $i_z = i_x$. (b) The negative conveyor CCII-, $i_z = -i_x$.

input. Therefore, this conveyor can easily be used to process both current and voltage signals unlike the first generation current-conveyor or the operational amplifier. A further enhancement to the second generation current-conveyor is that there are two types of conveyors: in the positive current-conveyor CCII+, the currents i_x and i_z have the same direction as in a current-mirror and in the negative current-conveyor CCII- the currents i_x and i_z have opposite direction as in a current buffer.

The second-generation current-conveyor is in principle a voltage-follower with a voltage input, Y, and a voltage output, X, and a current-follower (or current-inverter) with a current input X and a current output Z connected together (Figure 3.7). The negative second-generation current-conveyor CCII- can also be considered an idealised MOS-transistor, where the currents $i_y = i_g = 0$ and $i_z = i_d = -i_x = -i_s$ and the voltages $v_x = v_s = v_y = v_g$. An 'ideal MOS transistor' is one that has a zero threshold voltage V_t and zero channel length modulation parameter λ and operates in the saturation region regardless of the drain-source voltage (positive or negative).

Three simple MOS CCII realisations are presented in Figure 3.8. The first is an NMOS transistor biased with two current-sources I_{BX} and I_{BZ}. Then there is always a DC-voltage level shift of V_{GS1} from Y- to X-terminal, which can be avoided by an additional level-shifting transistor M_2 in the enhanced CCII- implementation of Figure 3.8b [12]. If these two negative MOS conveyor implementations are compared to the current buffer implementation of Figure 2.17, both are found very similar. However, the terminal names are different and also their usage is different because the terminal Y can be used apart from the biasing purposes as a signal input.

The class-A MOS implementation of the positive second generation current-conveyor CCII+ of Figure 3.8c is close to the simple class-A implementation of the first generation current-conveyor of Figure 3.2a. Only one transistor, which was used to mirror the current form X-terminal back to the Y-terminal, is missing. As a conse-

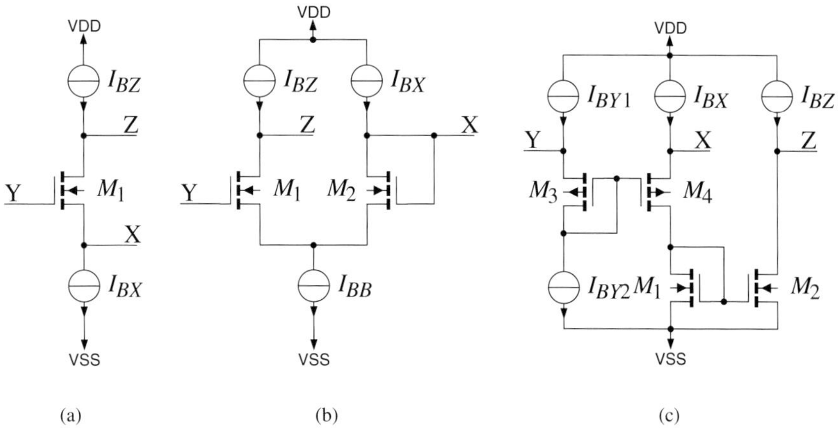

Figure 3.8 Simple class-A MOS implementations of the second generation current conveyor CCII. (a) NMOS transistor as a negative conveyor CCII- (b) An enhanced NMOS CCII-. (c) The positive conveyor CCII+.

quence of the lack of local feedback, the impedance level at the X-terminal of this CCII+ implementation is much higher than in the comparable CCI. The impedance level at the Y-terminal is as a result limited by the output conductances of the current sources I_{BY1} and I_{BY2}.

3.2.1 Linear nonidealities

To represent the linear nonidealities of the conveyor implementations of Figure 3.8, the matrix representation of a second generation current-conveyor of Equation (3.17) is rewritten because the matrix no longer contains uniquely zeroes and ones. The CCII matrix representation with linear nonidealities [13] is

$$\begin{bmatrix} i_y \\ v_x \\ i_z \end{bmatrix} = \begin{bmatrix} Y_y & A_{ir} & G_{mr} \\ A_{vf} & Z_x & A_{vr} \\ G_{mf} & A_{if} & Y_z \end{bmatrix} \begin{bmatrix} v_y \\ i_x \\ v_z \end{bmatrix}, \qquad (3.18)$$

where A_{vf} and A_{if} are the conveyor voltage and current forward gains close but not equal to one (or minus one) and A_{vr} and A_{ir} are the conveyor voltage and current reverse gains close but not equal to zero. Respectively, Y_y, Z_x, and Y_z are the limited terminal admittances and impedance. In certain applications, the transconductance G_{mf} from terminal Y to Z is similarly significant, although the reverse transconductance G_{mr} seldom has any effect on the circuit behaviour.

For the single NMOS transistor CCII- of Figure 3.8a, the nonideal matrix for low frequencies is calculated as

$$\begin{bmatrix} i_y \\ v_x \\ i_z \end{bmatrix} = \begin{bmatrix} 0 & 0 & 0 \\ \frac{g_m}{g'_x} & \frac{1}{g'_x} & \frac{g_{ds}}{g'_x} \\ \frac{g_m g_{bx}}{g'_x} & -1+\frac{g_{bx}}{g'_x} & g_{bz}+g_{bx}\frac{g_{ds}}{g'_x} \end{bmatrix} \begin{bmatrix} v_y \\ i_x \\ v_z \end{bmatrix}, \quad (3.19)$$

where

$$g'_x = g_m + g_{mb} + g_{ds} + g_{bx} \quad (3.20)$$

and g_{bx} and g_{bz} are the limited conductances of the current sources I_{BX} and I_{BZ}. Thus the voltage buffer of this conveyor is much more inaccurate as the current buffer essentially because of the bulk effect i.e. transconductance g_{mb}. The X-terminal impedance and Y- and Z-terminal admittances are expected but there is a significant forward transconductance term as well as a reverse voltage gain term. The forward transconductance is deriving from the nonzero output conductance g_{bx} of the current-source I_{BX}, and is the product of the forward voltage gain A_{vf}, the forward current gain A_{if} and the current source conductance g_{bx}

$$G_{mf} = \left.\frac{i_z}{v_y}\right|_{i_x=0} = A_{vf}\, g_{bx} A_{if} . \quad (3.21)$$

Similarly, the reverse voltage gain from Z to X A_{vr} is arising from the limited NMOS transistor drain-source conductance g_{ds}, and will reduce the X-terminal impedance if the impedance level at the Z-output is high.

As for the enhanced MOS CCII- implementation of Figure 3.8b, the matrix is approximated as

$$\begin{bmatrix} i_y \\ v_x \\ i_z \end{bmatrix} \approx \begin{bmatrix} 0 & 0 & 0 \\ 1-\frac{g_{ds}+g_{bb}}{g_m} & \frac{2}{g_m} & \frac{g_{ds}}{g_m} \\ g_{bb} & -1+\frac{g_{bb}+2g_{bx}}{g_m} & g_{bz}+\frac{g_{ds}(g_{bb}+g_{bx})}{g_m} \end{bmatrix} \begin{bmatrix} v_y \\ i_x \\ v_z \end{bmatrix}. \quad (3.22)$$

Consequently, the transistors M_1 and M_2 are assumed ideally matched so that the transconductance g_m and the drain-source conductance g_{ds} is identical for both transistors and g_{bb}, g_{bx} and g_{bz} are the output conductances of the current sources I_{BB}, I_{BX} and I_{BZ}.

This MOS CCII- implementation not only cancels the voltage level shift of one V_{GS}, but it also cancels the errors in the forward voltage gain A_{vf} deriving from the bulk effect. Additionally, the reverse voltage gain is approximately halved compared to the single NMOS CCII-. Unfortunately, all this happens at the expense of doubled X-terminal impedance Z_x and slightly increased gain error in the forward current gain A_{if}.

In the case of the simple MOS CCII+ implementation of Figure 3.8c, the nonideal conveyor matrix is approximately

$$\begin{bmatrix} i_y \\ v_x \\ i_z \end{bmatrix} \approx \begin{bmatrix} g_{by1} + g_{by2} & 0 & 0 \\ 1 - \frac{g_{dsp}+g_{bx}+g_{by2}}{g_{mp}} & \frac{1}{g_{mp}} & 0 \\ -g_{bx} & 1 - \frac{g_{bx}}{g_{mp}} - \frac{g_{dsn}}{g_{mn}} & g_{dsn} + g_{bz} \end{bmatrix} \begin{bmatrix} v_y \\ i_x \\ v_z \end{bmatrix}, \qquad (3.23)$$

where g_{mn} and g_{dsn} are the ideally matched transconductances and drain-source conductances of the NMOS transistors M_1 and M_2 and where g_{mp} and g_{dsp} are the corresponding parameters for the PMOS transistors M_3 and M_4. Respectively, g_{by1}, g_{by2}, g_{bx} and g_{bz} are the limited output conductances of the current sources in the circuit.

As the matrix shows, this CCII+ implementation entails no significant reverse gains. However, the Y-input admittance is in this case limited and, in order to ensure a low Y-input admittance and low forward voltage gain error, the current-source I_{BY2} should be realised as a cascode current source, which will limit the voltage range of the conveyor. Since the signal path for the forward current gain of this conveyor is a cascade of a current buffer (M_3) and a current-mirror (M_1 and M_2), the gain error is a sum of these two errors. In order to reach sufficient accuracy in the current gain, a cascode current-mirror should be used rather than the simple NMOS current mirror (M_1 and M_2). This additionally ensures high output impedance at the Z-output.

3.2.2 CCII macromodel

A current-conveyor macromodel generally includes only the gain error and the limited output impedance of the input voltage follower, and the current-mirror or current-follower as well as the limited Y-terminal impedance [14]. However, the nonideal matrix representations can be easily converted to a small-signal macromodel containing all significant linear nonidealities [15]. The macromodel for a CCII+ derived from equation (3.18) is presented in Figure 3.9. In this model, the forward transconductance G_{mf} is modelled by an admittance Y_x as the relation between the admittance and the forward transconductance is

$$G_{mf} = \left.\frac{i_z}{v_y}\right|_{i_x=0} = A'_{vf} A'_{if} \frac{Y_x}{1+Y_x Z_x}. \qquad (3.24)$$

However, in the model schematic the admittance Y_x is split into two admittances with a value of $Y_x/2$ between the X-terminal and VDD and between the X-terminal and VSS. The reason for this is best explained by comparing the test set-up for measuring the forward transconductance, presented in Figure 3.10a, and the set-up for measuring the power supply rejection ratio simultaneously for positive and negative supply, as pre-

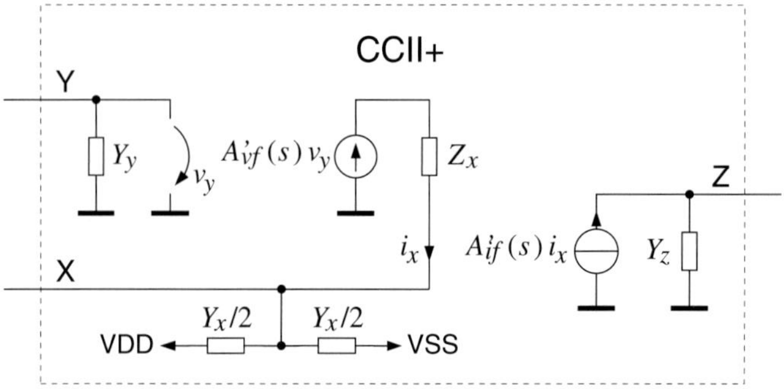

Figure 3.9 The linear macromodel for the positive second generation current-conveyor CCII+.

sented in Figure 3.10b. Moving the Y-input relative to the power supplies is equivalent to moving the power supplies relative to the Y-input and consequently the PSRR of the current-conveyor can be modelled if the admittance Y_x is tied rather than with the signal ground to power supplies. A better approximation for power supply rejection ratio would be obtained if positive and negative power supply rejection are measured separately and thus the admittance Y_x would be split into two unequal parts.

Since both the X-terminal impedance Z_x and the admittance Y_x affect the total forward voltage and current gains A_{vf} and A_{if} in the conveyor macromodel, these effects are excluded from the voltage and current gains A'_{vf} and A'_{if}. Therefore, the relation of the parameters A_{vf} and A'_{vf} is

$$A_{vf} = \frac{A'_{vf}}{1+Y_x Z_x}, \tag{3.25}$$

and the relation of A_{if} and A'_{if} is respectively

$$A_{if} = \frac{A'_{if}}{1+Y_x Z_x}. \tag{3.26}$$

Modelling a negative second-generation current-conveyor CCII- is almost identical to modelling a CCII+. However, for certain CCII- realisations such as the two discussed earlier it is necessary to model the reverse voltage gain A_{vr} by inserting an admittance between the X- and Z-terminals. The value of this admittance can be calculated by Z_x, Y_x and A_{vr} as

$$Y_{zx} = \frac{A_{vr}}{1-A_{vr}}\,\frac{1+Y_x Z_x}{Z_x} \approx \frac{A_{vr}}{Z_x}. \tag{3.27}$$

In the case of most conveyor applications the high frequency behaviour is modelled accurately enough by adding the parasitic capacitances to ground at all three current-

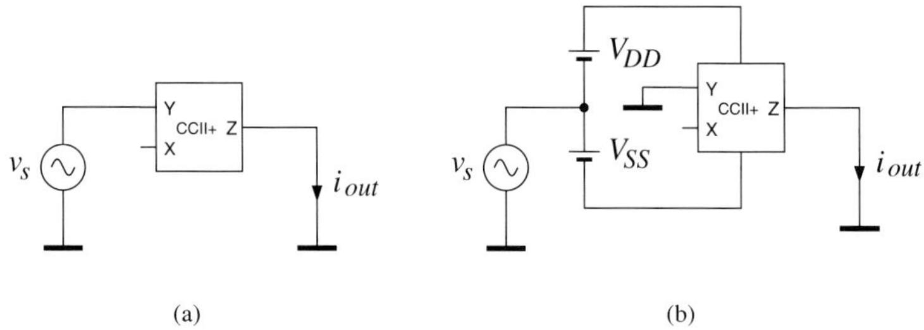

Figure 3.10 (a) The forward transconductance test set-up. (b) The test set-up for CCII PSRR.

conveyor terminals. Adding the forward transconductance in the conveyor macromodel by the admittance Y_x is similarly advantageous in this respect as the dependency of the forward transconductance on the parasitic capacitances at the X-terminal is similarly modelled correctly. For better accuracy, certain frequency dependencies could be added to the forward voltage and current gains, as depicted in the model. Commonly, a simple one-pole low-pass transfer function is sufficient for this purpose.

3.2.3 Applications of the CCII

Figure 3.11 shows typical current-conveyor circuits using the positive second-generation current-conveyor CCII+ [3]. Because of the separate voltage and current inputs both voltage and current amplifiers can easily be realised with the second-generation current-conveyors and the gain can be set by resistor ratios as in operational amplifier circuits. However, there is neither high voltage gain nor high current gain present in the current-conveyor and, as a consequence, signal processing in current-conveyor circuits is based on voltage-to-current and current-to-voltage conversions and on signal buffering by voltage and current buffers. Because there is typically no feedback in current-conveyor circuits, wide bandwidth operation without any slewing at large signal amplitudes is achieved. For the same reason, however, the accuracy of the current-conveyor circuits is affected by the voltage-follower and current-follower (or current-inverter) inaccuracies arising from the transistor process parameter deviation and device mismatches.

Using the current-conveyor macromodel of Figure 3.9, the unloaded voltage gain of the noninverting voltage amplifier, presented in Figure 3.11a, is expressed as

$$A_v(s) = -\frac{R_2}{R_1 + Z_x} \frac{A'_{vf}(s) A'_{if}(s) \left(1 + Y_x R_1\right)}{\left(1 + Y_x Z'_x\right)\left(1 + Y_z R_2\right)}, \tag{3.28}$$

where

$$Z'_x = Z_x \| R_1 = \frac{Z_x R_1}{Z_x + R_1}. \tag{3.29}$$

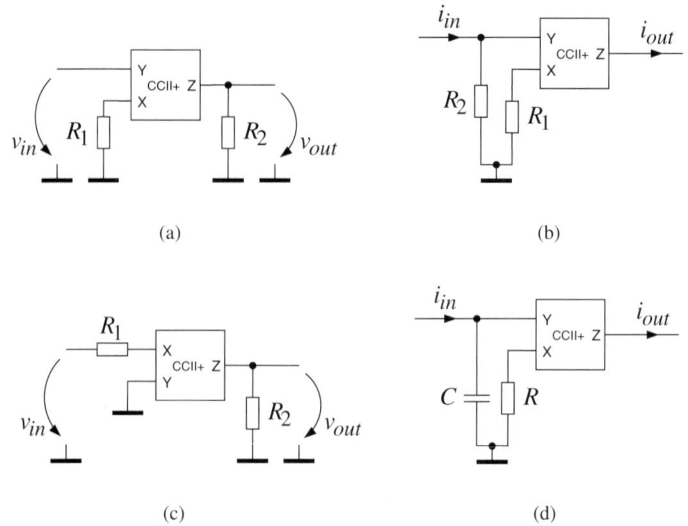

Figure 3.11 Application examples of a positive second generation current-conveyor CCII+. (a) Noninverting voltage amplifier, $A_v \approx R_2/R_1$. (b) Inverting current amplifier, $A_i \approx -R_2/R_1$. (c) Inverting voltage amplifier, $A_v \approx -R_2/R_1$. (d) Inverting current integrator, $i_{out} \approx -\frac{1}{RC}\int i_{in}dt$.

Because the forward current gain of the CCII+ is negative, i.e. $A'_{if}(0) \approx -1$, this voltage amplifier is noninverting. In order to minimise Y_z, a voltage buffer is quite often added to the output in order to isolate the conveyor Z-output from the load.

At low frequencies, the most significant gain error is arising from the X-terminal impedance Z_x. As the parasitic capacitances at the conveyor terminals are generally much larger than the parasitic capacitances of the internal nodes of the current-conveyor, the X- and Z-terminal parasitic capacitances C_x and C_z represent the most significant error sources at high frequencies and thus the equation reduces to

$$A_v(s) \approx \frac{R_2}{R_1 + Z_x} \frac{1 + sR_1C_x}{(1 + sZ'_xC_x)(1 + sR_2C_z)}. \tag{3.30}$$

At high voltage gains (high resistor ratios R_2/R_1), the bandwidth of the amplifier primarily depends on the output resistor R_2 and the Z-terminal parasitic capacitance C_z. In order to prevent the reduction of gain and bandwidth deriving from the load impedance, a voltage buffer is necessary. As voltage gain approaches one, the left half-plane zero arising from X-terminal parasitic capacitance begins to cancel the pole arising from the Z-terminal capacitance. In order to maximise the bandwidth, the resistances R_1 and R_2 should be maintained as small as possible and, as a result, it is important to minimise the conveyor X-terminal impedance Z_x, so that the gain error is similarly maintained at a tolerable level.

The current gain to a short circuit load of the conveyor current amplifier of Figure 3.11b is almost identical to the voltage gain given by Equation (3.28), except that the admittance Y_z is changed to Y_y:

$$A_i(s) = \frac{R_2}{R_1 + Z_x} \frac{A'_{vf}(s) A'_{if}(s)\,(1 + Y_x R_1)}{(1 + Y_x Z'_x)\,(1 + Y_y R_2)}. \tag{3.31}$$

Therefore, all such approximations at high and low frequencies are valid for this equation as well. Consequently, as $A'_{if}(0) \approx -1$ this current amplifier is inverting. Similarly, it frequently is necessary to preserve the high impedance at the Y-terminal by means of an additional input current buffer, particularly when large current gain is needed.

In all such current-conveyor applications, the amplifier type can be changed from noninverting to inverting or vice versa only by changing the conveyor from a CCII+ to a CCII-. However, an inverting voltage amplifier with a positive conveyor CCII+ can similarly be realised, as presented in Figure 3.11c. Since the Y-input is in this case grounded, the conveyor forward voltage gain $A'_{if}(s)$ has no effect on the voltage gain of this inverting amplifier:

$$A_v(s) = \frac{R_2}{R_1 + Z_x} \frac{A'_{if}(s)}{(1 + Y_x Z'_x)\,(1 + Y_z R_2)}. \tag{3.32}$$

Similarly, there is no left half-plane zero deriving from the parasitic capacitances at the X-terminal. Furthermore, the X-and Z-terminal parasitic capacitances dominate the high frequency behaviour as in the noninverting voltage amplifier case.

Additionally, current and voltage integration is possible in the case of second generation current conveyors and thus both current and voltage mode filters can be realised [2, 3, 17]. As illustration, a lossless current integrator is presented in Figure 3.11d and its nonideal transfer function is

$$A_i(s) = \frac{1}{(R + Z_x)\,(Y_y + sC)} \frac{A'_{vf}(s) A'_{if}(s)\,(1 + Y_x R)}{(1 + Y_x Z''_x)}, \tag{3.33}$$

where $Z''_x = R \| Z_x$. The equation shows that the integration time constant depends similarly on the conveyor X-terminal impedance Z_x. The capacitive part of the Y-input admittance Y_y adds on the integration capacitance C and therefore affects the integration time constant. Additionally, the resistive part of Y_y limits the DC-gain of the integrator. Furthermore, additional capacitance or conductance at the Y-input will affect the transfer function of the integrator and therefore an additional current buffer is often needed to isolate the signal source.

As $R > Z''_x$, the zero deriving from the parasitic capacitance at X-terminal is at a lower frequency than the pole deriving from the same capacitance and, as a conse-

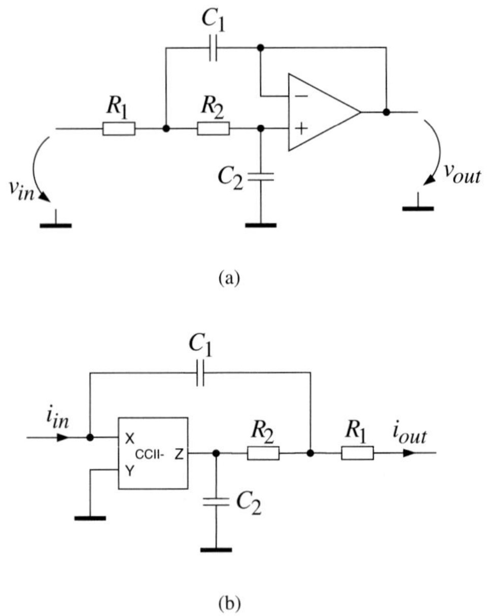

Figure 3.12 Sallen-Key low-pass filter implementations. (a) Voltage-mode filter using an operational amplifier. (b) Current-mode filter using a current-conveyor CCII-.

quence, the integrator phase error at high frequencies is actually phase lead instead of phase lag typical of most other integrator topologies. Thus most Q-enhancement techniques for high frequency filters cannot be used with filters using current-conveyor integrators.

Certain operational amplifier based filter topologies can be converted to current-conveyor based circuits by applying the adjoint principle, described in Section 1.4. An example of this is the Sallen-Key SAB (=single operational amplifier biquad) filter in Figure 3.12a, which is converted to a CCII- based current-mode filter [3, 16, 17]. However, in the voltage-mode circuit, the operational amplifier operates as a voltage follower and, as a result, its adjoint circuit element is a current follower. For this reason a second generation current-conveyor can by used as a replacement for the operational amplifier in a limited number of applications.

Perhaps the most useful current-conveyor application is, however, the instrumentation amplifier [18]: a differential amplifier with a high common-mode rejection ratio can be realised with two current-conveyors and two resistors without any matched components. Instrumentation amplifiers are typically realised with three operational amplifiers, as presented in Figure 3.13a. In order to reach a high common-mode rejection ratio (CMRR), all resistors but R_1 must be matched or the resistor ratios must

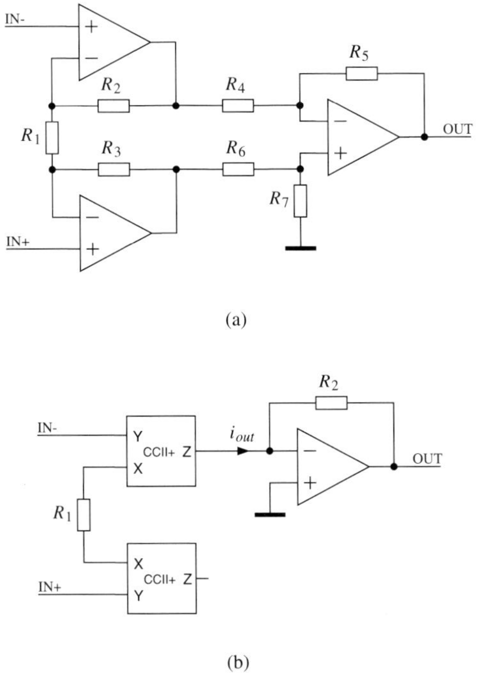

Figure 3.13 (a) A typical instrumentation amplifier using three operational amplifiers. (b) A typical current-mode instrumentation amplifier.

be trimmed. The CMRR is similarly dependent on the voltage gain of the input stage constructed of the first two operational amplifiers and resistors R_1, R_2 and R_3 and, as a consequence, a high CMRR can be realised only with a high voltage gain. Because of the gain-bandwidth product limitation of the voltage-mode operational amplifier, these high CMRR instrumentation amplifiers have generally a relatively low bandwidth. Furthermore, the CMRR of the operational amplifiers decreases with frequency, and is normally around 0 dB at the unity gain frequency and therefore good common-mode rejection is difficult to maintain at high frequencies.

The current-mode instrumentation amplifier in Figure 3.13b overcomes most of these difficulties [18]. In this circuit, there are only two resistors R_1 and R_2 which set the gain of the amplifier according to the resistor ratio R_2/R_1. The bandwidth of the conveyor instrumentation amplifier is similarly large with high voltage gains as current-conveyors operate in open loop without the gain bandwidth product limitation. The active current-to-voltage conversion, realised with the operational amplifier and the resistor R_2 also helps to extend the bandwidth at high values of R_2 as the effect of the parasitic capacitances at the conveyor Z-output is reduced by the operational amplifier open-loop gain $A_{ol}(s)$, so that

$$v_{out} = -\frac{i_{out}R_2}{1+\frac{1+R_2Y_z}{A_{ol}(s)}}. \tag{3.34}$$

In the case of a differential input voltage, when $v_{in+} = v_{dm}/2$ and $v_{in-} = -v_{dm}/2$, the current i_{out} feeding the output current-to-voltage converter can be calculated as

$$i_{out,dm} = v_{dm}A'_{vf}A'_{if}\frac{1+Y_xR_1/2}{R_1+2Z_x+2Y_xR_1Z_x}. \tag{3.35}$$

In this differential input signal case, the conveyor model parameters can be assumed to be identical for both conveyors. These equations show that the differential voltage gain is similar to the gain equation (3.28) of the noninverting voltage amplifier, in which case, the effect of the conveyor X-impedance Z_x is doubled because the input resistor R_1 is not grounded but floating between the two conveyor X-terminals. Therefore, for a differential input voltage, the input resistor acts as a grounded resistor with a value of $R_1/2$ and thus the zero arising from the X-terminal parasitic capacitance is similarly shifted up.

In the case of a common-mode input voltage, when $v_{in+} = v_{in-} = v_{cm}$, the mismatch of the current-conveyor parameters must additionally be taken into account. Therefore, in order to simplify calculations the common-mode output current is derived separately for each parameter mismatch. By introducing a conveyor forward voltage gain mismatch parameter $\Delta A'_{vf}$, which represents the difference between the two conveyor forward voltage gain parameters, the common-mode output current is

$$i_{out,cm}(\Delta A'_{vf}) \approx v_{cm}A'_{if}\frac{\Delta A'_{vf}+A'_{vf}Y_x(R_1+2Z_x)}{(1+Y_xZ_x)(R_1+2Z_x+Y_xR_1Z_x)}. \tag{3.36}$$

All other conveyor parameters are in this case assumed to be the average of the two current-conveyor parameters. Similarly, the common-mode output current as a function of the X-terminal admittance mismatch ΔY_x is

$$i_{out,cm}(\Delta Y_x) \approx v_{cm}A'_{if}A'_{vf}Y_x\frac{R_1\left(1+\frac{\Delta Y_x}{2Y_x}\right)+2Z_x}{R_1+2Z_x+2Y_xR_1Z_x} \tag{3.37}$$

$$\approx v_{cm}A'_{if}A'_{vf}Y_x. \tag{3.38}$$

For the last parameter mismatch ΔZ_x the common-mode current gain is

$$i_{out,cm}(\Delta Z_x) \approx v_{cm}A'_{if}A'_{vf}Y_x\frac{R_1+2Z_x\left(1+\frac{\Delta Z_x}{2Z_x}\right)}{R_1+2Z_x+2Y_xR_1Z_x} \tag{3.39}$$

$$\approx v_{cm}A'_{if}A'_{vf}Y_x. \tag{3.40}$$

As seen from the equations, the only mismatch parameter that produces a significant amount of common-mode output current is $\Delta A'_{vf}$, as already reported in [19], when current-conveyors were implemented with standard 741 operational amplifiers with power supply sensing. At that time, the voltage gain mismatch was deriving from the bandwidth mismatch of two discrete low speed operational amplifiers, whereas in the case of two integrated conveyors on the same chip, the gain mismatch is merely a fraction of the mismatch of two discrete devices. Additionally, in the case of integrated conveyors, the voltage follower bandwidth is commonly much greater than the amplifier bandwidth and, as a consequence, bandwidth mismatch of $A'_{vf}(s)$ has little effect.

Because the common-mode rejection above the amplifier -3 dB corner frequency is seldom of interest, the terms that affect only the frequency behaviour near or even above the amplifier corner frequency are neglected, so that the common-mode rejection ratio of the current-mode instrumentation amplifier is approximated as

$$CMRR \approx \frac{1}{Y_x(R_1+2Z_x)+\frac{\Delta A'_{vf}}{A'_{vf}}}. \tag{3.41}$$

Thus, there is no need to match resistors to reach high CMRR. However, in order to maximise the CMRR, the input resistor R_1 should be maintained as low as possible, thereby minimising the X-terminal impedance Z_x and the X-terminal admittance Y_x. The parasitic capacitance at the conveyor X-terminals in particular should be minimised in order to maintain an effective CMRR at high frequencies. Eventually the forward voltage gain mismatch $\Delta A'_{vf}$ will limit the CMRR if Z_x and Y_x are sufficiently small.

3.2.4 Nonlinearity of the class-A CCII

There are three main sources of nonlinearity in the current-conveyor, the non-linear forward voltage gain A_{VF}, the non-linear X-terminal impedance Z_X and the non-linear forward current gain A_{IF}. Because in most CMOS CCII topologies the linear input voltage range is quite limited, the nonlinearity of the A_{VF} is arising from clipping with large input voltage swings. Additionally, the bulk-effect may add distortion in simple voltage-follower implementations as in Figure 3.8a. Other nonlinearities in the input voltage follower are largely arising from the X-terminal impedance Z_X, which is a nonlinear function of i_X. Moreover, in most cases, all nonlinearities of the input voltage follower can be modelled in the non-linear Z_X.

In most current-conveyor applications, the non-linear impedance Z_X is in series with an external linear resistor R and thus the low-frequency distortion arising from Z_X is attenuated by the linear resistor R, as in source degenerated common source MOS-transistor amplifier stages (or emitter degenerated common-emitter bipolar transistor amplifier stages). As signal frequencies increase, this "X-degeneration" resistance is increasingly shunted by a parasitic capacitance and therefore distortion, arising from Z_X, increases with frequency.

In the simple class-A CMOS CCII- implementations of Figure 3.8a and b, the non-linearity of the forward current gain A_{IF} is quite weak and thus almost all distortion is generated in the non-linear X-terminal impedance Z_X. However, in the simple class-A CMOS CCII+ of Figure 3.8c the most significant source of low frequency distortion is the threshold voltage mismatch in the output current-mirror, and as the signal frequencies increase, the distortion deriving from the parasitic capacitances at the X-input and the current-mirror input node, begin to dominate.

3.2.5 Alternative class-A CCII topologies

The two simple implementations of the negative second generation current-conveyor CCII- discussed so far have similar nonidealities as the simple class-A current buffer implementations, with the addition of a voltage input and output. All of such implementations have a relatively high X-terminal impedance, which results in quite significant voltage and current gain errors in many applications. Unfortunately, this X-terminal impedance is further increased by the nonzero reverse voltage gain A_{vr}.

The CCII+ implementation discussed earlier suffers from similar problems. Consequently, X-terminal impedance is not increased by the reverse voltage gain, while the X-terminal impedance depends on the transconductance of a PMOS-transistor, which has typically 2-4 times lower transconductance with identical aspect ratio and bias current than an NMOS transistor. Therefore the discussed CCII+ has higher gain errors and narrower bandwidth than a NMOS current-mirror with comparable device sizes.

Furthermore, as in most current-conveyor applications there is no global feedback present to reduce the nonlinearity of the main amplifier, the current-conveyor should be as linear as possible. The X-terminal impedance is similarly important in this case since, in most conveyor circuits, this non-linear impedance is in series with an external gain setting resistor and consequently decreasing Z_X additionally decreases distortion if the gain resistance is maintained constant.

A straight-forward method for lowering the X-terminal impedance is to realise the input voltage follower with a more complex closed loop amplifier, as depicted in Figure 3.14. In Figure 3.14a, the transconductance of the source-follower transistor $M_1$is boosted with an additional transconductance amplifier A_1, providing external voltage gain $A_v = \frac{g_m}{g_o}$. In most cases, a simple differential amplifier stage provides enough additional amplification. The transconductance of the amplifier A_1 can be quite large without any stability problems. In addition to the lowered X-terminal impedance,

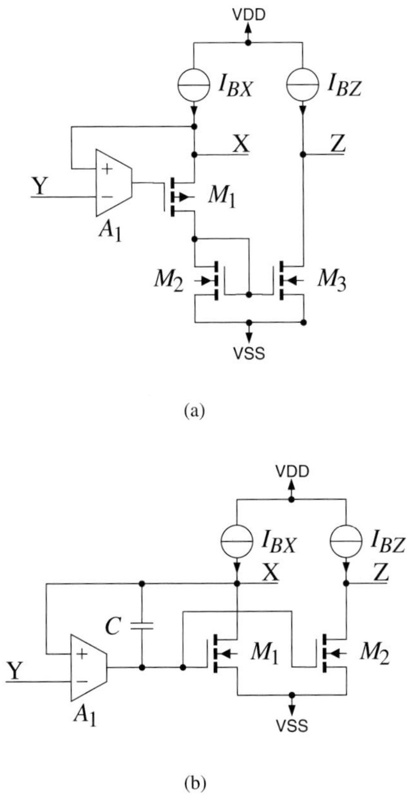

Figure 3.14 Lowered X-terminal impedance by additional loop gain. (a) Enhanced input voltage follower. (b) A class-A operational amplifier with a replica output.

the added loop gain additionally makes the voltage-follower less sensitive to the bulk effect.

An alternative way to realise a second generation current conveyor is to build the conveyor out of a Miller-compensated operational amplifier, as presented in Figure 3.14b. In this case, the Z-output is realised by a replica output. However, without any cascode devices, significant errors in the Z-output current arising from channel length modulation will result. The cascode transistors must be added to both X- and Z-terminals to ensure insensitivity to channel length modulation, which involves adding an extra pole to the input voltage follower feedback loop. As a consequence, a larger compensation capacitor C is needed, and a lower bandwidth in a normal Miller compensated operational amplifier will result. Thus, this conveyor topology would not be an ideal solution for high-speed applications.

The replica output operational amplifier approach has still the advantage that the voltage swing at the X-terminal is large, and if the input stage of the amplifier A_1 has a large input voltage swing, the Y-terminal voltage swing can similarly be increased. By using both NMOS and PMOS differential stages, a rail-to-rail input range amplifier can be constructed, as published in many papers [22–24]. Hence, using this technique in the current-conveyor a wide input voltage range both in Y- and X-terminals can be achieved [25]. Although such a rail-to-rail current-conveyor is complex and slow, it has applications in input voltage to current converters and current-mode instrumentation amplifiers.

A simple way to lower the X-terminal impedance is to add a CCI-like local feedback loop to the class-A CCII+ topology. When the diode connected PMOS transistor M_3 in the class-A MOS CCI of Figure 3.2 is changed to an NMOS source-follower and the obsolete current source I_{BY} is dropped off, the result is the circuit of Figure 3.15a, where the X-terminal impedance equations (3.2) and (3.4) similarly apply to this topology. However, in order to reach low X-terminal impedance, the transconductances of the NMOS transistor M_3 and the PMOS transistor M_4 should be well matched, which is impossible when relying on current CMOS technologies. Moreover, if the NMOS transistor transconductance g_{m3} is smaller than the PMOS transistor transconductance g_{m4} the X-terminal impedance may turn negative and thus cause instability. Since the mobility of the electrons may easily vary between two to four times the mobility of the holes, the NMOS transistor transconductance g_{m3} must be significantly higher than g_{m4} in order to avoid instability. Therefore, the X-terminal impedance cannot be designed as low as in the first generation current-conveyor and similarly the variation of the impedance is larger.

As a consequence of the uncorrelated threshold voltages and bulk effects of the NMOS and the PMOS transistors, there is a significant offset voltage and gain error

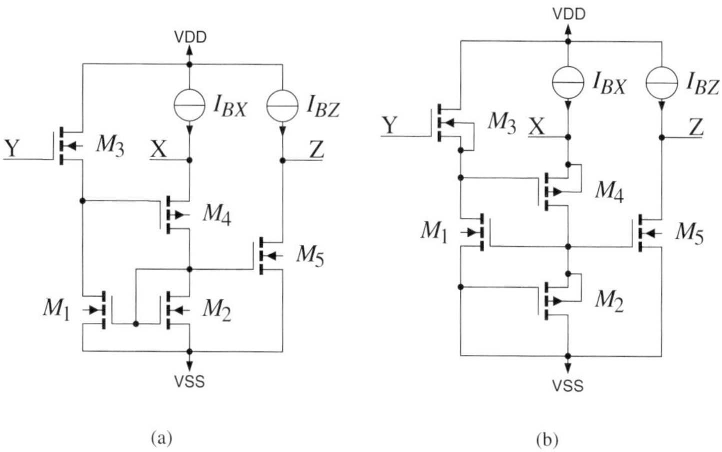

Figure 3.15 (a) A class-A CCII+ topology that lowers Z_x with a CCI-like local feedback [26]. (b) A similar topology, which lowers the X-impedance in the expense of the current gain accuracy [28].

present in the input voltage follower and both of these errors have large variations. Furthermore, if both M_3 and M_4 are realised without floating wells the input offset voltage and the X-terminal impedance is a function of the Y-terminals input voltage and thus additional distortion is generated. Unfortunately, realising only one of the transistors M_3 or M_4 as a floating well device serves only to make matters worse, since the bulk effect of the other transistor is not even partially canceled by the other transistor. Therefore, for the best result, a twin-well CMOS-process should be used to realise this conveyor.

The X-terminal impedance can be lowered if the input transistor M_2 in the NMOS current-mirror is changed to a PMOS transistor. In this case, the mirroring ratio of the current-mirror tracks the mobility ratio of NMOS and PMOS transistors and better matching of transconductances g_{m3} and g_{m4} is achieved. Furthermore, this method helps to reduce the input offset voltage between Y- and X-terminals. In order to reduce the inaccuracies arising from the bulk-effect, a floating p-well NMOS device (M_3) is needed in addition to the floating n-well PMOS devices (M_2 and M_4) and thus a twin-well CMOS-process is needed.

Unfortunately, as a result, the current gain from X to Y is a function of hole and electron mobilities. Thus, the bias current I_{BZ} cannot be equal to I_{BX} but it must be generated with a similar PMOS/NMOS hybrid current mirror as the Z-output current mirror is generated. However, no additional bias circuits are needed for a push-pull voltage follower realised with this principle [27]. In the case of a current-feedback operational amplifier [28], the uncertainty in the conveyor current gain does not represent as much a problem because then the second-generation current-conveyor forms the input stage of a high-gain feedback amplifier. Similar current-conveyor topologies, which similarly try to match the NMOS and PMOS threshold voltages and transconductances, have been published before [29, 30], but they require a twin well CMOS-process for accurate operation. Moreover, they are much more complicated.

3.2.6 Push-pull CCII topologies

Basic operation of a push-pull CCII+

A positive second generation CMOS current-conveyor in a push-pull amplifier configuration is presented in Figure 3.16. Unlike in the basic first generation push-pull conveyor of Figure 3.4a, in this conveyor the quiescent current is set directly to I_B by two current sources. In this topology, current flowing into the X-terminal is nonlinearly divided into signal paths through either an NMOS or a PMOS current-mirror which are summed up at the Z-output. Therefore, this current division must be analysed as mismatches in these signal paths may affect the conveyor linearity.

If one assumes that $\beta_3 = \beta_4 = \beta_p$, $\beta_7 = \beta_8 = \beta_n$, $V_{T3} = V_{T4} = V_{Tp}$ and $V_{T7} = V_{T8} = V_{Tn}$, the drain currents of the X-input transistor M_4 can be expressed as a function of the input voltage v_X:

$$i_{D4} = \begin{cases} I_B + v_X\sqrt{2\beta_p I_B} + \frac{1}{2}\beta_p v_X^2 & \text{if } v_X > -\sqrt{\frac{2I_B}{\beta_p}} \\ 0 & \text{if } v_X < -\sqrt{\frac{2I_B}{\beta_p}} \end{cases}, \tag{3.42}$$

and similarly for M_8:

$$i_{D8} = \begin{cases} I_B - v_X\sqrt{2\beta_n I_B} + \frac{1}{2}\beta_n v_X^2 & \text{if } v_X < \sqrt{\frac{2I_B}{\beta_n}} \\ 0 & \text{if } v_X > \sqrt{\frac{2I_B}{\beta_n}} \end{cases}. \tag{3.43}$$

Then the short transition to weak inversion before the transistors turn off completely is neglected, as this current does not normally contribute to a large error in the total current flowing into the X-terminal if the bias current I_B is sufficiently large. Therefore, the X-terminal current could be a linear function of the X-input voltage

$$i_X = i_{D4} - i_{D8} = v_X\sqrt{2I_B}\left(\sqrt{\beta_n} + \sqrt{\beta_p}\right) + \frac{1}{2}v_X^2\left(\beta_p - \beta_n\right), \tag{3.44}$$

if $\beta_n = \beta_p$ and $|i_X| < 4I_B$.

In Figure 3.17, the drain currents of the transistors M_4 and M_8 are represented graphically in the ideally matched case. This representation clearly shows that none of the transistors in the push-pull conveyor drops off the saturation region until the

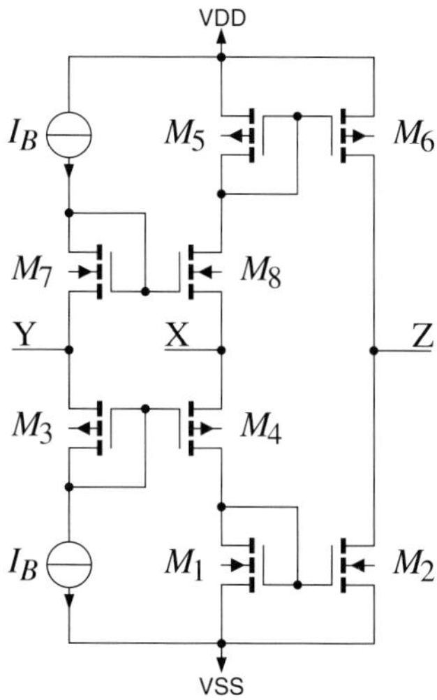

Figure 3.16 Push-pull MOS CCII+.

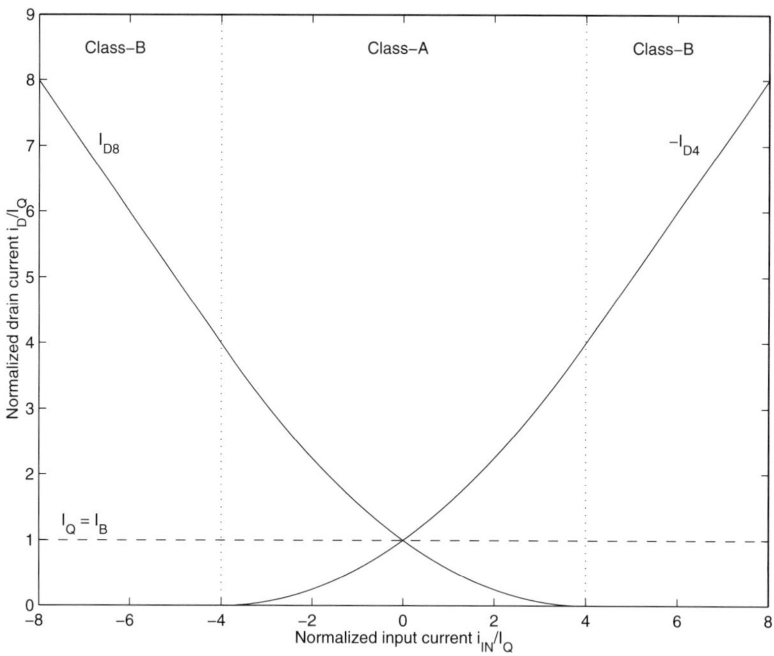

Figure 3.17 The division of the X-input current between ideally matched transistors M_4 and M_8.

X-input signal current is almost four times the bias current I_B, and thus the class-A operation range of the push-pull conveyor is almost four times greater than in the simple class-A CCII+. The push-pull conveyor additionally operates with larger current signal amplitudes than $4I_B$, but with increased nonlinearity.

When compared to a push-pull CCI with the corresponding quiescent current in Figure 3.5, the class-A region is doubled. Furthermore, in the push-pull CCII+ the transition from the class-A into the class-B region is smoother than in a push-pull CCI and consequently less cross-over distortion is expected in a push-pull CCII+ than in a push-pull CCI. The internal feedback of the CCI also results in a longer transient while one of the half-circuits turn on as the conveyor enters the class-A region once more, and consequently the cross-over distortion represents a more significant problem at high frequencies in a push-pull CCI than in a push-pull CCII+.

Basic operation of a push-pull CCII-

In addition, negative second-generation current-conveyors can be realised with a CMOS-technology, as the simple MOS implementation in Figure 3.18 reveals. It can operate with lower supply voltages than the positive push-pull conveyor of Figure 3.16 because, in this configuration, the source voltages are lower and hence the gate-source voltages are not increased as much as in the positive conveyor because of the bulk effect. However, this conveyor operates exclusively in the class-A region and, as a consequence, it is much more feasible to use differential structures rather than a push-pull negative conveyor as larger voltage swing with lower supply voltages can be achieved with better linearity.

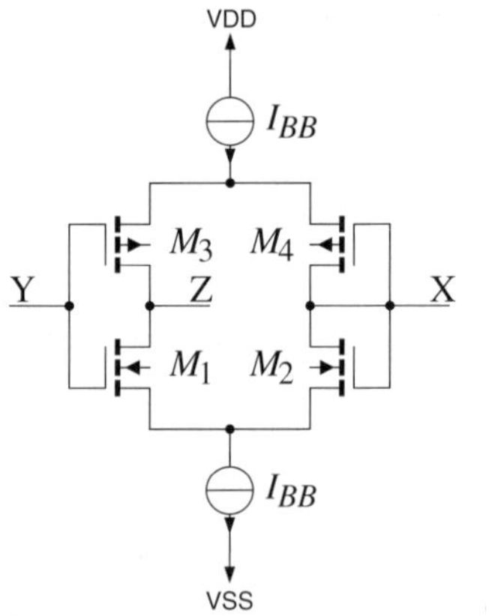

Figure 3.18 A push-pull MOS CCII-.

X-terminal impedance

As the equation (3.44) shows the linearity of the X-terminal impedance in the simple push-pull CCII+ depends strongly on the matching of the parameters β_n and β_p. Unfortunately, these parameters are uncorrelated and thus there is always second order nonlinearity present in the push-pull CCII X-terminal impedance. However, when compared to the class-A CCII+ where $i_X = i_{D4} - I_B$, the nonlinearity is attenuated by $\left|\frac{\beta_p - \beta_n}{\beta_p}\right|$. Thus, even in a worst case β mismatch situation, this second order term is typically three times lower in the push-pull CCII+ than in the simple class-A CCII+. Additionally, the X-terminal impedance is approximately two times lower than in the simple class-A CCII+:

$$Z_x = \frac{1}{\sqrt{2I_B}\left(\sqrt{\beta_n} + \sqrt{\beta_p}\right)}. \tag{3.45}$$

The required matching of the parameters β_n and β_p brings about a situation in which the PMOS input transistors M_3 and M_4 cannot be realised as floating well devices unless the NMOS input transistors M_7 and M_8 can similarly be realised as floating well devices. Unfortunately, this is possible only with relatively rare and high-cost twin-well CMOS-processes. Therefore with standard n-well CMOS-processes relative large supply voltages are required.

In most cases, the nonlinearity of the X-terminal contributes only indirectly to amplifier nonlinearity as in the case of current-mirrors. When the X-terminal is used as a current input, the nonlinearity of input impedance generates significant amount of distortion only at high frequencies. Fortunately, this distortion is significantly lower than in current-mirrors as the impedance is less non-linear because of the push-pull structure. However, as the NMOS and PMOS current-mirrors do not have equal bandwidths there is nevertheless significant high frequency distortion in the push-pull CCII+ arising from this unsymmetrical high frequency large signal behaviour.

When there is an external impedance, for example a resistor, connected to the X-terminal, as in the current-conveyor applications in Figure 3.11, 3.12 and 3.13, the nonlinearity of the X-terminal impedance exerts a stronger influence on the distortion. If as low distortion as possible is desired, the large mismatch between NMOS and PMOS transistors is no longer tolerable. A push-pull current amplifier that requires only one type of MOS-transistors for linearised X-terminal impedance was published in 1987 [31]. A version of this circuit topology, optimised for low-cost n-well CMOS-processes, is presented in Figure 3.19, where the input current is divided into push and pull currents with two PMOS transistors M_1 and M_5 of identical sizes. The NMOS current-mirror has very little effect on the high frequency behaviour of the circuit as the gate capacitances of the PMOS current-mirror and the bulk capacitance of the input transistor M_5 are connected to the X-input and thus totally dominate the high frequency

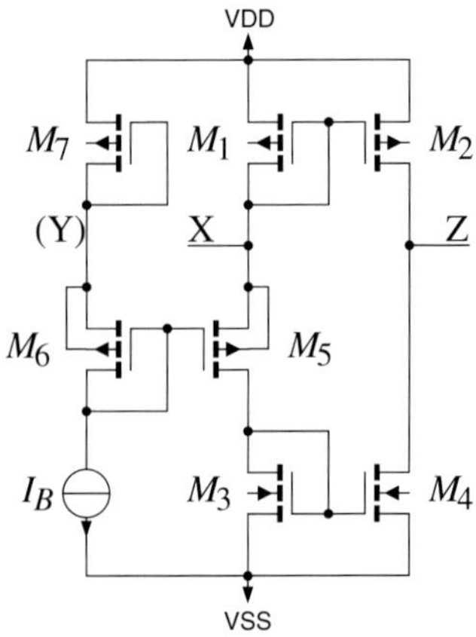

Figure 3.19 A push-pull current amplifier that does not require good matching between NMOS and PMOS transistors to maintain linear X-terminal input impedance. [31]

response. Unfortunately, this topology cannot be used as a general purpose CCII as the Y-terminal voltage controls the quiescent current of the push-pull amplifier and thus the Y-terminal cannot be used as a signal input.

Current gain nonlinearity

Generally. the most significant nonlinearities are involved in the current gain path from the X-input to the Z-output. The push-pull connection brings symmetry to the circuit and thus reduces the second order nonlinearities in a similar way to differential structures. Because of this symmetry, the distortion arising from the channel length modulation is attenuated in a push-pull conveyor, although it is not entirely cancelled out. Therefore, adequate distortion performance can be obtained by implementing cascode current-mirror topologies, which are sensitive to the channel length modulation. Thus, a push-pull CCII+ topology with simple cascodes at the output is used, as seen in Figure 3.20 to show experimentally the typical distortion performance of a push-pull conveyor. In order to minimise the gain error deriving from the low quality output cascode current-mirrors, the cascode transistors M_9 and M_{10} are biased so that main mirror transistors have identical drain voltages in both NMOS and PMOS current-mirrors. This limits the Z-output voltage swing but, as the voltage swing at Y- and X- terminals are very limited because of the bulk effect, the voltage swing of the Z-terminal is seldom fully utilised.

Two series of simulations are performed to show the overall distortion behaviour of the push-pull conveyor. Firstly, the distortion is simulated at a fixed and low frequency (1 Hz) as a function of the input current signal amplitude. Secondly, the distortion is simulated as a function of frequency with a moderate input current signal amplitude.

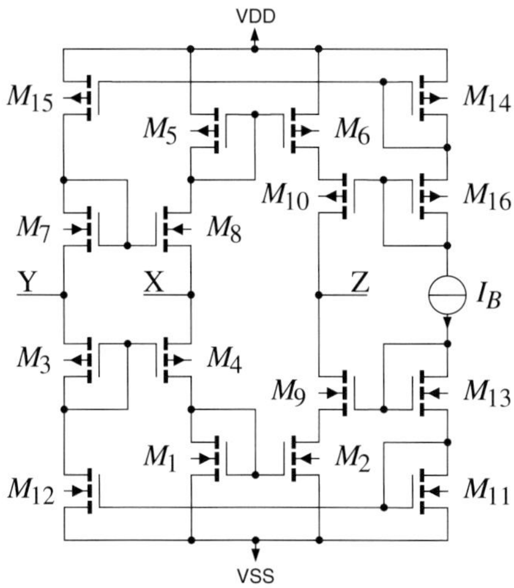

Figure 3.20 Push-pull MOS CCII+ with output cascodes.

Transistors	Aspect ratio / μm/μm
M_1, M_2, M_7, M_8 , M_{11} and M_{12}	100/5
M_3, M_4, M_5, M_6, M_{14} and M_{15}	300/5
M_9 and M_{13}	100/2
M_{10} and M_{12}	300/2

Table 3.1 Transistor dimensions of the cascode push-pull MOS CCII+ of Figure 3.20

In both series of simulations, the push-pull CCII+ is compared with the lower half-circuit and the upper half replaced with current sources. In the push-pull conveyor, the quiescent current is set to 25 μA, with the current source I_B, whereas the half-circuit is biased with 100 μA current sources. Thus, both the half circuit and the push-pull circuit have comparable class-A regions of operation. In all simulations, no threshold voltage mismatches are assumed in the circuit as this would have increased the simulation task considerably.

The simulation results of the low frequency distortion as a function of signal amplitude are presented in Figure 3.21. The results clearly show that the push-pull circuitry effectively reduces the second order distortion, which in this case is deriving from channel length modulation. However, the third order distortion is higher in the push-pull conveyor than in the half-circuits. However, the unsymmetrical second order nonlinearity of the half circuits changes to symmetrical third order nonlinearity in the push-pull conveyor. The level of distortion is quite moderate until clipping occurs. However, if the threshold voltage mismatch is included and worst case matching of NMOS and PMOS transistor parameters are assumed, the distortion increases quite significantly and thus it is doubtful whether using better cascode current-mirror

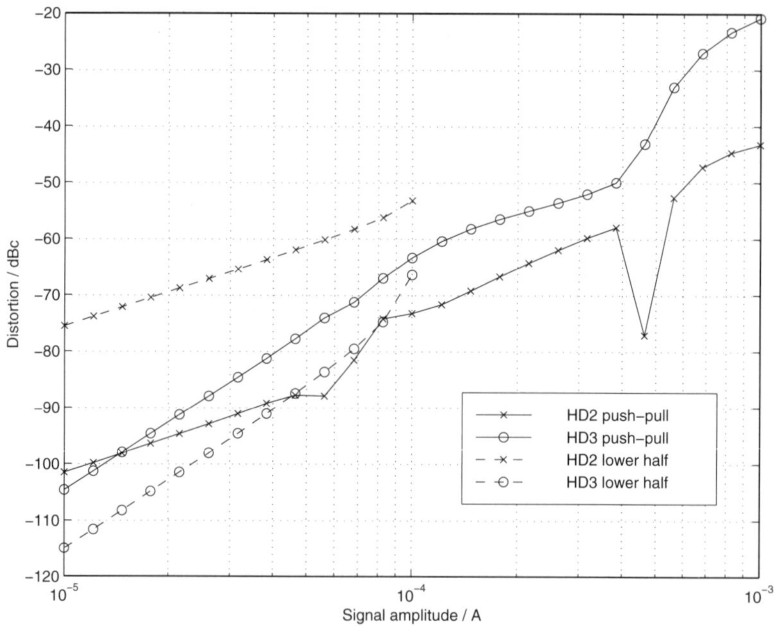

Figure 3.21 Simulated distortion of a push-pull CCII+ as a function of signal amplitude compared to the distortion of the lower half-circuit operating alone. The push-pull conveyor has a quiescent current of 25 μA and the the half-circuit is biased with four times larger currents i.e. 100 μA, so that both amplifiers have in theory equal class-A regions.

topologies could significantly enhance the distortion performance.

At low frequencies, the nonlinearity of the upper and lower half-circuits can be transformed to the nonlinearity of the push-pull amplifier by the methods described in Appendices A and B. The equations in Appendix A explain the reduction of the second order distortion but the third order distortion equation does not work as well. However, the lambda model based third order distortion equations were inaccurate, even for simple class-A current-mirrors. Additionally, the fact that the push-pull connection of the CMOS CCII+ extends the class-A operating range to four times the quiescent current is not included in Equations (B.8) and (B.9) of Appendix B.1 so certain differences can be expected. In effect, the third order distortion follows better Equation (B.17) of Appendix B.2 where the class-B operation is assumed.

Alternative methods for predicting the distortion has been published [32–35]. These methods assume that the class-A region is so small that it can be neglected and thus the resulting equations include large errors near the class-A region. However, the simulations show that the transistors in a CMOS push-pull conveyor easily fall off their proper operation regions with larger current signal amplitudes and thus the class-B region is often merely a narrow transition region between the class-A region and sig-

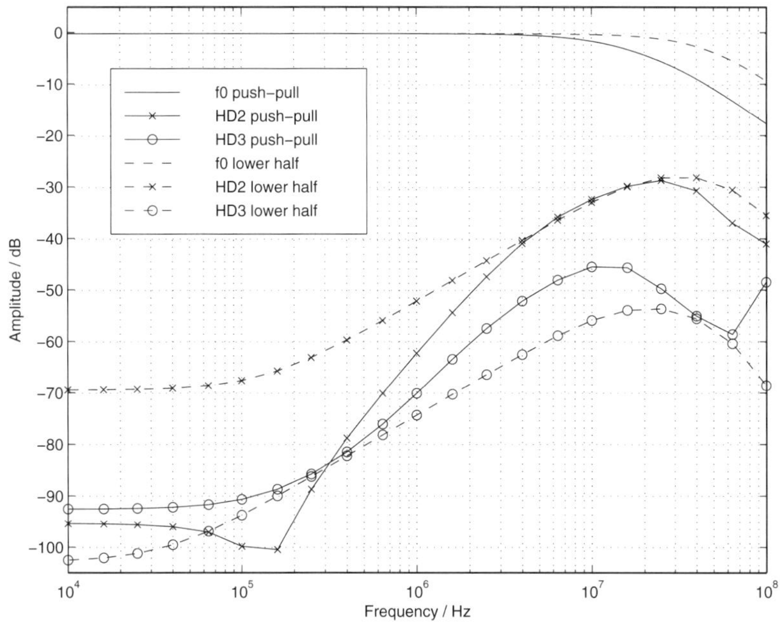

Figure 3.22 Simulated frequency response and distortion as a function of frequency of a push-pull CCII+ compared to the lower half-circuit operating alone with four times larger bias currents. The input signal peak amplitude is 20 μA so that the modulation index can be assumed as 0.2 in both circuits.

nal clipping. Therefore, in order to maximise the operation in the class-B region, the quiescent current must be very low and consequently the bandwidth of the amplifier remains limited.

The simulated distortion as a function of frequency with an input signal peak amplitude of 20 μA is presented in Figure 3.22. The simulation results show that the push-pull configuration can attenuate second order nonlinearities only at moderately low frequencies. The high frequency distortion is high, with signal amplitudes comparable to the quiescent current. Therefore, push-pull conveyors can adequately operate in the class-B region probably only with signal frequencies two decades lower than the conveyor -3 dB corner frequency.

The distortion calculation methods for push-pull amplifiers mentioned earlier are based on the nonlinearities in the DC transfer curve. Therefore, they cannot be used to describe the frequency dependent nonlinearities. Furthermore, the operation of the push-pull structure is too strongly non-linear for Volterra-analysis, and by using the harmonic balance technique only numerical results can be obtained. However, symbolic equations for the high frequency distortion of the push-pull CMOS conveyors are probably not needed. Simple simulations and a degree of intuition should be enough to discourage the use of push-pull techniques at high frequencies, at least in the case of the CMOS technology. Fortunately, most differential structures maintain good balance, even near the corner frequency of the amplifier, although they are limited to class-A operation.

3.3 Third generation current-conveyor CCIII

Yet another current-conveyor was proposed in 1995 [36]. The network of this third generation current-conveyor CCIII is formulated in a matrix form as follows

$$\begin{bmatrix} i_y \\ v_x \\ i_z \end{bmatrix} = \begin{bmatrix} 0 & -1 & 0 \\ 1 & 0 & 0 \\ 0 & 1 & 0 \end{bmatrix} \begin{bmatrix} v_y \\ i_x \\ v_z \end{bmatrix}. \tag{3.46}$$

The operation of the third generation current-conveyor CCIII is similar to that of the first order current-conveyor CCI, with the exception that the currents in ports X and Y flow in opposite directions.

A CMOS realisation on the CCIII was also proposed in the same year [37] and is presented in Figure 3.23. It is a push-pull conveyor built from four simple first generation current-conveyors. Thus, the X- and Y-terminal impedances are maintained comparably low. However, the impedance level is sensitive to threshold voltage mismatches. The quiescent current is similarly very sensitive to process and supply voltage variation and consequently the quiescent current control schemes [9] described in 3.1.4 must be added to the circuit.

This current-conveyor can be used as an active current probe. The typical current measurement set-up presented in Figure 3.24a, where the voltage drop over a small shunt resistor is amplified with a voltage amplifier, is problematic if a low shunt resistance is required. In such a case, a large voltage gain is needed to amplify this small voltage drop V_R across the resistor R, which limits the measurement bandwidth and makes the measurement more sensitive to offset voltage, noise, and RF-interference. By using the CCIII in preference to the shunt resistor, the voltage drop can be maintained small without other problems arising. However, as the circuit is relatively complex, its bandwidth cannot be very wide. Moreover, in the case of CMOS-technology very large currents cannot easily be measured.

As the input current flows into the Y-terminal and out from the X-terminal, one might think that a differential current input could be realised with this amplifier. However, the CCIII has a high input impedance with common-mode current signals, i.e. identical currents are fed both to Y- and X-terminals. Therefore common-mode currents can push the input terminals out from the proper operation range. Thus, unfortu-

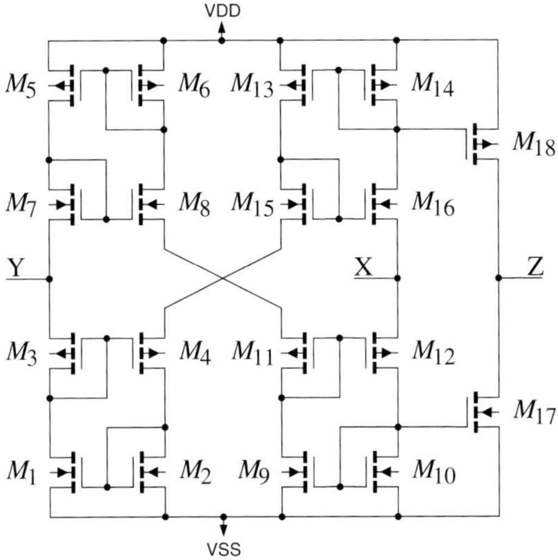

Figure 3.23 A CMOS implementation of the third generation current-conveyor CCIII [37].

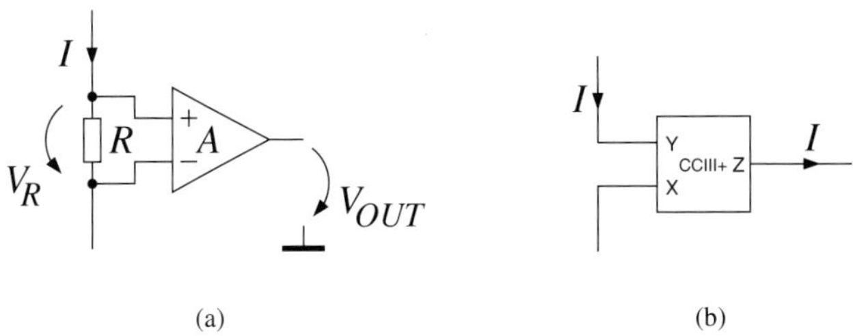

Figure 3.24 Current measurement (a) with a voltage amplifier and (b) with a third generation current-conveyor CCIII.

nately, this conveyor has not many applications other than the current probing.

References

[1] K. Smith, A. Sedra, "The current-conveyor - a new circuit building block," *IEEE Proc.,* vol. 56, pp. 1368-69, 1968.

[2] A. Sedra, K. Smith, "A second-generation current-conveyor and its applications," *IEEE Trans.,* vol. CT-17, pp. 132-134, 1970.

[3] C. Toumazou , F. J. Lidgey, D.G. Haigh (ed.), *Analogue IC design: the current-mode approach*, London, Peter Peregnirus Ltd, 1990, 646 p.

[4] P. Gray, R. Meyer, *Analysis and Design of Analog Integrated Circuits*, second edition, New York, John Wiley & Sons, 1984, 771 p.

[5] R. Brennan, T. Viswanathan, J. Hanson, "The CMOS negative impedance converter," *IEEE J. Solid State Circuits*, vol. 23, pp. 1272-1275, Oct. 1988.

[6] R. Kaunisto, D. Webster, D. Haigh, "Improved MMIC active filters based on passive LC resonators with active negative resistance circuits," *Proc. IEE Colloquium on Advanced Signal Processing for Microwave Applications*, London, UK, pp. 3/1-3/6, Oct. 1996.

[7] D. Python, C. Enz, "A 2V low-distortion CMOS biquadratic cell," *Proc. IEEE International Symposium on on Circuits and Systems* (ISCAS-98), Monterey, 1998, pp. 139 -142.

[8] Z. Wand, "Wideband class AB (push-pull) current amplifier in CMOS technology," *Electronics Letters*, vol. 26, pp. 543-545, April 1990.

[9] E. Bruun, "Class AB CMOS first-generation current conveyor," *Electronics Letters*, vol. 31, pp. 422-423, March 1995.

[10] G. Temes, W. Ki, "Fast CMOS current amplifier and buffer stage," *Electronics Letters*, vol. 23, pp. 696-697, June 1987.

[11] K. Chan, R. Duncan, A. Sedra, "High-performance current amplifier and current conveyor," in *Proc. IEEE Int. Symposium on Circuits and Systems* (ISCAS-91), Singapore, 1991, pp. 1825-1828.

[12] P. Racz, G. Roberts, M. Blostein, " BiCMOS current mode analog Nyquist filter at video rates," in *Proc. IEEE Int. Symposium on Circuits and Systems* (ISCAS-92), San Diego, 1992, pp. 2344-2347.

[13] K. Koli, K. Halonen ,"CMRR Enhancement Techniques for Current-Mode Instrumentation Amplifiers," *IEEE Trans. Circuits and Systems-I*, vol. CAS-47, pp. 622-632, May 2000.

[14] B. Wilson, "Performance analysis of current conveyors," *Electronic Letters*, vol. 25, pp. 1596-1598, Nov. 1989.

[15] K. Koli, K. Halonen, "Current-conveyor subscriber line interface circuits" in *Proc. IEEE Int. Symposium on Circuits and Systems* (ISCAS-92), San Diego, 1992, pp. 1384-1387.

[16] G. Roberts, "All current-mode frequency selective circuits," *Electronics Letters*, vol. 25, pp. 759-761, June 1989.

[17] G. Roberts, A. Sedra, "A general class of current amplifier-based biquadratic filter circuits," in *Proc. IEEE Int. Symposium on Circuits and Systems* (ISCAS-91), Singapore, 1991, pp. 1821-1824.

[18] C. Toumazou, F. Lidgey, "Novel current-mode instrumentation amplifier," *Electronics Letters*, vol. 25, pp. 228-230, Feb. 1989.

[19] B. Wilson, "Universal conveyor instrumentation amplifier," *Electronic Letters*, vol. 25, pp. 470-471, March 1989.

[20] W. Su, F. Lidgey, "Common-mode rejection in current-mode instrumentation amplifiers," *Int. J. Analog Integrated Circuit Design and Signal Processing*, vol. 7, pp. 257-260, May 1995.

[21] Q. Zhu, F. Lidgey, W. Su, "High CMRR, second generation current-mode instrumentation amplifiers," *Proc. IEEE Int. Symposium on Circuits and Systems* (ISCAS-93), Chicago, 1993 pp. 1326-1328.

[22] J. Huising, D. Linebarger, "Low-voltage operational amplifier with rail-to-rail input and output ranges," *IEEE J. Solid State Circuits*, vol. SC-20, pp. 1144-1150, Dec. 1985.

[23] J. Huising, R. Hogervorst, K. Langen, "Low-power low-voltage VLSI operational amplifier cells," *IEEE Trans. Circuits and Systems-I*, vol. CAS-42, pp. 841-852, Nov. 1995.

[24] S. Sakurai, M. Ismail, "Robust design of rail-to-rail CMOS operational amplifiers for a low power supply voltage," *IEEE J. Solid State Circuits*, vol. SC-31, pp. 144-156, Feb. 1996.

[25] H. Elwan, A. Soliman, "Low-voltage low-power CMOS current conveyors," *IEEE Trans. Circuits and Systems-I*, vol. CAS-44, pp. 828-835, Sep. 1997.

[26] H.-W. Cha, K.Watanabe, "Wideband CMOS current conveyor," *Electronics Letters*, vol. 32, pp. 1245-1246, July 1996.

[27] K. Manetakis, C. Toumazou, "A new high-frequency very low output impedance CMOS buffer," *Proc. IEEE Int. Symposium on Circuits and Systems* (ISCAS-96), Atlanta, 1996 pp. 485-488.

[28] K. Manetakis, C. Toumazou, "Current-feedback opamp suitable for VLSI Technology," *Electronics Letters*, vol. 32, pp. 1090-1092, June 1996.

[29] M. Cheng, C. Toumazou, "3V MOS Current-conveyor cell for VLSI Technology," *Electronics Letters*, vol. 29, pp.317-319, Feb. 1993.

[30] A. Piovaccari, S. Graffi, G. Masetti, "A low-voltage low-power CMOS current-conveyor," *European Conference on Circuits Design and Theory* (ECCTD'95), Istambul, Turkey, 1995, pp. 5-8.

[31] K. Bult, H. Wallinga, "A class of analog CMOS circuits based on the square-law characteristic of an MOS transistor in saturation," *IEEE J. Solid State Circuits*, vol. SC-22, pp. 357-365, June 1987.

[32] D. Pederson, K. Mayaram, *Analog Integrated Circuits for Communication (Principles, Simulation and Design),* Norwell, MA, Kluwer Academic Publishers, 1991, 568 p.

[33] G. Palmisano, G. Palumbo, S. Pennisi, "Harmonic distortion on class AB CMOS current output stages," *IEEE Trans. Circuits and Systems-II*, vol. CAS-45, pp. 243-250, Feb. 1998.

[34] E. Bruun, "Harmonic distortion in CMOS current-mirrors," *Proc. IEEE Int. Symposium on Circuits and Systems* (ISCAS-98), Monterey, 1998, pp. 567-570.

[35] E. Bruun, "Worst case estimate of mismatch induced distortion in complementary CMOS current mirrors," *Electronics Letters*, vol. 34, pp. 1625-1627, August 1998.

[36] A. Fabre, "Third-generation current conveyor: a new helpful active element," *Electronics Letters*, vol. 31, pp. 338-339, March 1995.

[37] A. Piovaccari, "CMOS integrated third-generation current conveyor," *Electronics Letters*, vol. 31, pp. 1228-1229, July 1995.

Chapter 4

Current-mode feedback amplifiers

4.1 Current-feedback operational amplifier

The current-feedback operational amplifier presented in Figure 4.1a is in effect a positive second generation current-conveyor CCII+ with an additional voltage buffer at the conveyor current output [1–3]. The current at the inverting input of the current-feedback operational amplifier is transferred to the high impedance current-conveyor output Z, causing a large change in output voltage. The current-feedback operational amplifier has a transresistance equal to the impedance level at the conveyor Z-output. Therefore, in the literature, the current-feedback operational amplifier is also referred to as a transimpedance amplifier.

Several semiconductor manufacturers have current-feedback operational amplifiers in their product range. They are typically realised with rather costly and complex complementary bipolar integration processes because a typical bipolar current-feedback operational amplifier, presented in Figure 4.1b, requires pnp-transistors with comparable f_T to npn-transistors for optimal operation. Both the input and output voltage followers are usually implemented with identical push-pull voltage follower structures but, since only the output buffer ($Q_{13}...Q_{16}$) must drive significant amount of current to the load, these buffers are biased differently. In certain current-feedback operational amplifiers, the input voltage follower ($Q_1...Q_4$) is alternatively realised with two current-mirrors, as in the CMOS push-pull CCII+ in Figure 3.16.

In order to reach high transimpedance gain the impedance level at the node TZ is maintained high by cascode current-mirrors, typically Wilson-type current mirrors such as the npn-mirror $Q_5...Q_7$ and the pnp-mirror $Q_9...Q_{11}$. The slew-rate of the current-feedback operational amplifier is large (typically over 1000 V/μs) because the push-pull current-conveyor can supply several (or tens of) milliamperes to the parasitic capacitances at the high impedance node TZ. Because the parasitic capacitance at the node TZ degrades the high frequency performance of the amplifier this node is usually

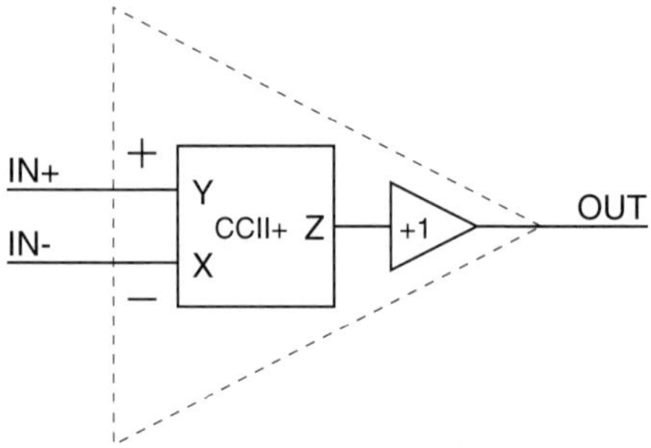

(a)

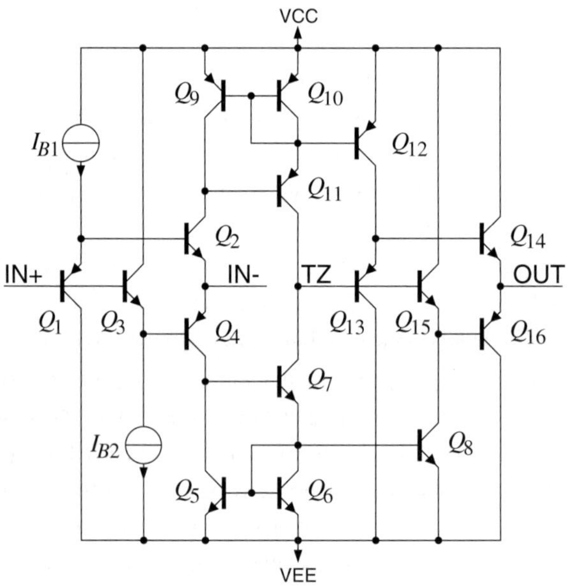

(b)

Figure 4.1 (a) The operating principle of the current-feedback operational amplifier. (b) The simplified schematic of a typical commercial current-feedback operational amplifier.

maintained as an internal node. However, there is at least one commercial current-feedback operational amplifier, the AD844 [4], where the user has access to the high impedance node TZ. This amplifier can also be utilised as a second generation current-conveyor and consequently many other applications are possible with this amplifier than with other commercial current-feedback operational amplifier. Additionally, there is also a very high-speed commercial current-feedback operational amplifier, in which the current-conveyor and the voltage buffer have separate inputs and outputs, namely the OPA 660 [5].

In most current-feedback operational amplifier implementations, the current-mirrors supply dynamic bias currents to the input transistors of the output voltage follower (Q_{13} and Q_{15}) by additional transistors Q_8 and Q_{12}. This helps the output buffer to supply larger currents to the load as the internal circuitry can now supply more base current to the output transistors.

The distortion is low in most current-feedback operational amplifiers because the used output stages, bipolar push-pull voltage followers, have intrinsically low distortion, which is reduced yet further by feedback in closed-loop configuration. However, with simpler voltage follower structures, more distortion is generated, particularly in the case of low impedance loads but even then the distortion performance of the current-feedback operational amplifier compares well to typical voltage-mode operational amplifiers as discussed in Appendix C.

4.1.1 Closed loop bandwidth

The current-feedback operational amplifier can be used as a traditional operational amplifier in inverting and non-inverting configurations if the feedback is resistive, as presented in Figures 4.2a and b. For the calculations, a simplified macromodel for the current-feedback operational amplifier was used (Figure 4.2c). In the internal current-conveyor model the X-terminal admittance Y_x is omitted as the low impedance feedback network dominates the impedance level at the noninverting input anyway. In addition, no high frequency nonidealities are included in the two voltage followers. If we assume a one pole frequency response for the current-feedback operational amplifier so that $A_i(s) = A_i$ and $Y_z = sC_z$ we can express the closed loop voltage gain for the noninverting amplifier circuit as

$$A_{vcl}(s) = \frac{1 + \frac{R_2}{R_1} + s\frac{C_z Z_o}{A_i}}{1 + s\frac{C_z}{A_i}\left(R_2 + Z_o\left(1 + \frac{Z_x}{R_1}\right) + Z_x\left(1 + \frac{R_2}{R_1}\right)\right)}. \tag{4.1}$$

Accordingly, for the inverting voltage amplifier the closed loop voltage gain is

$$A_{vcl}(s) = \frac{-\frac{R_2}{R_1} + s\frac{C_z Z_o Z_x}{A_i R_1}}{1 + s\frac{C_z}{A_i}\left(R_2 + Z_o\left(1 + \frac{Z_x}{R_1}\right) + Z_x\left(1 + \frac{R_2}{R_1}\right)\right)}. \tag{4.2}$$

It can clearly be seen that the frequency response is almost the same for both amplifier topologies, except that the DC-gain and the right half-plane zeroes differ slightly.

If both Z_x and Z_o are assumed far smaller than R_2, the corner frequency does not depend on the resistor ratio $\frac{R_2}{R_1}$ as in voltage-mode operational amplifiers but only on the feedback resistor R_2,

$$\omega_{0cl} = \frac{A_i}{C_z\left(R_2 + Z_o\left(1 + \frac{Z_x}{R_1}\right) + Z_x\left(1 + \frac{R_2}{R_1}\right)\right)} \approx \frac{A_i}{R_2 C_z}. \tag{4.3}$$

Therefore, there is no fixed gain-bandwidth product, as in voltage-mode operational amplifiers. However, there is a limit for the gain-bandwidth product

$$\lim_{R_1 \to 0} \omega_{0cl} \times A_{vcl}(0) = \frac{A_i}{Z_x C_z} \frac{R_2}{R_2 + Z_o} \approx \frac{A_i}{Z_x C_z} \tag{4.4}$$

A typical current-feedback amplifier integrated with a complementary bipolar process has a constant bandwidth of 10...100 MHz up to gains of 20...30 dB. Then the current gain A_i of the internal current-conveyor is one. The constant bandwidth gain range can be increased to 40 dB with additional circuit techniques that enable a slight increase in current gain of the bipolar current-mirrors without sacrificing speed [6]. Increasing the constant bandwidth range to yet higher gains is difficult as in this case the non-dominant poles of the amplifier deriving from the current-mirrors and voltage buffers will limit the bandwidth and cause stability problems. The constant bandwidth can be extended to higher closed loop gains if the feedback resistor R_2 is set to a higher value but this occurs at the expense of bandwidth.

Because of the low impedance level at the inverting input of the current-feedback operational amplifier, all operational amplifier circuits cannot be realised with current-feedback operational amplifiers. For example, in a voltage-follower configuration, two voltage-follower outputs, namely the internal current-conveyor voltage-follower output and the output of the whole amplifier, are connected together. Therefore, there is a minimum resistance which can be connected between the output and the inverting input to ensure stability and fast settling.

In the conventional realisation of a noninverting amplifier of Figure 4.2a, a voltage-mode operational amplifier rather than the current-feedback amplifier is used. In this

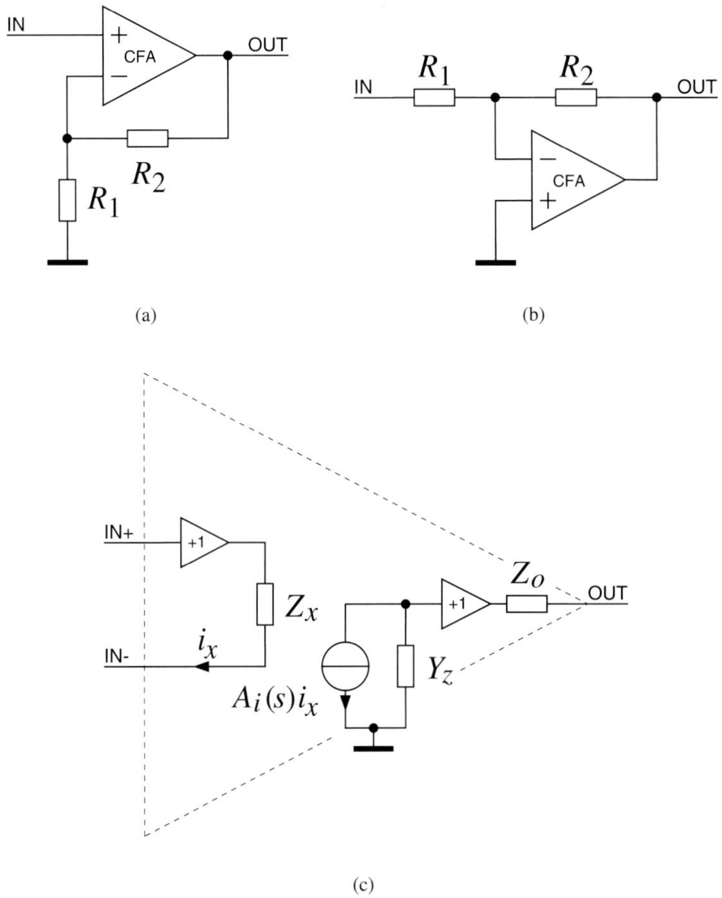

Figure 4.2 Voltage amplifier configurations of current-feedback operational amplifier. (a) Noninverting voltage amplifier. (b) Inverting voltage amplifier. (c) Simplified current-feedback operational amplifier macromodel for the calculations.

case, the closed loop gain becomes

$$A_{vcl}(s) = \frac{1+\frac{R_2}{R_1}}{1+s\frac{C_c}{g_m}\left(1+\frac{R_2}{R_1}\right)}, \tag{4.5}$$

if a one-pole model with infinite DC-gain and zero output impedance is assumed, so that the open loop gain of the voltage-mode operational amplifier is $A_{vol} = \frac{g_m}{sC_c}$. Thus, if the pole of the voltage-mode operational amplifier closed-loop gain is compared to Equation (4.3), a relation between the voltage-mode operational amplifier transconductance and the current-feedback amplifier circuit parameters results

$$g_m = \frac{A_i\left(1+\frac{R_2}{R_1}\right)}{R_2+Z_x\left(1+\frac{R_2}{R_1}\right)}, \tag{4.6}$$

if $C_c = C_z$ and $Z_o = 0$. Therefore, the current-feedback operational amplifier can be considered as a voltage-mode operational amplifier with a variable transconductance and thus with a variable gain-bandwidth product. This account for the high speed operation of the current-feedback operational amplifier as most conventional voltage-mode operational amplifiers are compensated to be unity gain stable, whereas a current-feedback operational amplifier can be left almost uncompensated because its gain-bandwidth product reduces approximately to $\frac{1}{C_zR_2}$ in the unity gain configuration [7]. The compensation of a discrete voltage-mode operational amplifier can be externally adjusted only with an off-chip capacitor and, since this adds more parasitic capacitance to the amplifiers signal path, there are only a few old operational amplifiers in the market with this feature. Therefore, the adjustment of the amount of compensation with the feedback resistor R_2 can be a useful feature in discrete operational amplifier circuits. However, when integrating a large analogue system into one chip this feature is no longer as useful as in this case the compensation capacitor of a voltage-mode operational amplifier is easily rescaled if less than 100% feedback is used.

4.1.2 Integrator implementations

If the feedback resistor R_2 in Figures 4.2b is replaced with a capacitor, an inverting lossless integrator will result as in Figure 4.3a with a transfer function

$$A_{vcl}(s) = \frac{sZ_xZ_oCY_z - A_i}{sC\left(RY_z\left(Z_x+Z_o\right)+RA_i+Z_xZ_oY_z\right)+Y_z\left(R+Z_x\right)}. \tag{4.7}$$

The terms A_i and Y_z are frequency dependent and this results in a third order transfer function. Moreover, since the minimum feedback resistance requirement is broken at high frequencies, it is very difficult to avoid stability problems with this integrator

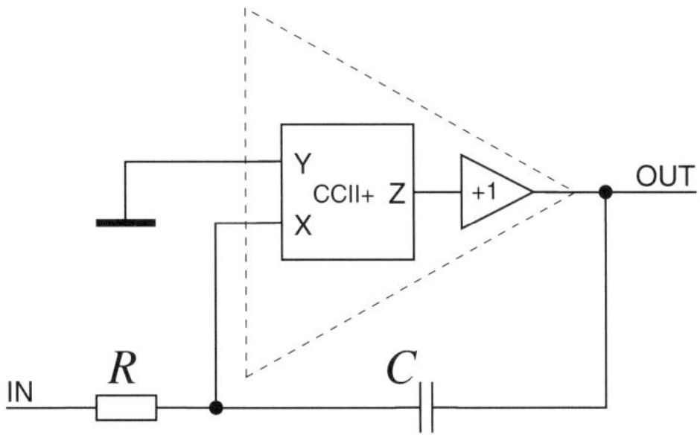

(a)

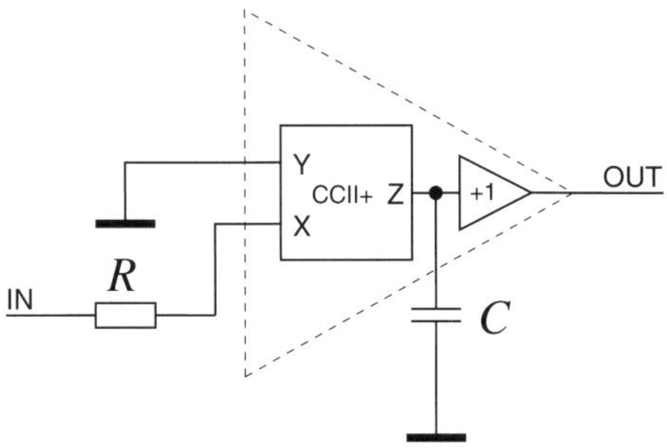

(b)

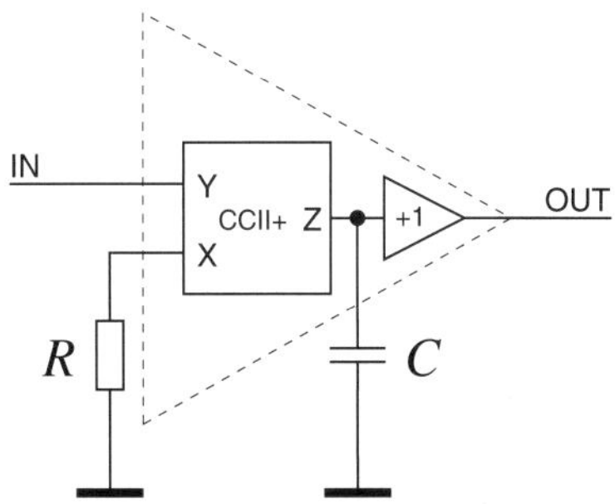

(c)

Figure 4.3 Lossless integrators implemented with current-feedback operational amplifiers. (a) Operational amplifier like inverting integrator. (b) An alternative inverting integrator realisation utilising the internal gain node of the current-feedback operational amplifier. (c) A similar noninverting integrator.

topology. However, it is possible to compensate the resulting integrator with techniques described in [8]. In practice this involves adding capacitance in parallel with Y_z. Unfortunately, this is possible only with a few commercial current-feedback operational amplifiers such as AD844 and OPA660 [4, 5] which has the internal current-conveyor output Z available as an additional pin.

The problems involved in the discussed integrators can be avoided by using an integrator realisation of Figure 4.3b. Then the resulting transfer function is significantly simpler

$$A_{vcl}(s) = -\frac{A_i}{s(R+Z_x)(C+C_z)}. \tag{4.8}$$

Since the dominant pole of the amplifier is now shifted down and used as the integrator time constant good phase response can be obtained even with relatively high frequencies. Furthermore, also noninverting integrators can be realised only by changing the signal input to the noninverting input as depicted in Figure 4.3c. Several filter realisations utilising this kind of integrators have been published [9–12]. However, as all these circuits use the internal current-conveyor output as the integration node these filters are actually current-conveyor filters. Moreover, these filters can be realised only with the AD844 or as full custom integrated circuits.

Furthermore, those commercial current-feedback operational amplifiers without access to the internal conveyor output node can be used to build continuous time filters utilising the active-R technique developed in the 1970's [14]. This technique uses the dominant pole of the voltage-mode operational amplifier as an integrating time constant. However, as the time constant depends on the transconductance of the differential input stage, the integrating time constant is sensitive to process and temperature variation. Moreover, the non-dominant poles of a voltage-mode operational amplifier can cause problems in active-R filters. When this filter technique is applied to current-feedback operational amplifiers the time constants are easily adjusted with external resistors and therefore sensitivities to process variation and parasitic poles can be significantly reduced [15].

4.1.3 Self-compensation of voltage followers

An interesting feature in the current-feedback operational amplifier design is the bandwidth optimisation of voltage followers. A typical voltage follower requires a resistor R rather than a short circuit between inverting input and amplifier output as seen in Figure 4.4a which results in a transfer function

$$A_{vcl}(s) = \frac{1+s\frac{C_z Z_o}{A_i}}{1+s\frac{C_z}{A_i}(R+Z_o+Z_x)}, \tag{4.9}$$

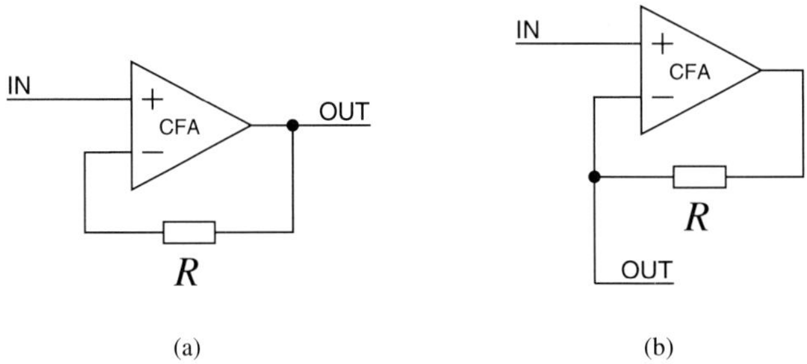

Figure 4.4 Voltage followers realized with current-feedback operational amplifiers. (a) A typical realisation. (b) A self-compensated voltage follower.

if $Y_z(s) = sC_z$ is assumed and the loading effects are neglected. However, if the output signal is taken from the noninverting input of the current-feedback operational amplifier, a circuit depicted in Figure 4.4b [16] results. Then the transfer function changes to

$$A_{vcl}(s) = \frac{1 + s\frac{C_z}{A_i}(R + Z_o)}{1 + s\frac{C_z}{A_i}(R + Z_o + Z_x)}. \tag{4.10}$$

Thus, almost perfect pole zero cancellation can be achieved if R is large enough or Z_x is very small. Unfortunately, the voltage drop across the resistor R can limit the output voltage swing when driving low impedance loads. Too large a feedback resistance can also shift down the pole deriving from the parasitic capacitances at the output of the current-feedback operational amplifier.

4.1.4 Common-mode rejection

The common-mode rejection of current-mode instrumentation amplifiers utilising second-generation current-conveyors was discussed earlier in Section 3.2.3. Because the input stage of the current-feedback operational amplifier is a second-generation current-conveyor similar common-mode behaviour could be expected. However, the common-mode behaviour of the current-mode instrumentation amplifier does not explain the common-mode gain mechanisms in the current feedback operational amplifier as this common-mode signal is inserted into the circuit in an entirely different way.

The common-mode gain of a current feedback operational amplifier is measured and calculated as if it were a normal voltage-mode operational amplifier, as depicted in the test set-up of Figure 4.5a. Therefore, the dominant source of the output common-mode voltage is not the conveyor X-terminal admittance Y_x but rather the gain error in the voltage gain from Y to X (A_{vf}). This gain error produces a Z-output current inversely proportional to the X-terminal impedance. This error is generated by the

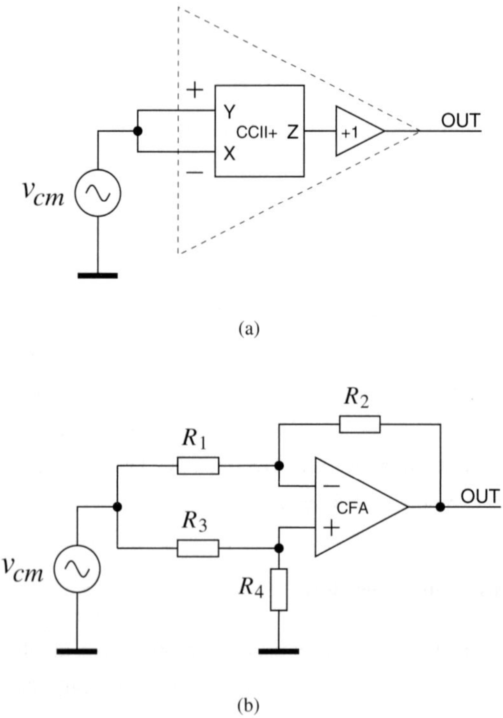

Figure 4.5 (a) CMRR test for a current-feedback operational amplifier. (b) CMRR test for a differential amplifier application.

limited output conductance of the input transistors [1, 17, 18]. Most techniques that improve the DC-accuracy of the current feedback operational amplifier similarly improve the CMRR of the amplifier. In certain current-feedback amplifiers, these errors are reduced by laser trimming.

Similarly, in the test set-up of Figure 4.5b, corresponding mechanisms produce the common-mode output voltage. In the case of voltage-mode operational amplifiers, maximum CMRR is reached with the test set-up of Figure 4.5a. Similar CMRR performance to the test set-up of Figure 4.5b can be reached with ideally matched resistor ratios. However, in the case of a current-feedback operational amplifier, the maximum CMRR is reached in the test set-up of Figure 4.5b. Thus, the resistor ratios are not ideally matched but suitably unbalanced in order to cancel out the input voltage follower gain error and other nonidealities [17].

4.1.5 CMOS implementations

CMOS implementations of current-feedback operational amplifiers have been reported [19–22]. However, CMOS integration technologies entail certain problems with regard to implementing current-feedback operational amplifiers. Using a class-A input conveyor, the input voltage range and the slew rate is comparable to conventional voltage-mode operational amplifiers. If high slew rate required with CMOS implementations, push-pull current-conveyors must be used. In this case, relatively large supply voltages are required and yet very limited input voltage range is achieved. The X-terminal impedance Z_x of the input conveyor plays an important role in maximizing the closed loop bandwidth. Since CMOS voltage followers have higher output impedances than bipolar voltage followers the maximum gain-bandwidth product is significantly lower in CMOS than in bipolar realisations. However, CMOS realisations have at least one advantage over the bipolar realisations: the impedance level at the conveyor output can easily be designed to a high value and thus large low frequency transimpedance gain is easily achieved in CMOS current-feedback operational amplifiers.

Perhaps the most promising CMOS current-feedback operational amplifier topology is presented in Figure 4.6 [20, 21]. This uses first generation current-conveyor like local feedback to lower voltage follower output impedance ,described earlier on page 71. The current-mirror structure used automatically matches the voltage-follower output transistor transconductances and thus optimal output impedance linearity is obtained. The current-gain of the input conveyor is not one but a ratio of NMOS and PMOS transistor process parameters. This uncertainty in the amplifier open loop gain does not present a problem since there is usually enough open loop gain. However, the variation in input conveyor current gain makes it unsuitable for most filter applications.

Recent test results of this amplifier topology realised with a standard n-well 0.6

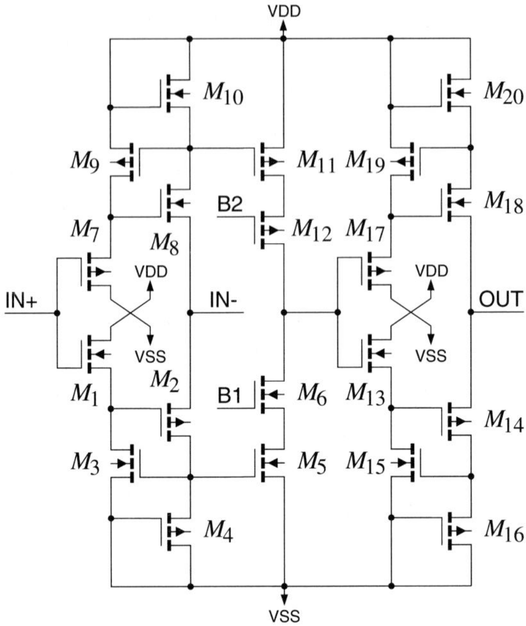

Figure 4.6 A CMOS current-feedback operational amplifier with low Z_x.

μm CMOS process exhibit 120 MHz bandwidth with a 12 mW power dissipation with a single 5 volt supply [21]. Although the performance of this CMOS current-feedback operational amplifier is comparable to bipolar implementations in many respects, the slew rate of the CMOS amplifiers falls more than one decade behind from bipolar implementations. That is partly caused by the mirror topology used. As a consequence of to the bulk effect, the threshold voltages of transistors M_4 and M_{10} are very large, which leads to very small aspect ratios for the main mirror transistors M_3, M_5, M_9 and M_{11} and limited current drive capabilities result. In the case of a twin-well CMOS process, the limitations deriving from the bulk effect can be reduced and in addition to slew rate also input and output voltage range and drive capabilities can be enhanced. Still bipolar transistors have so much better capabilities to handle large currents that the slew rate of bipolar current-feedback operational amplifiers cannot be reached in the near future.

The input voltage range can be increased by using class-A input conveyors and the output voltage range can be increased by using Miller compensated output stages, as reported in [22]. However, this leads to such slow settling and narrow bandwidth that most traditional voltage-mode operational amplifiers perform better. Alternatively, rail-to-rail voltage range operational amplifiers can be used to realise current-conveyors and voltage followers [23]. But, since this entails building one operational amplifier out of two operational amplifiers and additional circuitry, it would hardly appear worth the effort.

The main benefit of the current-feedback operational amplifier lies in obtaining the optimal bandwidth with any closed loop gain without changing the compensation capacitor solely by controlling the feedback resistance. This represents an advantage in discrete operational amplifiers because adding an external compensation pin will significantly lower the available bandwidth caused by increased parasitic capacitance in the signal path. In application specific integrated circuits, however, the compensation of each operational amplifier can easily be optimised. The bipolar implementations of current-feedback operational amplifiers are quite simple since the strong features of bipolar transistors are efficiently utilized whereas the CMOS implementations suffer from bulk effect and low transconductance, leading to complicated circuitry. Therefore, in CMOS integration technology, a traditional voltage-mode operational amplifier with application specific compensation will invariably exceed the performance of a current-feedback operational amplifier.

4.2 Operational floating conveyor

The operational floating conveyor (Figure 4.7) is a current-mode building block that combines the transmission properties of a current-conveyor and a current-feedback operational amplifier, and has an additional output current sensing capability [27]. The matrix representation of the operational floating conveyor is

$$\begin{bmatrix} v_x \\ i_y \\ v_w \\ i_z \end{bmatrix} = \begin{bmatrix} 0 & 1 & 0 & 0 \\ 0 & 0 & 0 & 0 \\ Z_t & 0 & 0 & 0 \\ 0 & 0 & -1 & 0 \end{bmatrix} \begin{bmatrix} i_x \\ v_y \\ i_w \\ v_z \end{bmatrix}, \tag{4.11}$$

where Z_t is the transimpedance of the internal current-feedback operational amplifier.

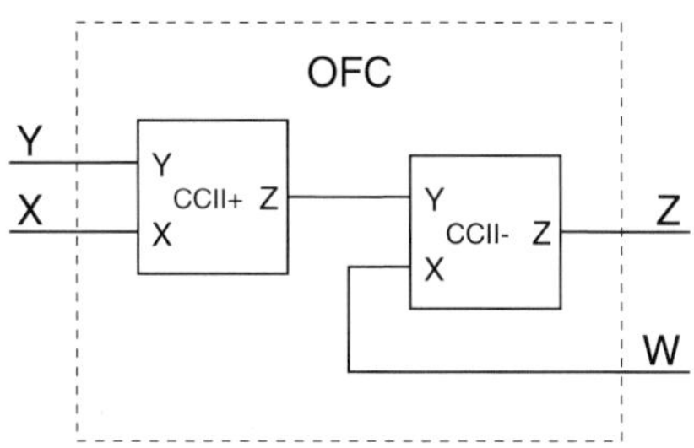

Figure 4.7 The operating floating conveyor constructed of two second generation current-conveyors.

If a current-conveyor is a voltage-follower with an additional output current-sensing circuit, the operational floating conveyor is a current-feedback operational amplifier with a similar output current-sensing circuit. Alternatively this conveyor can be constructed of two cascaded current-conveyors.

4.2.1 Applications

With this circuit, it is possible to realise all four types of amplifiers: voltage, current, transconductance, and transimpedance amplifiers, as presented in Figure 4.8. The voltage amplifier in Figure 4.8a operates identically to the current-feedback operational amplifier realization of the noninverting voltage amplifier in Figure 4.2a and thus the transfer function of Equation (4.1) similarly applies here. If the macromodel of the current-feedback operational amplifier, presented in Figure 4.2c, is extended by adding current gain $A_{io} = \frac{i_z}{i_w}$ and the current gain of the input conveyor is renamed A_{ii}, the closed loop current gain of the circuit in Figure 4.2b can be expressed as

$$A_{icl}(s) = A_{io} \frac{1 + \frac{R_2}{R_1} + s\frac{C_z Z_x}{A_{ii}}}{1 + s\frac{C_z}{A_{ii}}\left(R_2 + Z_o\left(1 + \frac{R_2}{R_1}\right) + Z_x\left(1 + \frac{Z_o}{R_1}\right)\right)}. \tag{4.12}$$

Accordingly, the transfer function of the transconductance amplifier is

$$G_{mcl}(s) = A_{io} \frac{-\frac{1}{R_1} + s\frac{C_z}{A_{ii}}}{1 + s\frac{C_z}{A_{ii}}\left(R_2 + Z_o\left(1 + \frac{R_2}{R_1}\right) + Z_x\left(1 + \frac{Z_o}{R_1}\right)\right)}. \tag{4.13}$$

The transimpedance amplifier similarly uses only the current-feedback operational amplifier part of the operational floating conveyor and the resulting transfer function is

$$R_{mcl}(s) = \frac{-R + s\frac{C_z}{A_{ii}} Z_x Z_o}{1 + s\frac{C_z}{A_{ii}}\left(R + Z_o + Z_x\right)}. \tag{4.14}$$

In the first three amplifier types, i.e. the voltage, current, and transconductance amplifiers, it is possible to maintain constant bandwidth at moderately large closed loop gains by keeping the resistor R_2 constant and adjusting the gain by the resistor R_1. However, in the transresistance amplifier, there is a similar gain-bandwidth product limitation to that in voltage-mode operational amplifier circuits.

The four amplifier types can also be realised with second generation current-conveyors as open loop amplifiers. However, when operational floating conveyor realisations are used, the amplifier gain is less sensitive to finite X-terminal impedance. Since the feedback reduces impedance levels at both X- and W-terminals the bandwidths of the amplifiers are less sensitive to parasitic capacitances. Furthermore, the

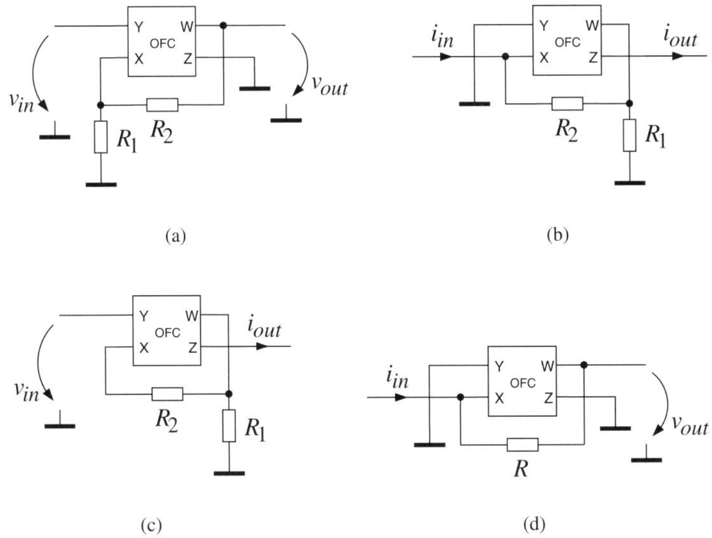

Figure 4.8 Basic amplifier types realised with operational floating conveyor. (a) Voltage amplifier $A_v \approx 1 + R_2/R_1$. (b) Current amplifier $A_i \approx 1 + R_2/R_1$. (c) Transconductance $G_m \approx 1/R_1$. (d) Transresistance $R_m \approx R$.

feedback also reduces distortion at low frequencies but still the current signal path from W- to Z-terminal remain outside the feedback loop and thus the nonlinearity remains unchanged in that part.

4.2.2 Composite conveyors

The operational floating conveyor can be also configured to form a high performance second generation current-conveyor as presented in Figure 4.9a. This is a useful technique for designing CMOS current-conveyors: with two poorly operating simple CMOS positive second generation conveyors, one positive conveyor with enhanced X-terminal impedance Z_x can be constructed. In the case of simple CMOS conveyors even the resistor R_F can generally be omitted as the X-terminal is high enough to prevent any stability and settling problems.

There is an alternative way to construct a composite conveyor which lowers the X-terminal impedance. This composite CCII- is presented in Figure 4.9b [28]. In this composite conveyor, the lower conveyor CC2 works as a negative impedance conveyor and consequently the X-terminal impedance of the composite CCII- is

$$Z_{x,composite} = Z_{x1} + A_{i2} Z_{x2} \approx Z_{x1} - Z_{x2}. \tag{4.15}$$

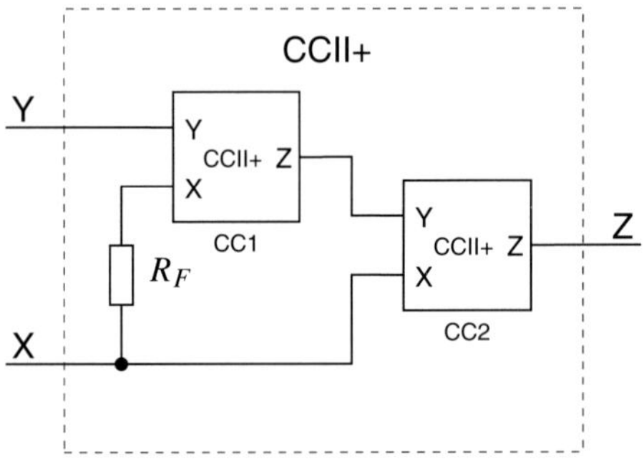

(a)

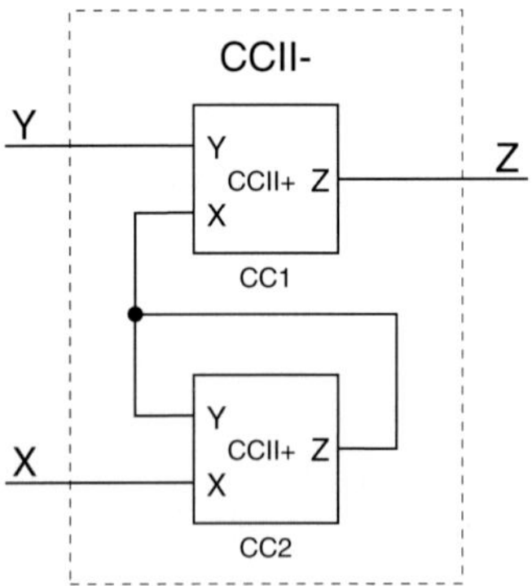

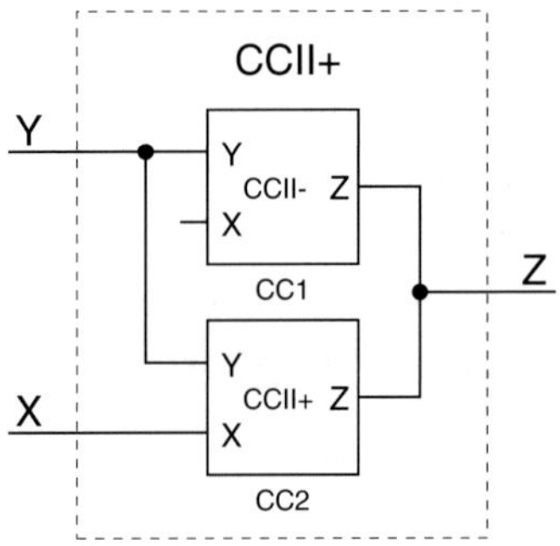

(b) (c)

Figure 4.9 Different composite conveyors. (a) A composite CCII+ with enhanced Z_x resembling an operational floating conveyor. (b) A composite CCII- with a different technique to lower Z_x. (c) A composite CCII+ with enhanced Y_x.

One can see clearly this technique is sensitive to mismatches and therefore the current gain A_{i2} should be designed slightly lower than one in order to prevent a negative X-terminal impedance for the composite conveyor. In addition, all even order nonlinearities in Z_{x1} and Z_{x2} are effectively summed together and hence X-terminal impedance nonlinearity is increased. Fortunately, in most cases the nonlinearity of the X-terminal impedance has little effect on the total amplifier distortion. Yet the overall performance of the current-conveyor build with the operational floating conveyor is better and less sensitive to process mismatches.

Composite techniques can be used also to enhance other current-conveyor parameters as the example in Figure 4.9c shows [29]. This composite conveyor reduces the unwanted X-terminal admittance by subtracting a replica of the error current deriving from Y_{x2} by using the additional conveyor CC1 with an opposite current gain polarity. However, maintaining an accurate replica of the error current is difficult at high frequencies with different types of conveyors, but in most cases performance enhancement at low frequencies is enough.

4.3 Current-mode operational amplifiers

Voltage-mode circuits can be converted into current-mode circuits by implementing the adjoint network transformation principle, as described in Section 1.4. According to this principle, a voltage-mode operational amplifier differential input is converted to a current-mode operational amplifier differential output and accordingly the voltage-mode output is converted to a current-mode single-ended input, as shown in Figure 4.10. Similarly, a typical voltage division feedback network in the voltage-mode circuit is converted to a current division feedback network in the current-mode circuit.

A CMOS implementation of the current-mode operational amplifier is presented in Figure 4.11 [30]. High gain is achieved in this circuit in the same way as in current-feedback operational amplifiers by driving the current-conveyor Z-output current to a high impedance load, which in this case is a push-pull CMOS differential stage.

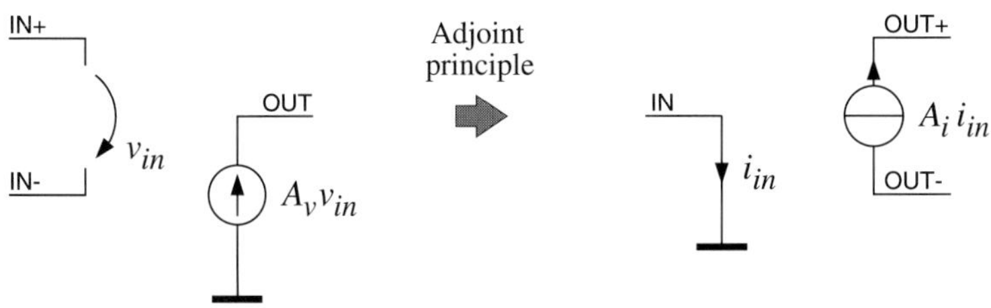

Figure 4.10 The voltage-mode operational amplifier and its current-mode counterpart derived by the adjoint network principle.

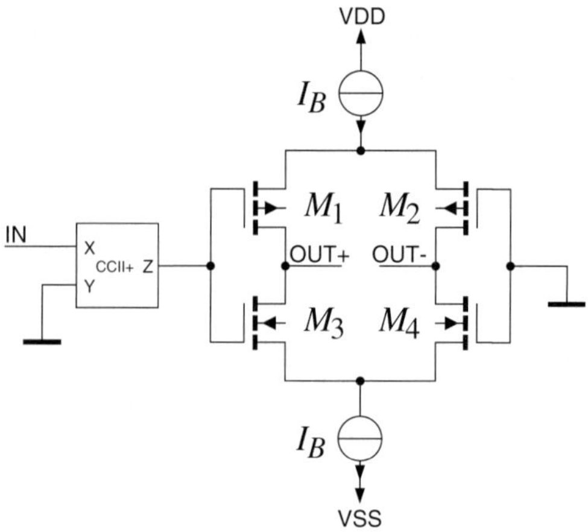

Figure 4.11 CMOS implementation of the current-mode operational amplifier.

In order to ensure high gain and slew rate, the positive conveyor driving the output stage should be a push-pull CMOS conveyor with cascodes at the output such as that presented in Figure 3.20.

The closed loop current gain of the current-mode operational amplifier is gain-bandwidth product limited as with the closed loop voltage gain in the voltage-mode operational amplifier. The unity gain frequency depends on the total transconductance of the output stage $g_{mo} = g_{m1} + g_{m3}$ and the parasitic capacitances at the conveyor Z-output C_z, so that

$$\omega_0 \approx \frac{g_{mo}}{C_z}. \tag{4.16}$$

The non-dominant poles of the amplifier are largely deriving from the current-mirrors in the input current-conveyor. Therefore, in order to reach sufficient phase margin, the bias current of the input conveyor cannot be aggressively scaled down.

4.3.1 Distortion

Because the current signal amplitude in the input conveyor is the same as the input current signal, the only dominant source of distortion in the amplifier is the output differential stage. Therefore, using similar assumptions as those used in the derivation of the distortion of the CMOS folded cascode OTA in Appendix C, the drain current of one of the four output transistors is

$$i_D = \frac{I_B}{2} + \frac{v_z}{4}\sqrt{\beta\left(4I_B - \beta v_z^2\right)}, \text{ if } |v_z| \leq \sqrt{\frac{2I_B}{\beta}}, \tag{4.17}$$

where v_z is the voltage signal at the current conveyor Z-output. As a consequence of the push-pull connection, the inverting output current is

$$i_{OUT-} = \frac{v_z}{2}\sqrt{\beta\left(4I_B - \beta v_z^2\right)}, \tag{4.18}$$

which is identical to the nonlinearity of normal differential stage of Equation (C.21). Similarly, the output current can be expressed as a function of input current as a Taylor-series

$$i_{OUT} = R_m i_{in}\sqrt{\beta I_B} - \frac{\beta R_m^3 i_{in}^3}{8}\sqrt{\frac{\beta}{I_B}} - \ldots, \tag{4.19}$$

where $R_m = \frac{v_z}{i_{in}}$ is the transimpedance gain of the input current-conveyor. Then, by using Equation (A.21) at low frequencies the third order distortion of the current-mode operational amplifier is

$$HD3(0) = \frac{1}{32}\left(\frac{\hat{i}_{out}}{I_B}\right)^2 \frac{f^2 A_{ol}^2}{(1+fA_{ol})^3} \approx \frac{1}{32 f A_{ol}}\left(\frac{\hat{i}_{out}}{I_B}\right)^2, \tag{4.20}$$

where $A_{ol} = R_m\sqrt{\beta I_B}$ and $\hat{i}_s = f\hat{i}_{out}$.

This equation is again very similar to the harmonic distortion Equation (C.24) for the folded cascode CMOS operational transconductance amplifier (OTA). However, the distortion of the OTA depends only indirectly on the signal amplitude since the distortion depends on the output current amplitude rather than output voltage. Thus, in the case of a high-impedance load, the distortion of the OTA is very low. However, the distortion of the OTA increases rapidly if the load impedance is decreased because of the load dependent open-loop voltage gain whereas the distortion of the current-mode operational amplifier does not depend on the load impedance.

At high frequencies, a one-pole open-loop transfer function $A_{ol} = \frac{\omega_0}{s}$ can be assumed in which the unity gain frequency is already defined in Equation (4.16). Then using Equation (A.39), the third order harmonic distortion can be described as a function of frequency

$$HD3(\omega) = \frac{3}{32}\left(\frac{\hat{i}_{out}}{I_B}\right)^2 \frac{\omega f^2 \omega_0^2}{\left(\omega^2 + f^2\omega_0^2\right)\sqrt{9\omega^2 + f^2\omega_0^2}}. \tag{4.21}$$

When the signal frequency is significantly lower than $\frac{f\omega_0}{3}$ the distortion equation reduces to

$$HD3(\omega)\Big|_{\omega \ll \frac{f\omega_0}{3}} = \frac{1}{32}\left(\frac{\hat{i}_{out}}{I_B}\right)^2 \frac{1}{fA_{ol}(3\omega)}. \tag{4.22}$$

These distortion equations are again very similar to the distortion equations of the

CMOS OTA. However, below $\frac{f\omega_0}{3}$ the distortion of the current-mode operational amplifier is proportional to frequency whereas the distortion of the OTA is proportional to the third power of frequency. Above ω_0 the distortion of the OTA is fixed and large whereas the distortion of the current-mode operational amplifier is inversely proportional to the second power of frequency.

The noninverting output of the current-mode operational amplifier is outside the feedback loop and therefore similar distortion occurs to that in simple current buffers discussed in Chapter 2.2. This distortion is frequency dependent and the distortion peak is usually above the unity gain frequency of the amplifier and thus no dramatic increase in distortion is expected.

4.3.2 Slew rate and full power bandwidth

In the case of a push-pull input conveyor, the slew-rate depends primarily on the current handling capabilities of the conveyor. The maximum output current of the conveyor depends on the used cascode current-mirror topology and on the biasing of the cascode transistors. The voltage swing at the conveyor output is not large and the nonlinearity of the input stage does not contribute a great deal to the total distortion of the amplifier. Thus, the used cascode structures can only be optimised for high current drive capability and consequently the slew rate is usually sufficiently large for most applications.

The slew rate of the CMOS current-mode operational amplifier is rather good even if a class-A CCII+ is used at the input. In a standard Miller-compensated CMOS voltage-mode operational amplifier, the slew rate depends on the compensation capacitor C_c and differential pair tail current I_{BD}, so that

$$SR_V = \left.\frac{dV}{dt}\right|_{max} = \frac{I_{BD}}{C_c}. \tag{4.23}$$

In the current-mode operational amplifier, slew rate limiting occurs at the Z-terminal output so that the maximum voltage change there is $\frac{I_{Zmax}}{C_z}$ and by using Equation (4.16) the resulted slew rate at the output is

$$SR_I = \left.\frac{dI}{dt}\right|_{max} = \frac{g_{mo} I_{Zmax}}{C_z} = \omega_0 I_{Zmax}. \tag{4.24}$$

Since the slew rate of the current-mode operational amplifier is actually a function of gain-bandwidth product, the equation for full power bandwidth becomes quite interesting:

$$\omega_M = \frac{SR_I}{I_{Omax}} = \omega_0 \frac{I_{Zmax}}{I_{Omax}}. \tag{4.25}$$

Therefore, the full power bandwidth can be set in respect to gain-bandwidth product directly by the ratio of the input and output stage bias currents. A similar ratio can be found also for voltage-mode operational amplifiers as

$$\frac{\omega_M}{\omega_0} = \frac{\frac{I_{BD}}{C_c V_{Omax}}}{\frac{g_{mi}}{C_c}} = \frac{I_{BD}}{g_{mi} V_{Omax}}. \tag{4.26}$$

For bipolar transistors the ratio of transconductance and collector current $\frac{g_m}{I_C}$ is an unscalable parameter $\frac{q}{nkT}$ and the same applies also to MOS-transistors operating in weak inversion. This leads to a full power bandwidth roughly ten times below gain-bandwidth product even with a moderate one volt output signal amplitude. Today when MOS channel lengths are scaled aggressively down the $\frac{g_m}{I_D}$ ratio is getting closer to $\frac{q}{nkT}$ even in strong inversion and thus the current-mode operational amplifier becomes more and more feasible at least when the full power bandwidth is considered.

4.3.3 Alternative topologies

However, the current-mode operational amplifier derived by the adjoint principle does entail certain problems. In voltage-mode circuits, output signals are easily shared among several inputs whereas in current-mode circuits, multiple outputs must be provided when driving several inputs or the output current must be converted to a voltage. The output voltage swing of this output structure is severely limited and if cascode transistors are added at the output, this voltage swing shrinks to almost nothing. Therefore, increasing the output impedance is difficult and only low impedance loads, such as other current-mode operational amplifiers, can be driven with this output structure. The feedback impedances must also be kept low since this increases voltage signal amplitude at the output.

The output voltage swing can be extended if the drain currents of the output transistors are fed to the output nodes via additional current mirrors [31]. However, this increases the circuit complexity yet further and adds additional poles to the transfer function. Since the noninverting output is outside the feedback loop, such additional mirrors add a significant amount of distortion to the signal. Alternatively, rather than mirroring, the drain currents can be folded to the output [32]. The resulting output stage is effectively a folded cascode operational transconductance amplifier (OTA), presented in most textbooks in analogue integrated circuit design [24–26]. The folded cascode OTA as an output stage does not add as much distortion as the output mirrors and the high frequency behaviour is more satisfactory.

An alternative current-mode operational amplifier has also been proposed [33] (Figure 4.12) that has a differential input and a single-ended output similar to a voltage-mode operational amplifier. In this structure it is easier to multiply output branches

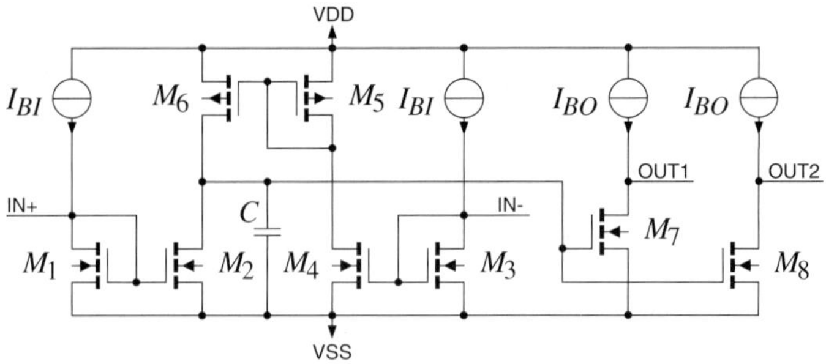

Figure 4.12 A CMOS current-mode operational amplifier with differential input and single-ended output and optional replica output.

because of the simple output structure. However, this structure is difficult to apply because of the lack of input symmetry: in closed-loop circuits the inverting input has a significantly lower input impedance than the non-inverting input. In voltage-mode CMOS operational amplifiers ,both inputs have very high impedance levels but the input impedance of a CMOS current-amplifier is not very low, even with feedback, which may represent a serious problem.

Yet more serious problems with this circuit involve the power efficiency and the distortion performance. By way of illustration, we might consider a unity gain noninverting current amplifier configuration where the output OUT1 is short-circuited to the inverting input and the replica output OUT2 is used to supply the output current signal to the load. In this case, the current-mirror transistors M_1 and M_2 at the noninverting input must handle signal amplitudes as large as those handled by the outputs and consequently the input stage bias current I_{BI} must be equal to the output stage bias current I_{BO}. Since one current mirror at the output and the replica output OUT2 are outside the feedback loop, the distortion performance of this current follower is comparable to a current follower realised with two cascaded NMOS current-mirrors with comparable aspect ratios and bias currents. Additionally, the current-mode operational amplifier must be compensated with the capacitor C which reduces the closed-loop bandwidth compared to two current-mirrors. Therefore, the current-follower based on this current-mode operational amplifier has at least 50% bigger area and power consumption and barely the same performance as two cascaded NMOS current-mirrors.

There are older current-mode operational amplifier realisations with differential inputs and outputs [34, 35]. However, this topology uses three current-conveyors and, as a consequence of the circuit complexity, it is rather slow. Furthermore, the differential inputs in this topology have the same problems as the amplifier in Figure 4.12 and it has a considerable amount of circuitry outside the feedback loop.

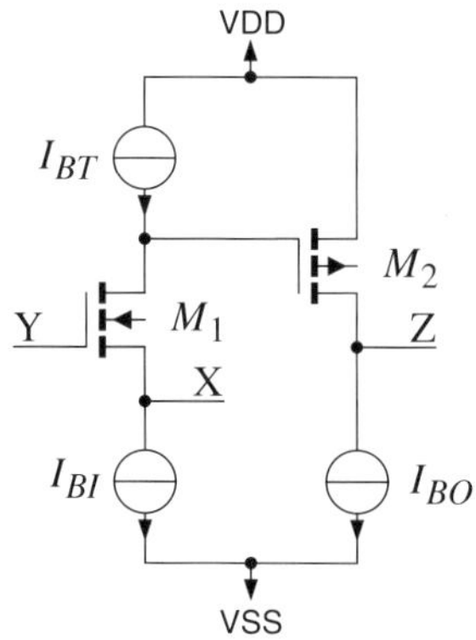

Figure 4.13 A simple class-A CMOS realisation of a high-gain current-conveyor CCII∞.

4.4 High-gain current-conveyor CCII∞

A voltage-mode circuit transformed to a current-mode circuit with the adjoint principle preserves all nonidealities of the circuit. Therefore, the used current-mode amplifier should exhibit better performance characteristics than the voltage-mode amplifier in order to justify the effort of using current-mode techniques. Unfortunately, most current-mode CMOS operational amplifiers discussed in the previous section are more complicated than commonly used voltage-mode operational amplifier topologies or have limited signal ranges or other problems. Thus, simpler circuit topologies should be found if a current-mode amplifier faster than voltage-mode amplifiers with otherwise comparable performance is desired.

A simple high-gain current-mode amplifier can be realised in CMOS technology with the circuit principle in Figure 4.13 [36–38]. This amplifier is similar to a second generation current-conveyor but it has a large current gain from X to Z rather than the unity gain of the standard CCII so to characterise it as a high-gain second generation current-conveyor CCII∞ would not be far fetched. This amplifier is constructed of a negative second-generation current-conveyor CCII- and a transconductance output buffer. The high gain is again achieved by the high impedance at the internal node achieved by a MOS cascode current-source I_{BT}.

The high-gain current-conveyor concept, using bipolar transistors, is published earlier, in [1,39]. However, the fully bipolar realisations such as the class-AB amplifier presented in Figure 4.14 [39] are more complicated than conventional bipolar voltage-mode operational amplifiers.

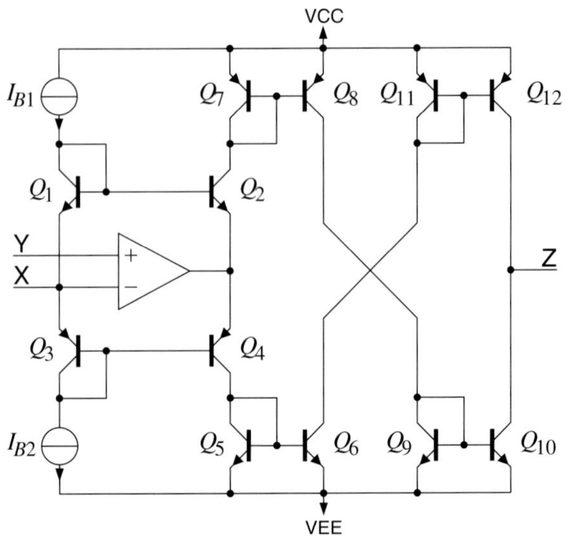

Figure 4.14 A simplified schematic of a bipolar class-AB high-gain current-conveyor CCII∞.

4.4.1 Linear nonidealities

Because the high-gain current-conveyor CCII∞ differs from regular second generation current-conveyors only in its current gain, the matrix representation of the current-conveyor nonidealities in Equation (3.18) can similarly be applied to this conveyor. The forward current gain of the CMOS conveyor in Figure 4.13 can then be derived as

$$\begin{aligned} A_{if}(s) &= \frac{(g_{m1}+g_{ds1})(sC_{gd2}-g_{m2})}{g_t(g_{m1}+g_{ds1}+Y_x)+g_{ds1}Y_x+s(C_t+C_{gd2})(g_{m1}+g_{ds1}+Y_x)} \\ &\approx \frac{g_{m1}(sC_{gd2}-g_{m2})}{(g_{m1}+sC_x)(g_t+sC_t)}, \end{aligned} \tag{4.27}$$

where $Y_x = g_{bi} + sC_x$ is the sum of the output conductance of the current source I_{BI} and all parasitic capacitances at the X-terminal. Similarly, $Y_t = g_t + sC_t$ is the sum of the output conductance of the current source I_{BT} and all parasitic capacitances at the internal high impedance node. Since the DC-gain of the amplifier $A_{if}(0)$ is approximately $\frac{g_{m2}}{g_t}$, high current gain can be achieved by using a cascode current source as the current source I_{BT}. The unity current gain frequency ω_0 depends on the output stage transconductance and is approximated as $\frac{g_{m2}}{C_t}$. Similarly, the nondominant pole depends on the input stage transconductance and is approximately $\frac{g_{m1}}{C_x}$. The zero deriving from the output transistor Miller-capacitance C_{gd2} is negligible in most cases as it is at significantly higher frequency than the nondominant pole.

The X-terminal impedance is quite high and frequency dependent:

$$
\begin{aligned}
Z_x(s) &= \frac{g_t + g_{ds1} + s\left(C_t + C_{gd2}\right)}{g_t\left(g_{m1} + g_{ds1} + Y_x\right) + g_{ds1}Y_x + s\left(C_t + C_{gd2}\right)\left(g_{m1} + g_{ds1} + Y_x\right)} \\
&\approx \frac{g_{ds1} + sC_t}{\left(g_{m1} + sC_x\right)\left(g_t + sC_t\right)}.
\end{aligned}
\tag{4.28}
$$

Fortunately, this impedance is high only at moderately low frequencies and any feedback in the amplifier efficiently lowers this impedance. However, with this circuit topology, as low input impedances cannot be reached as, for example, with the CMOS current-mode operational amplifier of Figure 4.11. At high frequencies the X-terminal impedance reaches the typical value $\frac{1}{g_{m1}}$ and consequently the pole deriving from the X-terminal parasitic capacitance remains approximately $\frac{g_{m1}}{C_x}$ regardless of the frequency dependencies in Z_x. The moderately high X-terminal impedance can still be undesirable in certain applications, in which case increasing the channel length of M_1 or using other techniques described in Section 2.2, such as regulated cascodes, can be used to lower Z_x.

There is also a nonzero reverse voltage gain present in this circuit, approximated as

$$
\begin{aligned}
A_{vr}(s) &= \frac{sg_{ds1}C_{gd2}}{g_t\left(g_{m1} + g_{ds1} + Y_x\right) + g_{ds1}Y_x + s\left(C_t + C_{gd2}\right)\left(g_{m1} + g_{ds1} + Y_x\right)} \\
&\approx \frac{sg_{ds1}C_{gd2}}{\left(g_{m1} + sC_x\right)\left(g_t + sC_t\right)}.
\end{aligned}
\tag{4.29}
$$

This reverse voltage gain can be converted into an admittance between X- and Z-terminals by using Equation (3.27)

$$
Y_{zx}(s) \approx \frac{A_{vr}}{Z_x} = \frac{sg_{ds1}C_{gd2}}{g_{ds1} + g_t + s\left(C_t + C_{gd2}\right)} \approx \frac{sg_{ds1}C_m}{g_{ds1} + sC_t}.
\tag{4.30}
$$

This admittance is in effect a series connection of a capacitance C_{gd2} and a conductance $g_{ds1}\frac{C_{gd2}}{C_t}$. Therefore, the effect of the reverse gain can be decreased by decreasing g_{ds1} or C_{gd2} or increasing C_t. Because this admittance is parallel to any feedback admittance between X- and Z-terminals, one can clearly see whether the reverse gain has an effect in the closed loop circuit or not. Regardless, in the case of a low impedance feedback network, the reverse gain is negligible and, in the case of very high feedback impedances, adding a cascode transistor at the drain of M_2 will efficiently attenuate the reverse gain.

Other nonidealities have quite similar mechanisms to other simple class-A second

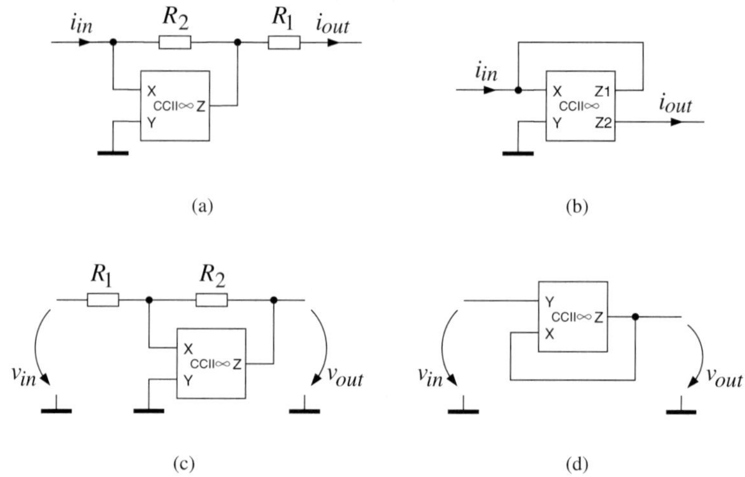

Figure 4.15 Closed-loop amplifiers realised with a CCII∞. a) Inverting current amplifier. b) Positive second generation current-conveyor CCII+ c) Inverting voltage amplifier. d) Voltage follower.

generation current-conveyors. The forward transconductance

$$\begin{aligned} G_{mf}(s) &= \frac{g_{m1} Y_x (sC_{gd2} - g_{m2})}{g_t (g_{m1} + g_{ds1} + Y_x) + g_{ds1} Y_x + s(C_t + C_{gd2})(g_{m1} + g_{ds1} + Y_x)} \\ &\approx Y_x A_{if} \end{aligned} \quad (4.31)$$

is arising from the output conductance of the current source I_{BI} and any parasitic capacitance at the X-terminal. Similarly, the derivation of the X- and Z-terminal admittances is straightforward.

Choosing a NMOS-transistor as the input transistor M_1 is feasible for many reasons. A large transconductance is easily achieved with a NMOS-transistor even with a low bias current and thus also low g_t is achieved to ensure high DC-gain, and the non-dominant pole remain at a sufficiently high frequency. Similarly, the channel length of the input transistor M_1 should be longer than the minimum length because decreasing g_{ds1} lowers both the X-terminal impedance and the reverse voltage gain. Similarly, the output common-source stage is feasible to realise with a PMOS-transistor since a large output current can be achieved without an excessive transconductance, which would degrade the phase margin of the amplifier.

4.4.2 Applications

In Figure 4.15a an inverting closed-loop current amplifier, realised with the CCII∞, is presented, in which case neglecting reverse gains and X-terminal admittance causes the closed-loop current gain to be derived as

$$A_{icl}(s) = \frac{Z_x - A_{if}(s)R_2}{(1 + A_{if}(s))R_1 + (1 + Y_z R_1)(R_2 + Z_x)}. \quad (4.32)$$

If one assumes a one-pole approximation for the forward current gain so that $A_{if}(s) = \frac{\omega_0}{s}$, the corner frequency of the closed-loop current amplifier is

$$\begin{aligned} \omega_{0cl} &= \frac{\omega_0}{1 + \frac{R_2 + Z_x}{R_1}\left(1 + \frac{Y_z}{R_1}\right)} \\ &\approx \frac{\omega_0}{1 + \frac{R_2}{R_1}}. \end{aligned} \quad (4.33)$$

Thus, the gain-bandwidth product of the closed-loop current amplifier is fixed as in normal operational amplifier circuits.

If two identical current outputs are provided in the CCII∞, a CCII+ can also be realised as in Figure 4.15b is presented. Consequently, using Equations (4.27) and (4.28), the closed-loop X-terminal impedance is approximated below the corner frequency as

$$Z_{xcl}(s) \approx \frac{Z_x}{1 + A_{if}(s)} \approx \frac{g_{ds1} + sC_t}{g_{m1}g_{m2}}. \quad (4.34)$$

Typically, the closed-loop X-terminal impedance at low frequencies is approximately ten ohms and is thus comparable to the input impedance of the class-A CMOS CCI. However, this input impedance is still higher than in the composite current-conveyor of Figure 4.9a or in the CMOS current-mode operational amplifier of Figure 4.11.

Because adding an external current output to the CCII∞ involves adding only two MOS-transistors, scaling of these transistor is also possible. Thus, closed-loop inverting current amplifiers with arbitrary current gains can be realised without resistors. This is a useful feature in analogue integrated circuits, as in most integration processes, integrated high quality resistors are either available as an expensive option or are not available at all.

The closed-loop voltage gain of the inverting closed-loop voltage amplifier of Figure 4.15c is derived as

$$A_{vcl}(s) = -\frac{R_2}{R_1}\frac{A_{if}(s) - \frac{Z_x}{R_1}}{1 + \frac{Z_x}{R_1} + A_{if}(s) + Y_z\left(R_2 + Z_x\left(1 + \frac{R_2}{R_1}\right)\right)}. \quad (4.35)$$

Generally, R_1 is significantly larger than Z_x and thus the closed-loop voltage gain can be approximated as

$$A_{vcl}(s) \approx -\frac{R_2}{R_1}\frac{A_{if}(s) - \frac{Z_x}{R_1}}{1 + A_{if}(s) + Y_z R_2}. \quad (4.36)$$

If $A_{if}(s) = \frac{\omega_0}{s}$ is again assumed, the corner frequency of the closed-loop voltage amplifier is

$$\omega_{0cl} = \frac{\omega_0}{\left(1+\frac{Z_x}{R_1}\right)(1+Y_zR_2)+Y_zZ_x} \approx \frac{\omega_0}{1+Y_zR_2}. \tag{4.37}$$

Therefore, the closed loop corner frequency is independent of closed loop gain if the input resistor R_1 is large compared to the X-terminal open-loop input impedance and if the feedback resistor R_2 is small compared to the Z-terminal open-loop output impedance.

In most cases, the impedance level at the Z-output is higher in the closed-loop voltage amplifier than in the closed-loop current amplifier. Moreover, when both amplifiers are set to a gain of -1, so that $R_1 = R_2$ in the voltage amplifier the entire Z-output current is fed to X-input, resulting in a 100% feedback, whereas in the current amplifier only half of the Z-output current is fed to X-input resulting in a 50% feedback. Therefore, the closed-loop voltage amplifier is more sensitive to parasitic capacitances at the Z-output and additional compensation capacitance must be added to the gate of the output transistor in order to reach for the same phase margin than with the closed-loop current amplifier with the same closed-loop gain.

The example of the inverting voltage amplifier reveals that the high-gain current-conveyor CCII∞ can be used as a replacement for the current-feedback operational amplifier. Unlike the current-feedback amplifier this conveyor is stable with all feedback impedances. Therefore, in the voltage follower of Figure 4.15d, there is no resistor required in the feedback path. Because of the high-impedance output of the high-gain current-conveyor, the output impedance of closed-loop voltage amplifiers is higher than with a current-feedback operational amplifier. For the voltage follower this output impedance can be derived as

$$Z_{ocl}(s) \approx \frac{Z_x}{1+A_{if}}. \tag{4.38}$$

Although this output impedance may seem to be high when compared with a current-feedback operational amplifier, it remains comparable to the output impedance of unbuffered CMOS operational amplifiers such as the amplifier in Figure C.1.

Since the high gain current-conveyor is stable with all feedback impedances, capacitive feedback is also possible and most active filter topologies using voltage-mode operation amplifiers can be converted with the adjoint principle to high gain current-conveyor circuits. However, the CCII∞ can additionally be used as a direct replacement for the voltage-mode operational amplifier in most cases. The input structure of the Y-terminal must then be modified as, with the simple CCII∞ realisation of Figure 4.13, a large DC-voltage difference occurs between Y- and X-terminals. This offset voltage can be reduced by adding a level shifter to the Y-input, resulting for example in an input structure, as in the class-A CMOS CCII+ of Figure 3.8c.

Integrators can also be realised with high gain current-conveyors by using the internal gain node capacitance C_t as the integrating capacitor as with current-feedback operational amplifiers. Thus, in the current gain Equation (4.27), the dominant pole is shifted down while the nondominant pole and the zero remain unchanged. Moreover, increasing C_t also lowers X-terminal impedance at low frequencies in Equation (4.28) and the reverse admittance Y_{zx} at high frequencies in Equation (4.30). Therefore, a better high frequency performance can be achieved with this type of integrator. However, all this happens at the expense of linearity, as explained in the following section.

4.4.3 Distortion

As in high gain amplifiers in general the nonlinearity of the high gain current-conveyor is arising from the nonlinearity of the output stage. This nonlinearity is caused by the non-linear output current of the output transistor M_2 and thus the equation for the output current is

$$i_{OUT} = v_{gs2}\sqrt{2\beta_2 I_{BO}} + \frac{1}{2}\beta_2 v_{gs2}^2. \tag{4.39}$$

At low frequencies, the gate voltage of the output transistor can be expressed as a function of the input current

$$v_{gs2} = R_m i_{in} \approx \frac{i_{in}}{g_t}, \tag{4.40}$$

where R_m is the transresistance gain of the common-gate input stage.

By using the Equation (A.20) and extracting the required coefficients from the second order output current Equation (4.39), the low frequency second order harmonic distortion of the high gain current-conveyor can be expressed as

$$HD2(0) = \frac{1}{8}\frac{\hat{i}_{out}}{I_{BO}}\frac{fA_{ol}}{(1+fA_{ol})^2} \approx \frac{1}{8fA_{ol}}\frac{\hat{i}_{out}}{I_{BO}}, \tag{4.41}$$

where $A_{ol} = \frac{1}{g_t}\sqrt{2\beta_2 I_{BO}}$ is the open-loop current gain at low frequencies. Similarly, using the Equation (A.21) the third order harmonic distortion can be expressed as

$$HD3(0) = \frac{1}{32}\left(\frac{\hat{i}_{out}}{I_{BO}}\right)^2 \frac{f^3A_{ol}^3}{(1+fA_{ol})^4} \approx \frac{1}{32fA_{ol}}\left(\frac{\hat{i}_{out}}{I_{BO}}\right)^2. \tag{4.42}$$

These two harmonic distortion equations are similar to the low frequency harmonic distortion Equations (C.5) and (C.6) of the unbuffered Miller-compensated operational amplifier. However, the closed-loop distortion of the Miller-compensated operational amplifier is more dependent on the output load impedance than the high gain current-conveyor. Therefore, if an inverting voltage amplifier realised with the CCII∞ (Figure 4.15c) is compared to an inverting voltage amplifier realised with the Miller-compensated CMOS operational amplifier, the current-conveyor amplifier performs better especially with low impedance levels.

In open-loop configuration, the CCII∞ can be assumed to be an integrator for a wide frequency range. Since the dominant source of nonlinearity in this amplifier is the output stage, equations (A.38) and (A.39), derived in Appendix A, can be used to depict the high frequency distortion also in this case. Therefore, letting $b_1 = \sqrt{2\beta_2 I_{BO}}$, $b_2 = \frac{1}{2}\beta_2$ and $b_3 = 0$, the second-order high frequency harmonic distortion on the CCII∞ as a function of the output current amplitude is

$$HD2(\omega) = \frac{1}{4}\frac{\hat{i}_{out}}{I_{BO}} \frac{f\omega\omega_0}{\sqrt{\omega^2 + f^2\omega_0^2}\sqrt{4\omega^2 + f^2\omega_0^2}}. \tag{4.43}$$

Similarly, the third-order high frequency distortion is

$$HD3(\omega) = \frac{3}{32}\left(\frac{\hat{i}_{out}}{I_{BO}}\right)^2 \frac{\omega f^3\omega_0^3}{\left(\omega^2 + f^2\omega_0^2\right)\sqrt{4\omega^2 + f^2\omega_0^2}\sqrt{9\omega^2 + f^2\omega_0^2}}. \tag{4.44}$$

Comparing these equations to the high frequency harmonic distortion Equations (C.14) and (C.16) of the unbuffered Miller-compensated operational amplifier reveals that the CCII∞ has larger distortion with moderately low frequencies and high impedance loads. Since the Miller-compensation provides local feedback for the output stage at high-frequencies, it is to be expected that the CCII∞ will have larger high-frequency distortion. Near the corner frequency, the distortion of both amplifiers are still quite comparable as the feedback can no longer reduce the distortion. Similarly, low impedance levels decrease this local feedback and thus increase the distortion of the Miller-compensated operational amplifier, whereas the distortion of the CCII∞ remains unchanged.

The high-frequency distortion of the CCII∞ is similar to the high frequency distortion of a simple MOS current-mirror. The diode-connected input transistor of a simple current-mirror can be thought as a high-gain current amplifier in a closed-loop configuration with the feedback factor f equal to one. Similarly, the mirror output transistor can be thought as a replica current output. The CCII∞ resembles a cascode current-mirror, where the circuit is arranged so that the input signal is fed to the source of the cascode transistor rather than to the current-mirror transistor gates.

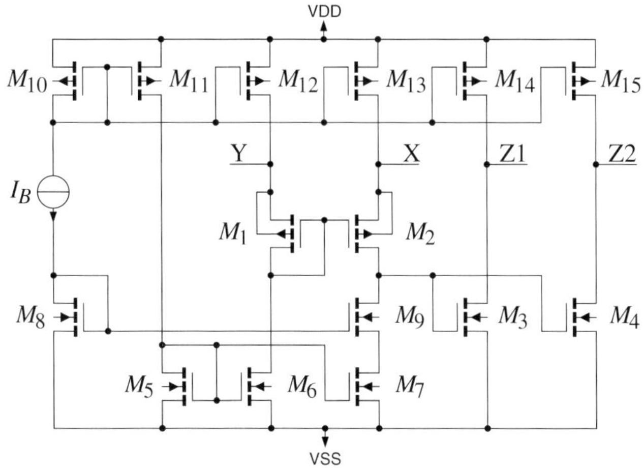

Figure 4.16 A CMOS high-gain current conveyor.

Transistors	Aspect ratio / μm/μm
M_1 and M_2	200/1.2
M_3 and M_4	100/5
M_5, M_6 and M_7	60/3
M_8	10/3
M_9	60/1.2
M_{10}	50/5
M_{11}, M_{12}, M_{13}, M_{14} and M_{15}	100/5

Table 4.1 The transistor dimensions of the CMOS CCII∞ in Figure 4.16. The bias current I_B is 50 μA resulting in 100 μA DC-currents in the main amplifier.

4.4.4 Design example

To further illustrate the behaviour of high-gain current-conveyor a complete CCII∞ implementation with a n-well CMOS-process is shown in Figure 4.16. This conveyor uses an input voltage follower structure typical to class-A second generation positive current-conveyors. The input voltage follower is implemented with PMOS-transistors M_1 and M_2 because then the input voltage swing can be maximised by using floating n-wells for these transistors. Since the offset voltage between Y- and X-terminals remains minimal this high-gain conveyor can be used as a drop-in replacement for voltage-mode operational amplifiers in most applications. In certain applications, the input impedance may be too low. However, then the performance can be improved by using cascode current sources rather than the transistors M_6, M_{12} and M_{13}.

This high-gain conveyor has two current outputs Z1 and Z2. Therefore, a CCII+ can also be constructed from this amplifier by connecting one of the current outputs to X-terminal. When only one current output is needed, the output current swing can

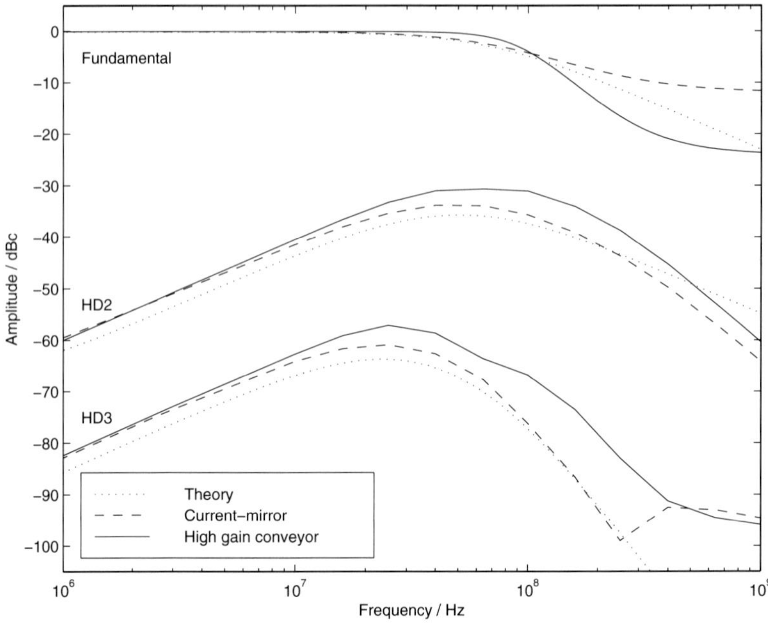

Figure 4.17 Simulated and calculated frequency response and distortion of the CMOS CCII∞ in Figure 4.16 compared to the distortion of a simple current-mirror with comparable device sizes. Signal amplitude in all distortion analyses is 20 μA and thus the modulation index m is 0.2.

be doubled by joining the two outputs together. However, this may degrade the phase margin with low closed-loop current gains. The output transistors M_3 and M_4 are NMOS-transistors with the same dimensions (100/5) as the transistors M_1 and M_2 in the current-mirror of Figure 2.1. Thus, the high frequency distortion performance of the high-gain current-conveyor can be compared to the simple current-mirror.

The distortion of the CMOS CCII∞ is simulated in the closed-loop configuration of Figure 4.15b, working as a CCII+. The simulation results are presented in Figure 4.17. These results are compared with the distortion Equations (4.43) and (4.44). The results are similarly compared with the simulated distortion of a simple CMOS current-mirror with the same dimension as the output transistors M_3 and M_4 of the CCII∞, as earlier presented in Section 2.1.3 (Figure 2.6). The input signal in these simulations is 20 μA resulting in the same modulation index 0.2 as before in the current-mirror simulations.

The simulated distortion of the CCII∞ is relatively similar to the distortion of the simple current-mirror and the calculated distortion agrees also well with both simulated distortions. At frequencies below the corner frequency, the simulated distortion of the CCII∞ is almost too close to the distortion of the current-mirror because the capacitances causing the frequency dependent distortion are not identical and the cir-

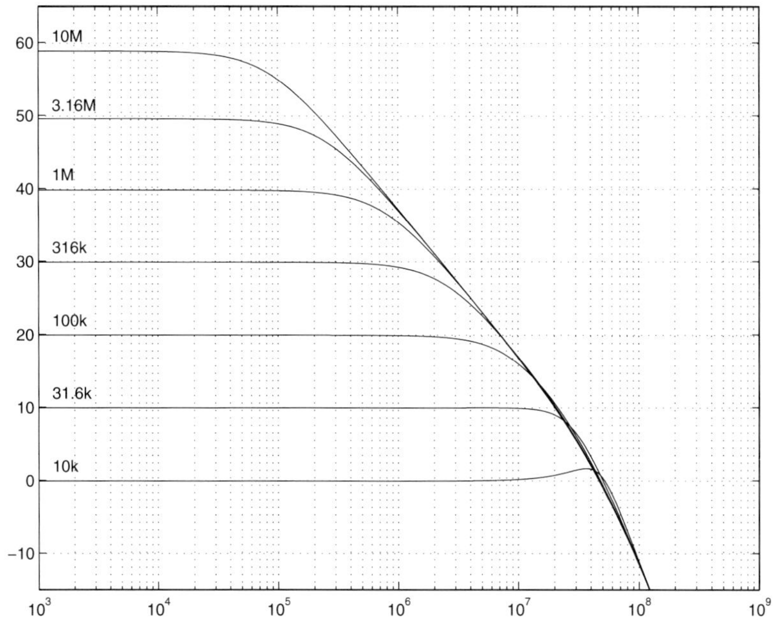

Figure 4.18 The closed-loop frequency response of the inverting current amplifier as a function of feedback resistor. This feedback resistor R_2 is varied from 10 kΩ to 10 MΩ in 10 dB steps while the load resistor R_1 is maintained at a constant value of 10 kΩ.

cuit topology differs significantly. When the distortion of the CCII∞ is compared to the simulated distortion of the cascode current-mirrors also presented earlier in Figure 2.6, the high-gain current-conveyor performs well because, in the cascode current-mirrors, there is a higher peak in the distortion.

This CMOS CCII∞ is additionally simulated with resistive feedback in the inverting current and voltage amplifier configurations, as depicted in Figure 4.15a and c. The simulation results of the inverting current amplifier are presented in Figure 4.18. In the simulations both Z-outputs Z1 and Z2 are connected together and the feedback resistor R_2 is varied from 10 kΩ to 10 MΩ in 10 dB steps, while the load resistor R_1 is maintained at a constant value of 10 kΩ. In the simulations, a large inductance is connected in parallel with the feedback resistor R_2 in order to minimise the effects of current and voltage offsets at the highest gains. The simulations show clear gain-bandwidth product limited operations as theory predicts, while there is a slight peak in the frequency response with the lowest gain. This result is relatively satisfactory because this conveyor implementation was not targeted for an optimal high frequency performance with resistive feedback but rather for comparing the distortion of the amplifier to a NMOS current-mirror.

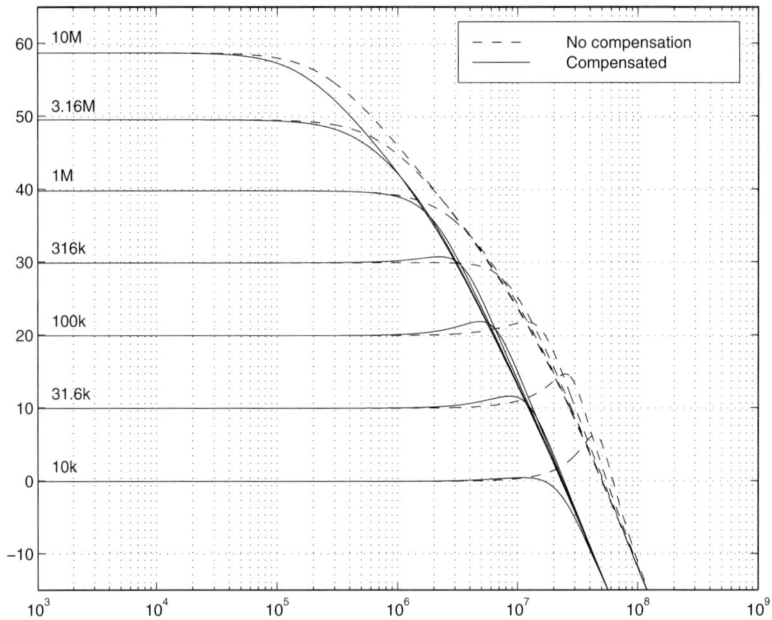

Figure 4.19 The closed-loop frequency response of the inverting current amplifier as a function of the feedback resistor R_2 which is varied from 10 kΩ to 10 MΩ in 10 dB steps while the input resistor R_1 is maintained at a constant value of 10 kΩ. Optionally, an external 5 pF grounded capacitance is added to the gate node of the output transistors M_3 and M_4.

The simulation results of the high gain conveyor in the inverting voltage amplifier configuration are presented in Figure 4.19. As in the case of the current amplifier, both Z-outputs are tied together and the input resistor R_1 is maintained at a constant value of 10 kΩ while the feedback resistor R_2 is varied from 10 kΩ to 10 MΩ. Since significant peaking in the frequency response occurs at low closed-loop gains, the simulations are repeated with an additional 5 pF compensation capacitor, which is connected between the gate node of the output transistors M_3 and M_4 and ground.

The peaking in the frequency response is deriving from the higher impedance level at the Z-output. If a 10 kΩ load resistor were added to the output, similar frequency response to that occurring with the inverting current amplifier would result. Similarly, if only one of the current outputs were used, less peaking would occur and the gain-bandwidth product would once more be constant. Even without the resistive load at the output, the bandwidth depends strongly on the closed-loop gain although the gain-bandwidth product is no longer constant. The reduction of the bandwidth at high gains is because of the Miller-capacitance of the output transistors M_3 and M_4. When cascode transistors are added to the drains of the output transistors, fixed bandwidth is reached with a wide closed-loop gain range [37, 38], in which case additional compensation capacitance is needed as the cascode transistors add poles to the transfer function.

In order to reach maximum bandwidth with a sufficient phase margin, a PMOS output transistor and NMOS input transistor should be selected, as depicted in Figure 4.13. The bandwidth can be further increased with BiCMOS technology by replacing the NMOS input transistor operating in the common-gate configuration to a npn-type bipolar transistor. Consequently, the bandwidth of the amplifier can be extended to radio frequencies [36]. The X-terminal impedance is radically lowered by the use of bipolar transistor and thus lower feedback impedances can be used. In the case of lower feedback impedances, the Miller effect in the output transistor is reduced and the bandwidth is further increased.

In addition, the floating wells in the PMOS input transistor M_1 and M_2 in Figure 4.16 add more than one picofarad of parasitic capacitance to the Y- and X-terminals. Therefore, a better high frequency performance can be obtained at the expense of the input voltage range .

References

[1] C. Toumazou , F. J. Lidgey, D.G. Haigh (ed.), *Analogue IC design: the current-mode approach*, London, Peter Peregnirus Ltd, 1990, 646 p.

[2] Comlinear Corporation, A new approach to op-amp design, Application Note 300-1, March 1985.

[3] F. J. Lidgey, K. Hayatleh, "Current-feedback operational amplifiers and applications," *Electronics & Communications Engineering Journal*, vol. 9, pp. 176-182, Aug. 1997.

[4] Analog Devices, Linear Products Databook 1990/91.

[5] Burr-Brown, OPA660 Datasheet, PO Box 11400, Tuscon, AZ 85734, 1990.

[6] T. Vanishri, C. Toumazou, "Wideband and high gain current-feedback opamp," *Electronics Letters*, vol. 28, pp. 1705-1707, Aug. 1992.

[7] E. Bruun, "Feedback analysis of transimpedance operational amplifier circuits," *IEEE Trans. Circuits and Systems-I*, vol. CAS-40, pp. 275-278, April 1993.

[8] J. Mahattanakul, C. Toumazou, "A theoretical sstudy of the stability of high frequency current feedback op-amp integrators," *IEEE Trans. Circuits and Systems-I*, vol. CAS-43, pp. 2-12, Jan. 1996.

[9] A. Fabre, "Gyrator implementation from commercially available transimpedance operational amplifier," *Electronics Letters*, vol. 28, pp. 263-264, Jan. 1992.

[10] A. Fabre, "Insensitive voltage-mode and current-mode filters from commercially available transimpedance opamps," *IEE Proceedings G*, vol. 140, pp. 319-321, Oct. 1993.

[11] S.-J. Liu, Y.-S. Hwang, "Realisation of R-L and C-D impedances using a current feedback amplifier and its applications," *Electronics Letters*, vol. 30, pp. 380-381, March 1994.

[12] C.-M. Chang, C.-S. Hwang, S.-H. Tu, "Voltage-mode notch, lowpass and bandpass filter using current-feedback amplifiers," *Electronics Letters*, vol. 30, pp. 2022-2023, Nov. 1994.

[13] A. Soliman, "Applications of the current feedback operational amplifiers," *Analog Integrated Circuits and Signal Processing*, vol. 11, pp. 265-302, 1996.

[14] J. Brand, R. Schaumann, "Active R filters: review of theory and practice," *IEE Electronic Circuits and Systems*, vol. 2, pp. 89-101, July 1978.

[15] C. Toumazou, A. Payne, S. Pookaiyaudom, "The active-R filter technique applied to current-feedback op-amps," in *Proc. IEEE Int. Symposium on Circuits and Systems* (ISCAS-95), Seattle, 1995 pp. 1203-1206.

[16] A. Payne, C. Toumazou, "High frequency self-compensation of current-feedback devices," in *Proc. IEEE Int. Symposium on Circuits and Systems* (ISCAS-93), San Diego, 1992 pp. 1376-1376.

[17] S. Smith, "The good, the bad and the ugly: current feedback - technical contributions and limitations," in *Proc. IEEE Int. Symposium on Circuits and Systems* (ISCAS-93), Chicago, 1993 pp. 1058-1061.

[18] B. Harvey, "Current feedback opamp limitations: a state-of-the-art review," in *Proc. IEEE Int. Symposium on Circuits and Systems* (ISCAS-93), Chicago, 1993 pp. 1066-1069.

[19] E. Bruun, "CMOS technology and current-feedback op-amps," in *Proc. IEEE Int. Symposium on Circuits and Systems* (ISCAS-93), Chicago, 1993 pp. 1062-1065.

[20] K. Manetakis, C. Toumazou, "Current-feedback opamp suitable for VLSI technology," *Electronics Letters*, vol. 32, pp. 1090-1092, June 1996.

[21] K. Manetakis, C. Toumazou, C. Papavassiliou, "A 120 MHz, 12 mW CMOS current feedback opamp," in *Proc. IEEE* Custom Integrated Circuits Conference (CICC-98), Santa Clara, May 1998, pp. 365-368.

[22] G. Giustolisi, G. Palmisano, G. Palumbo, S. Pennisi, "High-Drive CMOS Current-Feedback Opamp," in *Proc. IEEE Int. Symposium on Circuits and Systems* (ISCAS-97), Hong Kong, June 1997 pp. 229-232.

[23] H. Elwan, A. Soliman, "Low-voltage low-power CMOS current conveyors," *IEEE Trans. Circuits and Systems-I*, vol. CAS-44, pp. 828-835, Sep. 1997.

[24] P. Allen, D. Holberg, *CMOS Analog Circuits Design*, New York, Holt, Rinehart and Winston, Inc, 1987, 701 p.

[25] P. Gray, R. Meyer, *Analysis and Design of Analog Integrated Circuits*, second edition, New York, John Wiley & Sons, 1984, 771 p.

[26] K. Laker, W. Sansen, *Design of Analog Integrated Circuits and Systems*, New York, McGraw-Hill, Inc, 1994,898 p.

[27] C. Toumazou, A. Payne, F. Lidgey, "Operational floating conveyor," *Electronic Letters*, vol. 27, pp. 651-652, April 1991.

[28] A. Fabre, H. Barthelemy, "Composite second-generation current conveyor with reduced parasitic resistance," *Electronic Letters*, vol. 30, pp. 377-378, March 1994.

[29] K. Koli, K. Halonen ,"CMRR enhancement techniques for current-mode instrumentation amplifiers," *IEEE Trans. Circuits and Systems-I*, vol. CAS-47, pp. 622-632, May 2000.

[30] E. Bruun, "Bandwidth optimization of a low power, high speed CMOS current op amp," *Analog Integrated Circuits and Signal Processing*, vol. 7, pp. 11-19, Jan. 1995.

[31] I. Mucha, "Current operational amplifiers: basic architecture, properties, exploitation and future," *Analog Integrated Circuits and Signal Processing*, vol. 7, pp. 243-255, May 1995.

[32] E. Abou-Allam, E. El-Masry, "Design of wideband CMOS current-mode operational amplifiers," in *Proc. IEEE Int. Symposium on Circuits and Systems* (ISCAS-95), Seattle, 1995 pp. 1556-1559.

[33] R. Zele, S. Lee, D. Allstot, "A high gain current-mode operational amplifier," in *Proc. IEEE Int. Symposium on Circuits and Systems* (ISCAS-92), San Diego, 1992, pp. 2852-2855.

[34] E. Bruun, "Constant-bandwidth current mode operational amplifier," *Electronic Letters*, vol. 27, pp. 1673-1674, Aug. 1991.

[35] T. Kaulberg, "A CMOS current-mode operational amplifier," *IEEE J. Solid State Circuits*, vol. SC-28, pp. 849-852, July 1993.

[36] K. Koli, K. Halonen, "Inverting transimpedance amplifier for current- and voltage-mode applications," *Proceedings of the 11th European Conference on Circuit Theory and Design* (ECCTD-93), Davos , 30.8-3.9.1993, pp. 1583-1588.

[37] K. Koli, K. Halonen, "BiCMOS current-feedback operational amplifier with 60 dB constant bandwidth range," *Proceedings of the 1994 IEEE International Symposium on Circuits and Systems* (ISCAS-94), London, 30.5-2.6.1994, pp. 5.525-528.

[38] K. Koli, K. Halonen, "Temperature Compensated Logarithmic Amplifier," *Analog Integrated Circuits and Signal Processing*, vol. 9, pp. 131-140, March 1996.

[39] C. Toumazou, F. Lidgey, M. Yang, "Translinear class AB current amplifier," *Electronic Letters*, vol. 25, pp. 873-874, June 1989.

Chapter 5

System aspects of current-mode circuits

5.1 Input voltage-to-current conversion

Since signals are normally represented as voltages, a voltage-to-current conversion is required at the input of most current-mode integrated circuits. A simple example of this is presented in Figure 5.1a, where a simple NMOS current-mirror is used as the input stage of an integrated circuit. Because the DC-voltage at the current-mirror input is sensitive to process and temperature variation, a DC-decoupling capacitor C is added in series with the off-chip resistor R.

Because of the simplicity of the input structure there are several nonidealities present in the circuit. Since the input impedance of the NMOS current-mirror is generally approximately one kilo-ohm, a large off-chip resistance R is required for accurate voltage-to-current conversion. The parasitic capacitance C_p, including the mirror input capacitance and parasitic capacitances deriving from the pad and wiring, amount to several picofarads and, since this capacitance is in parallel with the input impedance of the current-mirror, the bandwidth is relatively limited. Moreover, this capacitance additionally increases high-frequency distortion.

If the DC-decoupling capacitor C is omitted, a current amplifier with a well defined input voltage must be used such as the second generation current-conveyor as the input stage. Since typical CMOS implementations of second generation current-conveyors have an input impedance as high as that of CMOS current-mirrors, similar problems with bandwidth and distortion will occur. Therefore, current-amplifiers with lower input impedance, such as first generation current-conveyors or high gain current-amplifiers in closed-loop configuration, perform more effectively as an input voltage-to-current converter.

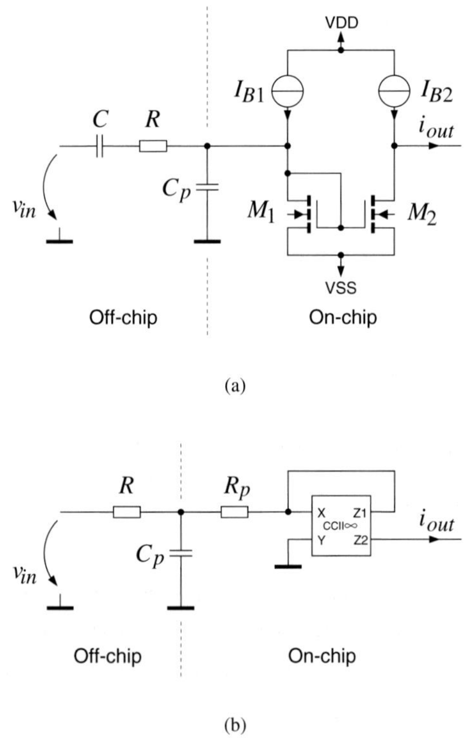

Figure 5.1 Input voltage-to-current conversion with off-chip resistor. a) A simple NMOS current-mirror as the input stage. b) A high gain current-conveyor CCII∞ operating as a CCII+ as the input stage.

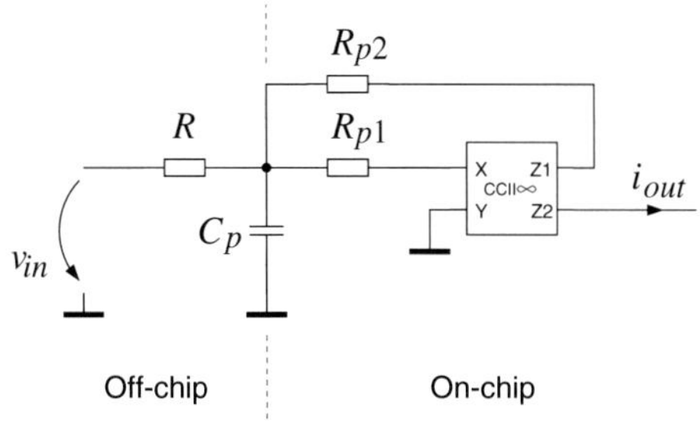

(a)

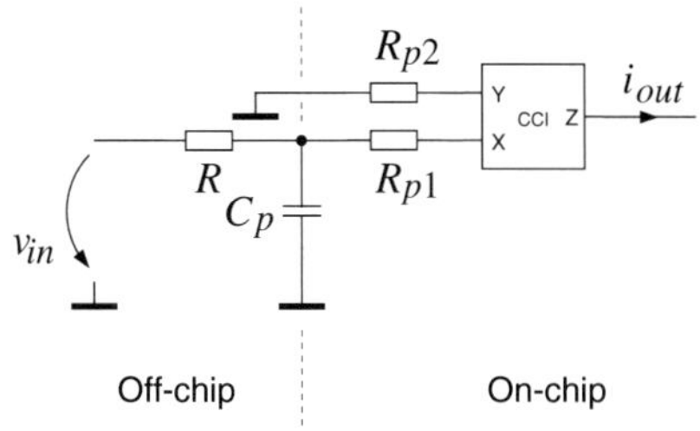

(b)

Figure 5.2 Input voltage-to-current conversion insensitive to resistance in input protection devices. a) A high gain current-conveyor CCII∞ operating as a CCII+ as the input stage. b) A first generation current-conveyor CCI as the input stage.

An example of an input voltage-to-current conversion using a current-amplifier with low input impedance is presented in Figure 5.1b, where a high gain current-conveyor CCII∞, operating as a CCII+, is used. As a consequence of the closed-loop operation, large input capacitance may cause instability. However, if this capacitance is in series with a resistance R_p of a few hundred ohms, the phase lag deriving from the parasitic capacitance C_p is significantly reduced. This series resistance cannot be avoided since it is included in the protection devices of the input pad. The resistor R_p is normally realised as a diffusion resistor, and so is weakly non-linear and temperature dependent. However, because this resistance is significantly smaller than the input impedance of the NMOS current-mirror, its contribution to the nonlinearity and accuracy of the voltage-to-current conversion is similarly minor.

In order to reduce the effects of the input pad diffusion resistor, the circuit can be modified so that the current output Z1 is connected directly to the input pad rather than the conveyor X-input, as depicted in Figure 5.2a. Since the resistor R_{p1} is in this case inside the feedback path, the input impedance seen at the input pad is reduced to roughly ten ohms. The parasitic capacitance C_p reduces feedback at high frequencies and thus stability problems seldom occur. It may not be possible to connect the conveyor Z-output directly to the input pad without protecting devices, so there is an additional resistance R_{p2} between the input pad and the high impedance Z-output. However, this resistance has virtually no effect on the circuit performance as it is in series with a current output.

This technique can also be used with such high-gain current amplifiers as current-mode operational amplifiers, in which case a lower input impedance and lower distortion can be achieved at the expense of reduced bandwidth. However, because the low input impedance of the first generation current-conveyor CCI is arising from a local feedback loop inside the amplifier, different techniques must be used to lower the input impedance. Such a technique is depicted in Figure 5.2b. As the impedance at the Y-terminal of the first generation current-conveyor is seen as a negative impedance at the X-terminal, the pad resistance R_{p1} can be cancelled out with an other pad resistance R_{p2}, as explained in Chapter 3.1.3. However, as in this case a negative resistance is used to cancel out nonidealities, instability may occur if R_{p2} is larger than R_{p1}.

The techniques described involving current-mode feedback amplifiers are similar to techniques used in voltage-mode operational amplifiers driving large capacitive loads. A large output voltage swing leads to large current drive requirements for the voltage amplifier, whereas the input voltage swing at the current amplifier input is always much smaller. Thus, a simpler circuitry with a lower power consumption can be used. It is therefore feasible to convert a voltage-mode system to its current-mode adjoint circuit when we compare the voltage output to its adjoint.

Most of the problems involved in voltage-to-current conversion at the low impedance current input can be avoided by using a voltage input for the system in conjunction with an on-chip voltage-to-current converter. Because almost all CMOS current-conveyors have a relatively poor input voltage swing, high-impedance voltage inputs are not feasible in low voltage current-mode systems unless the input signal amplitudes are relatively small. Another way to implement a voltage input with a large signal swing is to use a voltage-mode operational amplifier with rail-to-rail input and output swing. However, they are relatively complex circuits and commonly exhibit a moderate bandwidth and lengthy settling time.

5.2 Output current-to-voltage conversion

It is easier to transfer signals out of the chip as currents than as voltages. Thus, the series resistance in the output pads do not degrade the performance, and the parasitic capacitances at the output do not cause stability problems to the circuitry in the integrated circuit. Furthermore, the current to voltage conversion i.e. I/V-conversion can be realised easily with a voltage-mode operational amplifier as depicted in Figure 5.3. Since the input impedance of the I/V-converter is low, there is virtually no distortion due to channel length modulation generated in the current-output transistors. However, there may be some gain errors and output offset currents due to the same reason.

In the case of this circuit, it is straightforward to set the DC-voltage of the output simply by setting the voltage at the operational amplifier noninverting terminal. The output signal can be scaled relatively easily since the operational amplifier operates with 100% feedback and thus the closed-loop bandwidth is degraded only by extremely high feedback impedances. Thus, the parasitic capacitance C_p deriving from pad and wiring capacitances degrades the settling behaviour, as depicted in the small-signal transfer function of the current-to-voltage converter

$$\frac{v_{out}}{i_{out}} = -\frac{RA_{ol}(s)}{1 + A_{ol}(s) + sRC_p}. \tag{5.1}$$

Therefore, the parasitic capacitance C_p degrades the phase margin of the closed-loop amplifier if the inverse of the time constant RC_p is comparable to the gain-bandwidth product of the amplifier and thus settling times become long. This can be avoided by reducing the resistance R or by using an overcompensated operational amplifier. The latter is difficult to realise with discrete operational amplifiers as there are currently extremely few operational amplifiers available with an option for an external compensation capacitor.

However, adding a small capacitance in parallel with the resistor R will enhance settling behaviour as it lowers the impedance level of the feedback network at high frequencies. This capacitance can be as small as the capacitance C_p resulting in only slightly decreased bandwidth while the settling becomes significantly faster. Alternatively, a current-feedback operational amplifier can be used in preference to the

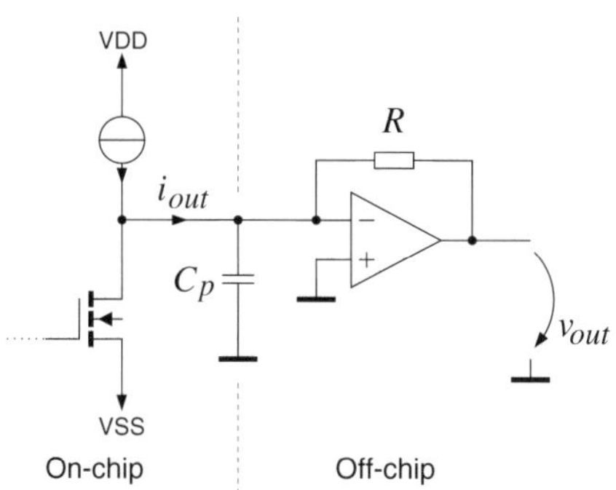

Figure 5.3 Typical output current to voltage conversion.

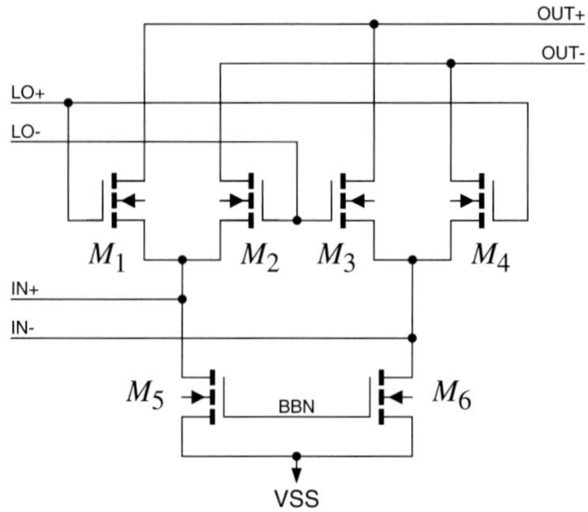

Figure 5.4 A balanced frequency mixer suitable for output interfaces using current signals.

voltage-mode operational amplifier in the current-to-voltage converter. Since the closed-loop bandwidth of the current-feedback operational amplifier is inversely proportional to the feedback resistance R, in the case of large feedback impedances the amplifier is overcompensated, and thus relatively insensitive to the extra phase lag arising from C_p. Additionally, adding a small capacitance in parallel with the resistor R will enhance settling behaviour with current-feedback operational amplifiers if this capacitance is kept sufficiently low [1].

Additional functions are straightforward to add to the current output. The CMOS implementation of the widely used balanced frequency mixer [2], for example, can be simplified to the circuit shown in Figure 5.4 as in this case the input voltage-to-current conversion can be omitted if the baseband input signal is a current [3]. The carrier signal in the ports LO+ and LO- is a differential voltage with a common-mode bias voltage, ensuring accurate operation for the circuit. The mixed differential output is convenient to realise as an open-drain output, so that pull-up resistors or a LC-resonator can be added outside the chip for an optimal termination.

Similar circuit techniques can also be used to realise variable gain control for the system either with discrete or continuous control circuitry. Moreover, if a current-mode system has both the input voltage-to-current and output current-to-voltage conversion realised with discrete resistors, the overall gain and impedance levels can be adjusted for a very wide range of applications, which is not the case with most voltage-mode circuits.

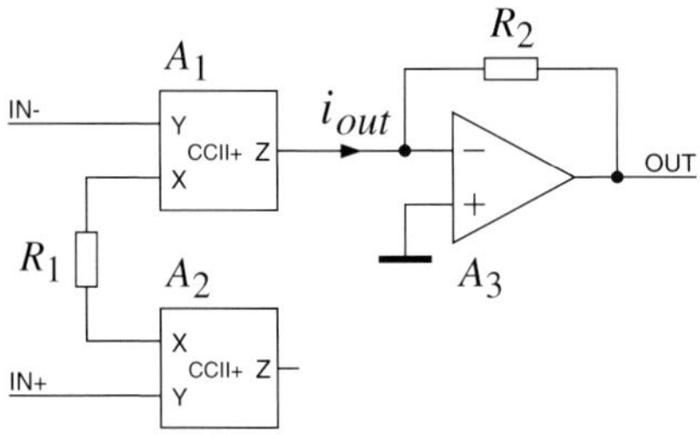

Figure 5.5 Current-mode instrumentation amplifier [4].

5.3 Differential voltage input structures

Although differential signals were used throughout the system, differential input structures usually need to reject unwanted common-mode signals out of the desired differential signal in order to maximise dynamic range and minimise distortion. Common-mode rejection of voltage signals can be easily realised with second generation current-conveyors by using the current-mode instrumentation amplifier illustrated in Figure 5.5, previously discussed in Chapter 3.1.3. This circuit does not require matched resistors to reach a high CMRR. Unlike typical voltage mode instrumentation amplifier topologies, the CMRR of the current-mode instrumentation amplifier is independent of gain. Therefore, the current-conveyor based instrumentation amplifier is an advantageous choice for wide bandwidth and low gain applications.

The common-mode rejection of the current-mode operational amplifier was derived in Chapter 3.1.3, resulting in Equation 3.41, reprinted here

$$CMRR \approx \frac{1}{Y_x(R_1 + 2Z_x) + \frac{\Delta A'_{vf}}{A'_{vf}}}. \tag{5.2}$$

This equation shows, that in order to maximise the CMRR, the input resistor R_1 should be maintained as low as possible which involves minimising the X-terminal impedance Z_x and admittance Y_x. The parasitic capacitance at the conveyor X-terminals, in particular, should be minimised in order to maintain high CMRR at high frequencies. Eventually, the mismatch $\Delta A'_{vf}$ in the forward voltage gains of the two current-conveyors will limit the CMRR if Z_x and Y_x are small enough.

As has been established, most MOS implementations of the second generation current-conveyor exhibit a relatively large Z_x and therefore the input resistor R_1 must be commonly in the range of tens of kilo-ohms in order to maintain the gain error and distortion low, which results in a relatively poor CMRR. Therefore, either the X-terminal impedance Z_x or the X-terminal admittance Y_x should be improved with special circuit techniques in CMOS implementations. On the other hand, the bipolar

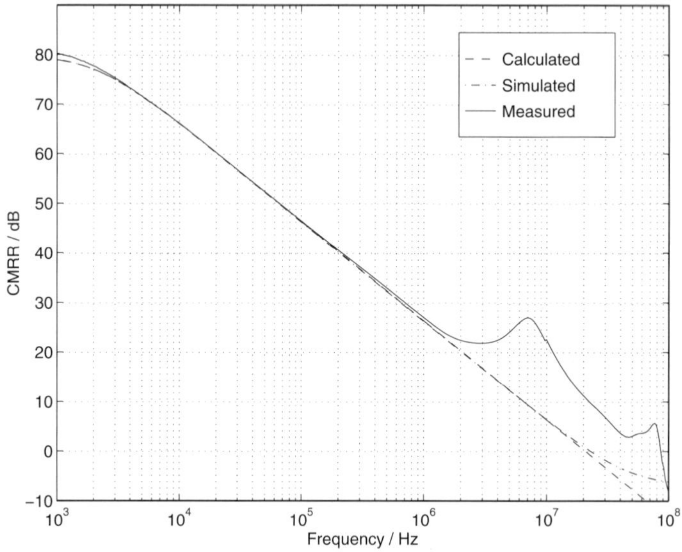

Figure 5.6 Calculated, simulated and measured CMRR of the current-mode instrumentation amplifier. The amplifier resistors R_1 and R_2 are both 1 kΩ.

voltage followers have low output impedances. Furthermore, the output impedance of bipolar push-pull voltage followers is relatively linear and therefore low values of R_1 can be used resulting in a CMRR of 70 dB or better.

The current-mode instrumentation amplifier of Figure 5.5 was built with the second generation current-conveyors realised with AD844 current-feedback operational amplifiers [5]. This circuit was also measured without the resistor R_1 in order to extract values for Y_x. The extracted values for the admittance were 90.9 nS for the conductance part and $\omega \times 6.9$ pS for the susceptance part. Approximately 5 pF from this 6.9 pF X-terminal capacitance was assumed to be deriving from the circuit board. Furthermore, the X-terminal impedance was set to 50 Ω based on the information on the datasheet of the device. As we can see from Figure 5.6, the measured CMRR of the amplifier agrees well with Equation (3.41) and with the vendor's SPICE-model. The voltage follower gain mismatch $\Delta A'_{vf}$ is assumed to be zero in the calculations.

5.3.1 CMRR enhancement techniques

Many voltage-mode instrumentation amplifiers can easily reach a CMRR above 100 dB at low frequencies and high gains whereas current-mode instrumentation amplifies using the circuit topology of Figure 5.5 seldom reach a CMRR above 80 dB even with high quality bipolar push-pull conveyors. Since with a CMOS conveyor the common-mode performance is even worse, additional circuit techniques are required in order to reach a very high CMRR with a bipolar technology or an moderate CMRR with CMOS technology. Therefore, in the following pages different circuit techniques to improve the CMRR of current-conveyors based instrumentation amplifiers are discussed.

Common-mode bootstrapping

As explained in Chapter 3, in second generation current-conveyors, both common-mode rejection and power supply rejection are based on the same mechanism. Therefore, the CMRR of the current-mode instrumentation amplifier can be reduced by forcing the conveyor supply voltages to follow the common-mode voltage [6–8]. This common-mode bootstrapping technique limits the common-mode input voltage range since the conveyors have to operate with lower supply voltages. Furthermore, the minimum supply voltage of a complementary current-conveyor is relatively high and thus this technique is not suited to low voltage applications even with bipolar conveyors. Additionally, the bulk effect in CMOS-processes makes this technique even less applicable for low voltage CMOS circuits.

Output current subtraction

At high frequencies the common-mode gain of the discussed amplifier solely depends on the capacitance at the X-terminals. By a careful circuit board design both conveyor X-terminal capacitances can be made to match very well and therefore subtracting the output current of the second conveyor from the output current of the first will effectively improve the CMRR of the instrumentation amplifier. Similarly, with a careful layout design the parasitic capacitances at conveyor X-terminals can be made to match well. The low frequency common-mode current is a function of several random variations so only a slight improvement in the range of 10-20 dB can be expected with discrete conveyors but, since the low frequency CMRR is relatively high, a large improvement is not needed. Because the differential gain is increased by 6 dB, further improvement on the CMRR can be expected. If the whole instrumentation amplifier is implemented as an integrated circuit, the low frequency CMRR can be enhanced with this technique since there is adequate matching between the two input conveyors.

Subtraction by an operational amplifier The simplest way to realise the output current subtraction is to terminate the second conveyor Z-terminal with a resistor R_3 and to connect it to the noninverting input of the operational amplifier, as depicted in Figure 5.7 [9]. However, current transfer errors arising from the finite Z-terminal admittance Y_z are different because the impedance levels at these two Z-terminals differ significantly from each other. The impedance at the first conveyor Z-terminal is inductive whereas the impedance at the first conveyor Z-terminal is almost equal to R_3.

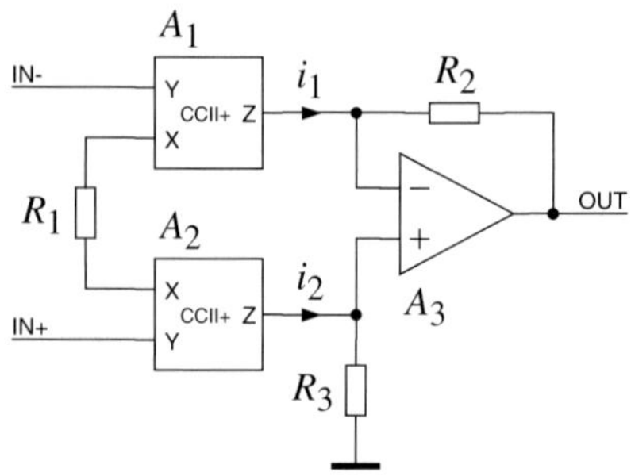

Figure 5.7 A current-mode instrumentation amplifier with common-mode current cancellation by an operational amplifier.

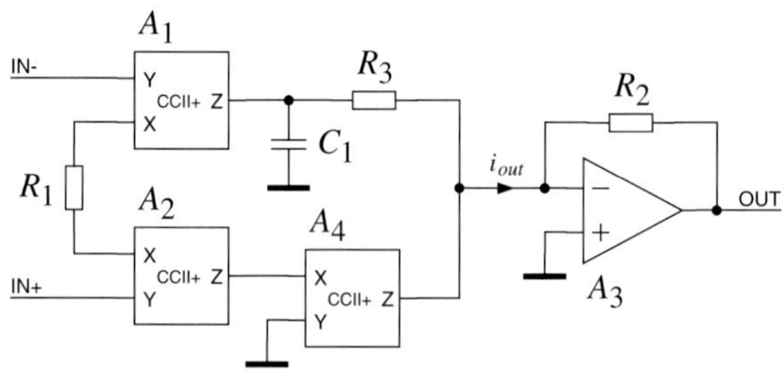

Figure 5.8 A current-mode instrumentation amplifier using an additional CCII+ for current inversion.

Furthermore, the CMRR of the operational amplifier is typically poor at high frequencies. The CMRR can be further improved by adjusting manually the value of R_3 but because of these frequency dependent nonidealities this trimming works exclusively at a narrow frequency range. In the case of low supply voltages, the input voltage swing of the operational amplifier A_3 is normally large and thus rail-to-rail input and output voltage swing topologies are required for the operational amplifiers.

Current inversion by a current conveyor A wide bandwidth current subtraction can be realised by inverting the second conveyor output current with an additional positive second generation current conveyor CCII+ and then adding it to the first conveyor output current, as presented in Figure 5.8 [10]. The accurate summing range can be further extended by adding an RC-network to the first conveyor output. The main purpose of this network is to compensate for the high frequency phase shift of the added conveyor, but the resistance R_3 can additionally compensate for the systematic current transfer error arising from the finite input and output impedances of the third conveyor. The values of C_1 and R_3 must be determined experimentally because the phase shift of the conveyor depends strongly on the circuit board parasitic capacitances, but once these values are found, there is no need for a component value trimming after the first prototype. Similarly, in integrated circuits, the parasitic capacitances must be modelled accurately in order to predict this compensation time constant adequately.

If all the current gain errors of the three conveyors are combined into one current gain mismatch parameter $\Delta A'_{if}$, neglecting the effect of the limited bandwidth of the third conveyor, the CMRR of this instrumentation amplifier can be approximated as

$$CMRR \approx \frac{1}{\frac{\Delta A'_{vf}}{A'_{vf}} + \frac{1}{2} Y'_x R_1} \tag{5.3}$$

where

$$Y'_x = \Delta Y_x + Y_x \left(\frac{2Z_x}{R_1} \frac{\Delta Z_x}{Z_x} + \frac{R_1 + 2Z_x}{R_1} \frac{\Delta A'_{if}}{A'_{if}} \right). \tag{5.4}$$

All other mismatch parameters in addition to $\Delta A'_{if}$ are defined as A'_{vf} and $\Delta A'_{vf}$ in Equation (3.41). Equation (5.3) shows that the sensitivity to the input voltage follower gain mismatch $\Delta A'_{vf}$ is not reduced, but the sensitivity to all other mismatches is nevertheless efficiently reduced. In most cases, the input resistor R_1 is significantly larger than Z_x in order to ensure sufficient linearity and gain accuracy. Thus, the voltage follower output resistance mismatch ΔZ_x can be neglected so that the CMRR equation reduces to

$$CMRR \approx \frac{1}{\frac{\Delta A'_{vf}}{A'_{vf}} + \frac{1}{2} R_1 \left(\Delta Y_x + Y_x \frac{\Delta A'_{if}}{A'_{if}} \right)}. \tag{5.5}$$

The current-mode instrumentation amplifier of Figure 5.8 is realised using AD844 current feedback operational amplifiers. This amplifier is measured with and without the conveyor phase shift compensation RC-network and it is compared with the conventional two conveyor current-mode instrumentation amplifier. The differential gain measurements of Figure 5.9a show that gain flatness of the new instrumentation amplifier is even better than in the conventional one. However, the conventional amplifier may suffer from circuit board crosstalk from the unused Z-output of the second conveyor A_2.

The bandwidth of the new amplifier is only slightly reduced when compared to the conventional topology, and the phase shift compensation extends the bandwidth further. The differential gains of the new topology were scaled down 6 dB for more convenient comparison. The gain setting resistors R_1 and R_2 are both 1 kΩ in all measured circuits and the gain error of 1 dB is deriving from the limited output impedance R_O of the conveyor input voltage followers. The measured CMRR of Figure 5.9b shows an improvement of 40 dB at high frequencies compared to the conventional topology. Furthermore, the phase shift compensation improves both the low and high frequency CMRR , which is above 40 dB even at the -3 dB corner frequency of the amplifier.

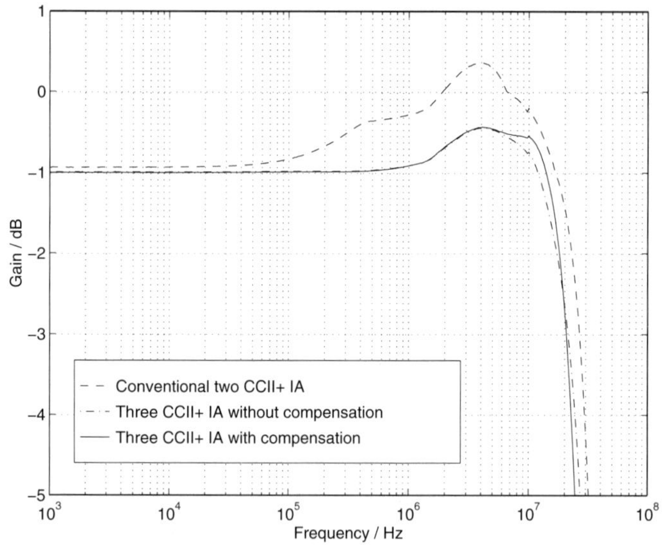

(a)

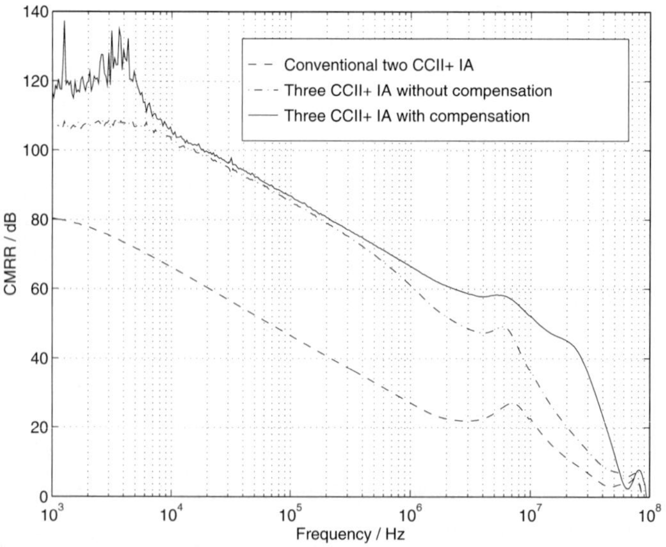

(b)

Figure 5.9 Comparison of current-mode instrumentation amplifier measurement results. a) The differential gains. b) The CMRR. The differential gains of the three CCII+ instrumentation amplifiers are scaled 6 dB lower for more convenient comparison.

Composite conveyors

As was discussed in Chapter 4.2.2, high performance current-conveyors can be realised by constructing a composite conveyor out of two or more conveyors. Most composite conveyors reduce the X-terminal impedance [11,12] and thus the CMRR of the current-mode instrumentation amplifier can be reduced by using lower resistance values. However, composite techniques can additionally be used to lower the forward transconductance G_{mf} of the CCII+ [10]. This can be realised by cancelling the error current arising from Y_x by generating a replica current and subtracting it from the initial error current. A simple way of cancelling the error current of a CCII+ is to use a CCII- with a floating X-terminal, as depicted in Figure 5.10a. Because the conveyors are of an opposite type, their high frequency performances do not match well and thus the forward transconductance reduction is efficient only at relatively low frequencies. Nevertheless, this remains sufficient for many instrumentation applications.

The circuit realisation of this composite conveyor can be very straightforward, as the example in Figure 5.10b shows. There, the negative conveyor CCII- shares the output structure with the positive conveyor CCII+ and thus only four transistors (M_7, M_8 and a PMOS cascode current source) are added to a simple MOS CCII+ to form a composite conveyor.

To illustrate this, a conventional current-mode instrumentation amplifier with class-A MOS conveyors is simulated. Since the MOS conveyors have relatively large Z_x, the resistors R_1 and R_2 are each 10 kΩ and there remains a differential gain error of approximately 1.5 dB. In the case of normal positive MOS conveyors, the CMRR of the amplifier is slightly over 50 dB. However, by replacing the first conveyor A_1 with a composite conveyor, almost identical performance is reached as when both conveyors are replaced with a composite conveyor. This can be explained by referring to Equation (3.37) in Chapter 3.2.3, which shows that the conventional current-mode instrumentation amplifier is not sensitive to the X-terminal admittance mismatch.

In practice, a dramatic improvement to the amplifier CMRR was not expected since there were no random variations in the simulations. If this amplifier is realised as an integrated circuit with integrated resistors, an additional dummy resistor should be added to the floating dummy X-terminal (DX) of the composite conveyor in order to compensate for the parasitic capacitances of the resistors and interconnections. Even with a careful layout design, a 40 dB improvement on the low frequency CMRR can be expected. Thus, adding four MOS transistors and a dummy resistor is very efficient.

In order to maximise input voltage swing the input transistors M_1, M_2 and M_7 should be implemented with floating n-wells. Such floating wells can add approximately one picofarad of parasitic capacitance to the conveyor input terminal, significantly degrading the CMRR even at moderately low frequencies. Additionally, be-

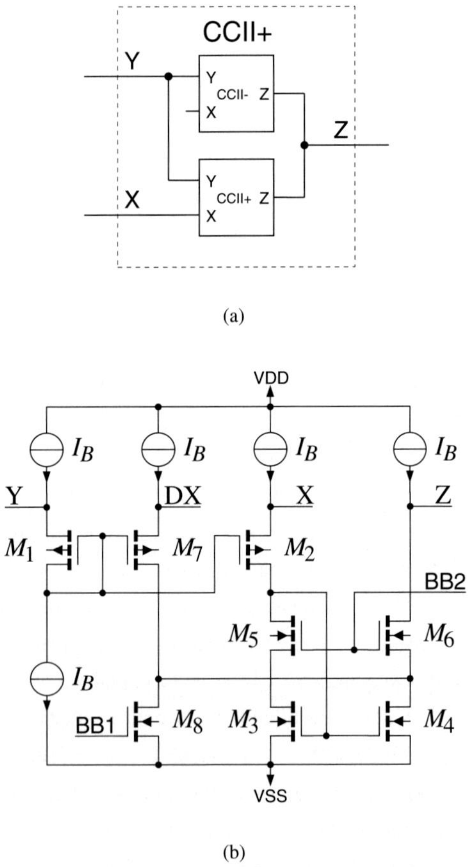

Figure 5.10 a) The principle of a composite conveyor lowering G_{mf}. b) A simple class-A CMOS implementation of this composite conveyor principle.

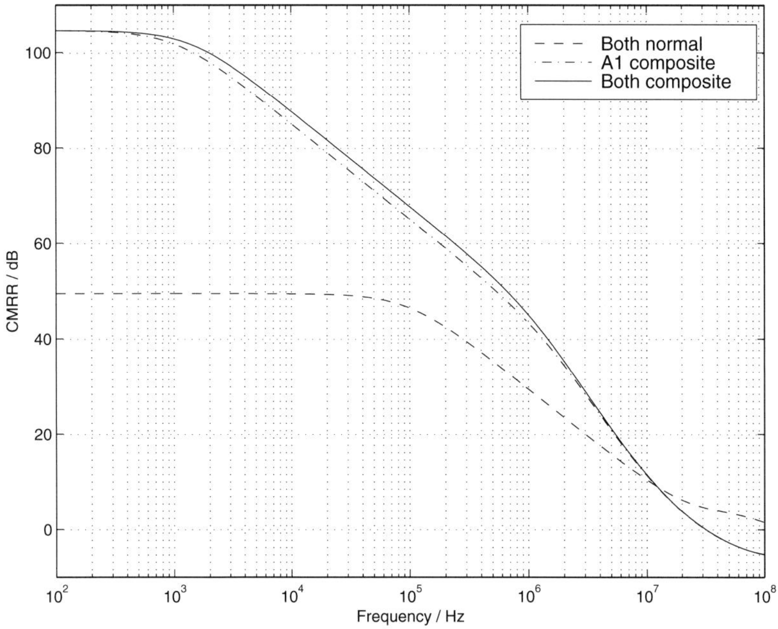

Figure 5.11 Simulated CMRR of the conventional current-mode instrumentation amplifier using either normal or composite MOS conveyors.

cause a low X-terminal admittance is achieved now without PMOS cascode current sources, the voltage swing can be increased by a few hundreds of millivolts compared to a conventional simple class-A CMOS CCII+.

5.4 Differential current input structures

Current signals frequently have a DC-components that may be different to the desired DC-current levels within the differential current-mode system. This DC-component can be removed from differential current signals for example by using the common-mode feedforward technique [3, 13] of Figure 5.12. There, the two input currents i_{ip} and i_{im} are mirrored by PMOS transistors M_1, M_2, M_3 and M_4 to the differential outputs i_{op} and i_{om}, while the transistors M_5 and M_6 are used to generate a current i_{cm} which is the average of the two input currents i_{ip} and i_{im}. This average current is then mirrored to both outputs with transistors M_7, M_8 and M_9. Thus, the DC-component is cancelled from the input currents and the even order distortion is attenuated.

For an optimal high frequency distortion performance, the two input mirrors should be significantly slower than the mirror inverting the average common-mode current. Consequently, the even order distortion is cancelled accurately even at high frequen-

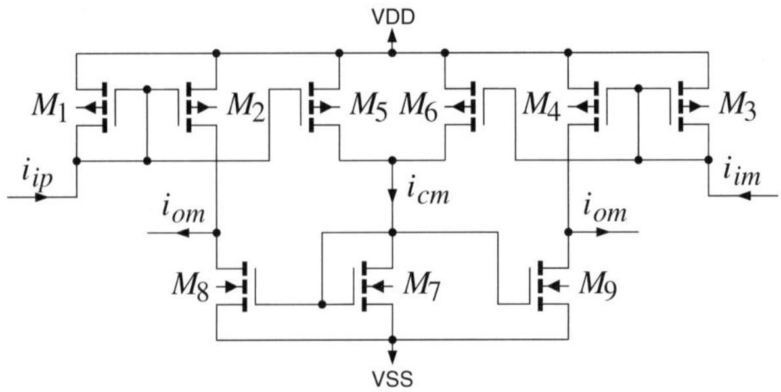

Figure 5.12 Cancelling DC-component from differential input current by common-mode feed-forward.

cies. Therefore, it is a good design practice to set an identical aspect ratio for all six PMOS-transistors and realise M_7 with an aspect ratio twice that of M_8 and M_9. Thus, the generation of the average current is accurate and the NMOS-mirror is fast.

Even if the NMOS-mirror is faster than the two PMOS-mirrors, it still generates distortion itself. However, since the time varying content of the input signal i_{cm} contains primarily the even order distortion components of the differential signal, the fundamental frequency of the NMOS-current mirror input signal is twice that of the differential signal. Therefore, the NMOS-mirror add almost exclusively even order distortion to the differential signal, which is in most cases efficiently rejected in differential systems.

5.5 Single-ended to differential conversion

The single-ended to differential conversion with current-conveyors is very similar to the differential to single-ended conversion as shown in Figure 5.13a. However, in this case the two outputs are less symmetrical than in the instrumentation amplifier case, because in this case there is a significant voltage swing only at the input terminals of the conveyor A_1. The two output currents can be calculated with the current-conveyor model as

$$i_{out+} = v_{in} A_{if} A_{vf} \frac{1 + Y_x (R + 2Z_x + Y_x Z_x R)}{(1 + Y_x Z_x)(R + 2Z_x + Y_x Z_x R)}, \tag{5.6}$$

$$i_{out-} = -\frac{v_{in} A_{if} A_{vf}}{(1 + Y_x Z_x)(R + 2Z_x + Y_x Z_x R)}, \tag{5.7}$$

if the two conveyors are assumed ideally matched. As can be seen there is an additional zero in the positive output current equation deriving from the parasitic capacitance at the X-terminal of the conveyor A_1.

If all parasitic capacitances at the X-terminal are assumed to be connected to

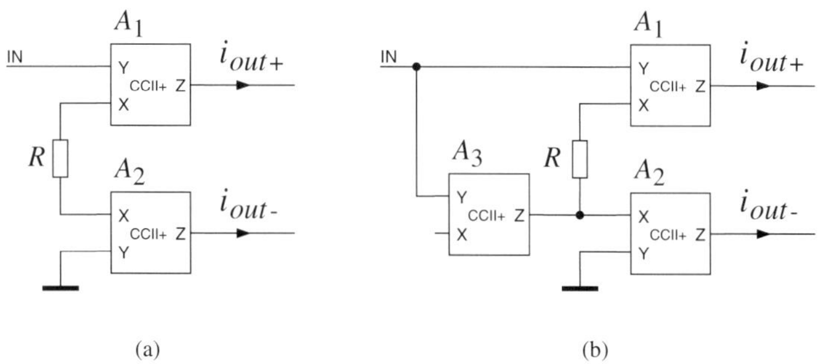

Figure 5.13 Single-ended voltage to differential current conversion realised with current conveyors. a) Conventional. b) Forward transconductance compensated.

ground there are typically significant gain and phase errors one decade below the -3 dB corner frequency. This assumption holds true with a fully integrated implementation of the circuit. Moreover, the normally large parasitic capacitance in integrated resistors increases this effect by adding more capacitance to Y_x. When this circuit is built with discrete components, there is also a significant parasitic capacitance in parallel with the resistor R, which effectively lowers the effect of grounded capacitances and reduces the high frequency gain and phase error.

For better high frequency gain and phase accuracy, an additional conveyor with a floating X-terminal can be used to add similar gain peaking deriving from the forward transconductance G_{mf} to the negative output current, as shown in Figure 5.13b. Nevertheless, in most cases the performance of the single-ended to differential converter is adequate without compensating techniques. This can be seen by comparing the measurement results of the compensated and uncompensated circuits in Figure 5.14. Both circuits are realised with AD844 current feedback operational amplifiers, as in the case of the current-mode instrumentation amplifier examples.

Because the distortion occurring in the single-ended to differential conversion can no longer be rejected in the differential system, the converter should be as linear as possible. The single-ended to differential converters discussed rely on the relatively low distortion of the bipolar push-pull conveyors fabricated with advanced complementary bipolar processes. However, in standard CMOS-processes, other circuit techniques are needed for low distortion. One interesting candidate for a low distortion single-ended to differential converter for CMOS-technology is presented in Figure 5.15, where a current-mode operational amplifier is used as a single-ended to differential converter. In effect, this circuit is the adjoint circuit of the well-known differential amplifier circuit based on a single voltage-mode operational amplifier presented in Figure 4.5b.

In an ideally matched case, the common-mode rejection of the voltage-mode dif-

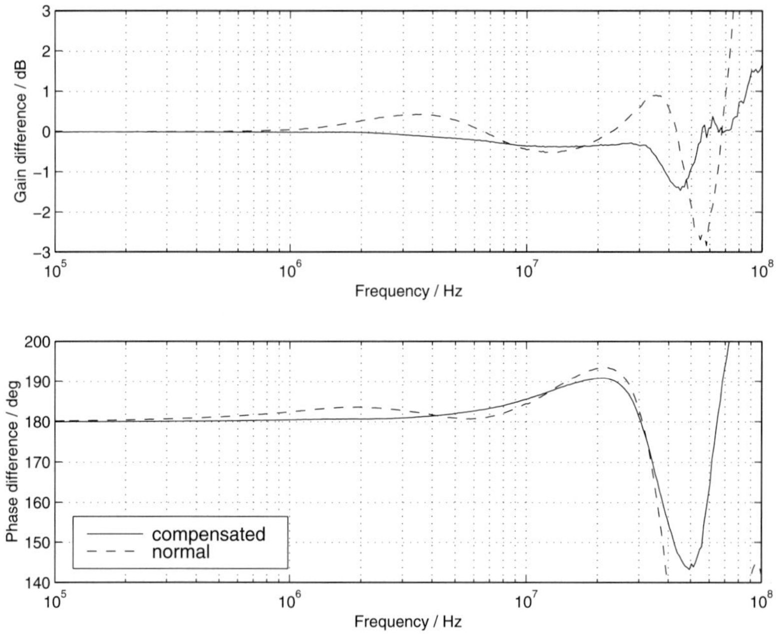

Figure 5.14 Measured gain and phase differences of the conventional and compensated single-ended to differential converter.

ferential amplifier is eventually deriving from the limited output conductance of the input differential pair tail current source. In the current-mode single-ended to differential converter, the limited output conductance of the output differential pair tail current source generates only a small common-mode component to the differential signal. This common-mode component is normally suppressed in the differential system and only a fraction of it leaks back to the differential signal. Therefore, different nonidealities dominate the accuracy of the current-mode operational amplifier based single-ended to differential converter.

If the nonidealities of the current-mode operational amplifier are assumed to be

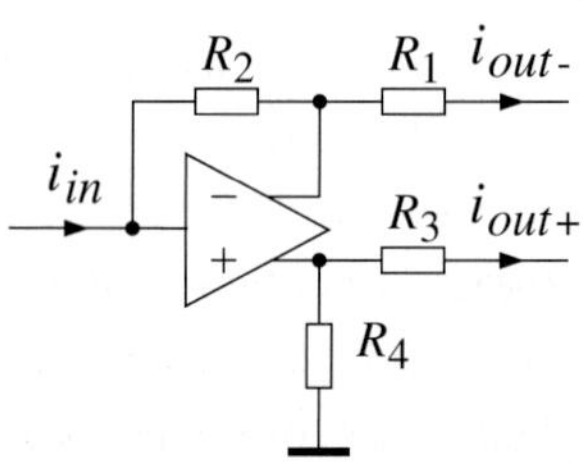

Figure 5.15 Single-ended to differential converter realised with a current-mode operational amplifier.

limited to an open-loop input impedance Z_{in} and an open-loop current gain $A_i(s) = \frac{\omega_0}{s}$, the two output currents can be derived by simple small-signal analysis as

$$i_{out+} = i_{in} \frac{1}{1 + s\frac{2R+Z_{in}}{\omega_0 R}}, \tag{5.8}$$

$$i_{out-} = -i_{in} \frac{1 - s\frac{Z_{in}}{\omega_0 R}}{1 + s\frac{2R+Z_{in}}{\omega_0 R}}. \tag{5.9}$$

In these equations, DC-current gains ± 1 with ideally matched resistors are assumed, so that $R_1 = R_2 = R_3 = R_4 = R$. The equations clearly shows that, provided the impedance level of the resistor network is significantly higher than the open-loop input impedance of the current-mode operational amplifier, the two output currents are well balanced, even at the corner frequency.

In these calculations, all high frequency losses between the inverting and noninverting outputs of the current-mode operational amplifier deriving from parasitic capacitances at the output differential pair are neglected. However, these nonidealities begin to dominate the behaviour of the circuit only well above the corner frequency.

5.6 Noise in current-mode circuits

5.6.1 Class-A CMOS CCII+

The noise analysis of current-conveyors and other current-mode amplifiers is not as straight-forward as with voltage-mode operational amplifiers. Because the low gain of the first and second generation current-conveyors almost all transistors in the circuits contribute to the total output noise and therefore referring the noise to the input may not be as straightforward as with traditional operational amplifiers. The output noise of the second generation current-conveyor depends on the impedance level at the Y-terminal differently to the way it depends on the impedance level at the X-terminal, which further complicates the noise modelling.

The general noise test set-up of the second generation current-conveyor is presented in Figure 5.16a, where two resistors R_{sy} and R_{sx} are connected to the conveyor input terminals. As an example, the output noise current $\overline{di_z^2}$ is derived in the case of the simple class-A MOS CCII+, depicted in Figure 5.16b. For simplicity, the case where $R_{sy} = 0$ and $R_{sx} = \infty$, i.e. a simple current amplifier, is considered. Consequently, neglecting high frequency effects, the output noise current is

$$\left.\overline{di_z^2}\right|_{R_{sy}=0, R_{sx}=\infty} = \overline{di_1^2} + \overline{di_2^2} + \overline{di_{B1}^2} + \overline{di_{B2}^2}, \tag{5.10}$$

if the forward current gain is assumed to be exactly one. Similarly, if $R_{sx} = 0$ the

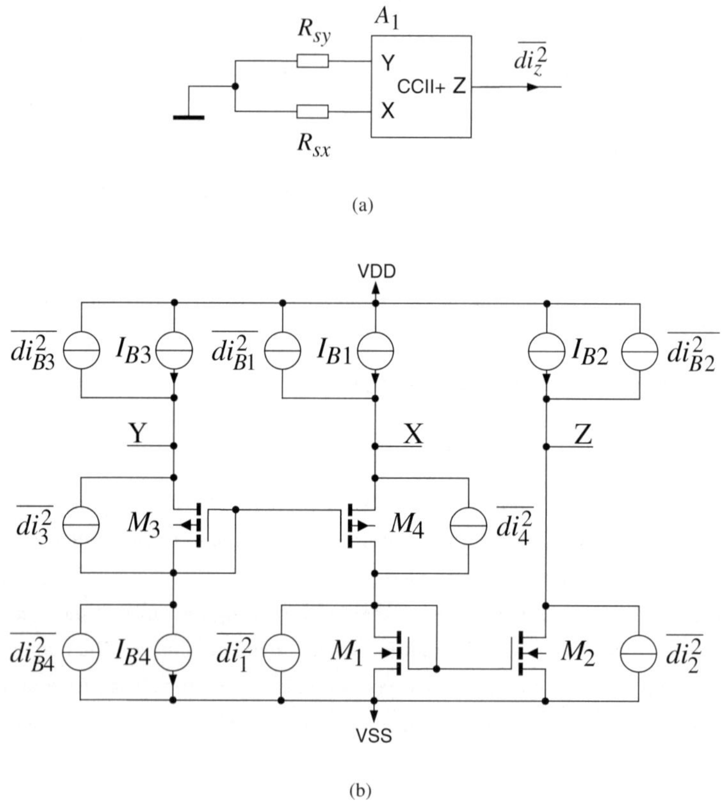

Figure 5.16 a) Noise test set-up for a second generation current-conveyor. b) Noise in simple class-AB MOS CCII+.

resulting output noise current is

$$\left.\overline{di_z^2}\right|_{R_{sy}=0,R_{sx}=0} = \overline{di_1^2} + \overline{di_2^2} + \overline{di_{B2}^2} + \overline{di_3^2} + \overline{di_4^2} + \overline{di_{B4}^2}, \tag{5.11}$$

A comparison of these two equations reveals that the noise arising from the NMOS current-mirror and the current source I_{B2} is present at the output regardless of the impedance level at the X-terminal. Therefore, these noise sources can be combined into an equivalent current noise source $\overline{di_{zeq}^2}$ at the Z-output as

$$\overline{di_{zeq}^2} = \overline{di_1^2} + \overline{di_2^2} + \overline{di_{B2}^2}. \tag{5.12}$$

Similarly, the noise term arising from the current source I_{B1} is missing in Equation 5.11 and consequently it can be expressed as an equivalent current noise source $\overline{di_{xeq}^2}$ at the X-input as

$$\overline{di_{xeq}^2} = \overline{di_{B1}^2}. \tag{5.13}$$

The remaining noise sources in Equation 5.11 can be collected to an equivalent voltage noise source $\overline{dv_{xeq}^2}$ at the X-terminal as

$$\overline{dv_{xeq}^2} = \frac{\overline{di_3^2} + \overline{di_4^2} + \overline{di_{B4}^2}}{g_{m4}^2}. \tag{5.14}$$

Since the voltage gain from Y- to X-terminal is very close to one, this equivalent noise source can also be moved to the Y-terminal as $\overline{dv_{yeq}^2}$.

The noise model for the second generation current-conveyor, using the equivalent noise sources discussed, is shown in Figure 5.17a [14]. However, there is one equivalent current noise source present in the model that has not been discussed. Equations (5.10) and (5.11) both assume that Y-terminal is shorted to ground. However, with high Y-terminal impedance levels, current noise in this terminal contributes to the total output noise, resulting in an equivalent current noise source $\overline{di_{yeq}^2}$ the Y-terminal as

$$\overline{di_{yeq}^2} = \overline{di_{B3}^2} + \overline{di_{B4}^2}. \tag{5.15}$$

The total output noise of the test set-up of Figure 5.16a, according to the conveyor noise model, can be expressed as

$$\begin{aligned}\overline{di_{out}^2} &= \overline{di_{zeq}^2} + \left(\frac{R_{sx}}{R_{sx}+Z_x}\right)^2 \left(\overline{di_{xeq}^2} + \frac{4kT}{R_{sx}}df\right) \\ &+ \frac{\overline{dv_{yeq}^2} + R_{sy}^2\overline{di_{yeq}^2} + 4kTR_{sy}df}{R_{sx}^2},\end{aligned} \tag{5.16}$$

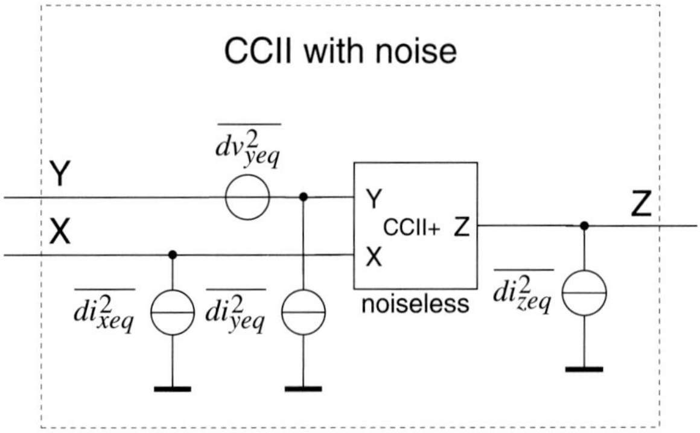

(a)

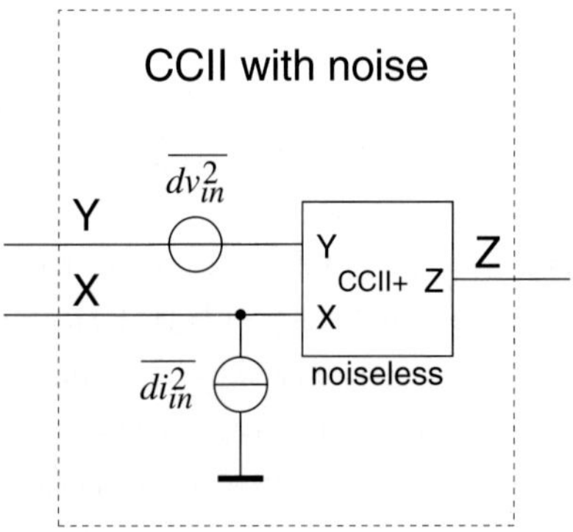

(b)

Figure 5.17 Equivalent noise sources of a second generation current conveyor. a) An accurate model. b) A simpler and less accurate model.

where $Z_x = \frac{1}{g_{m4}}$. The equation clearly shows that the lowest noise is achieved by maximising R_{sx} and minimising R_{sy}. Consequently, the output noise approaches the noise of a simple current-mirror, thus minimising the noise of transistors M_1 and M_2 as well as the current sources I_{B1} and I_{B2}, proves to be an efficient way of restricting the noise level, as Equation (2.32) for the current noise MOS-transistors clearly shows.

In most applications $R_{sx} \gg Z_x$ and thus the equivalent current noise source $\overline{di^2_{zeq}}$ can be moved to the X-terminal without significant errors although it is not theoretically correct [15]. Similarly, in most cases $R_{sx} \gg R_{sy}$ so that $\overline{di^2_{yeq}}$ can be omitted. Therefore, the noise model can be simplified to an equivalent circuit of Figure 5.17b, where

$$\overline{di^2_{in}} = \overline{di^2_{xeq}} + \overline{di^2_{zeq}}, \quad (5.17)$$

$$\overline{dv^2_{in}} = \overline{dv^2_{yeq}}. \quad (5.18)$$

5.6.2 Other low-gain conveyor topologies

With the method described, four equivalent noise sources can also be derived for other conveyor topologies. In a simple class-A MOS CCII-, as in Figures 3.8a and b, the current noise at Y-terminal $\overline{di^2_{yeq}}$ can be omitted since the Y-input node is a MOS-transistor gate. Similar assumption holds true for certain MOS CCII+ topologies such as those in Figures 3.14 and 3.15.

The symmetrical topology of push-pull conveyors eliminates the bias current source at the X-terminal and thus there is no current noise source $\overline{di^2_{xeq}}$. Because the noise in MOS-transistors depends on the transistor transconductance, lowering quiescent current reduces noise without sacrificing the maximum output current, resulting in an increased dynamic range. This occurs at the expense of speed. Additionally, when the signal amplitude exceeds the quiescent current, transistor transconductance and noise become signal dependent.

The noise model discussed can also be used to depict the noise behaviour of the first generation current-conveyors [14]. When the Y-terminal is grounded, the noise behaviour is almost identical to the second generation current-conveyor. However, because the X-terminal input impedance is low, combining both X- and Z-terminal current noise sources into one equivalent input-reduced current noise source is possible. Because of the internal feedback in this conveyor, additional current noise is generated to the Y-terminal.

5.6.3 High-gain current-conveyor

The input stage of almost all current-mode feedback amplifiers discussed in Chapter 4 is either a positive or a negative second generation current-conveyor. Therefore, the conveyor noise models of Figure 5.17 can be used to simplify the noise calculations

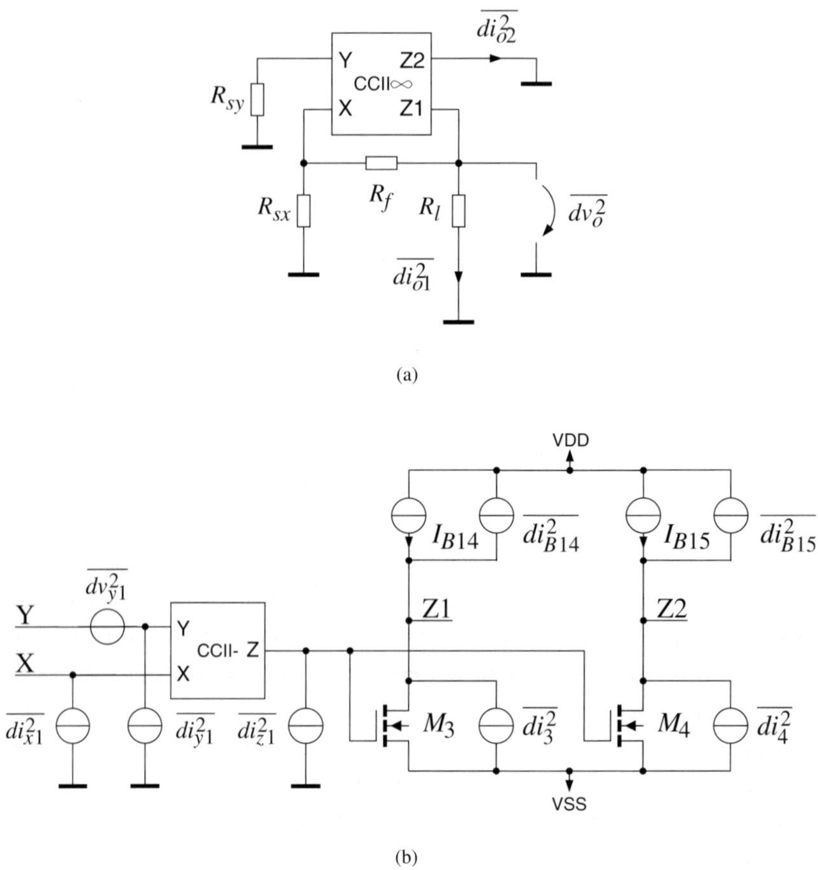

Figure 5.18 a) The noise test set-up for a dual-output high gain current-conveyor. b) The noise sources in a dual output MOS CCII∞.

of current-mode feedback amplifiers. To illustrate this, the high gain conveyor design example of Figure 4.16 is used. Thus, in the resulting equivalent circuit, the only transistors not included in the conveyor noise model are the output stage transistors M_3, M_4, M_{14} and M_{15}, as depicted in Figure 5.18b.

By using the circuit configuration of Figure 5.18a, the noise of most high-gain current-conveyor applications can be evaluated. Since the closed-loop input impedance at the X-terminal is in most cases significantly lower than R_{sx}, the output noise voltage $\overline{dv_o^2}$ at the Z1-output can be approximated at low frequencies as

$$\begin{aligned}\overline{dv_o^2} &\approx \left(1+\frac{R_f}{R_s}\right)^2\left(\overline{dv_{y1}^2}+R_{sy}^2\overline{di_{y1}^2}+4kTR_{sy}df\right)\\ &+R_f^2\left(\overline{di_{x1}^2}+\overline{di_{z1}^2}+\frac{4kT}{R_{sx}\|R_f}df\right)\\ &+\frac{R_f^2}{\left(1+A_i+\frac{R_f}{R_l}\right)^2}\left(\overline{di_3^2}+\overline{di_{B14}^2}+\frac{4kT}{R_f\|R_l}df\right).\end{aligned} \tag{5.19}$$

Because the open-loop current gain A_i is high, the noise contribution of output transistors M_3 and M_{14} can be neglected unless very high frequencies are considered. If the output current noise is required, it can be obtained by letting

$$\overline{di_{o1}^2}=\frac{\overline{dv_o^2}}{R_l^2}. \tag{5.20}$$

Since the replica output is outside the feedback loop, the noise contribution of transistors M_4 and M_{15} cannot be neglected, and thus the output current noise is

$$\overline{di_{o2}^2}=\frac{\overline{dv_o^2}}{(Rf\|R_l)^2}+\overline{di_4^2}+\overline{di_{B15}^2}. \tag{5.21}$$

However, the replica output is typically used in the closed-loop current-conveyor configuration of Figure 4.15b without resistive feedback. Consequently, the output current noise can be expressed as

$$\overline{di_{o2}^2}\approx A_{icl}^2\left(\overline{di_{x1}^2}+\overline{di_{z1}^2}+\overline{di_3^2}+\overline{di_{B14}^2}\right)+\overline{di_4^2}+\overline{di_{B15}^2}, \tag{5.22}$$

where A_{icl} is the closed-loop current gain depending on the aspect ratios of the output transistors M_3, M_4, M_{14}, and M_{15}. Therefore, the noise contribution of the output transistors becomes more significant whereas the resistor noise and the conveyor input voltage noise is omitted.

The examples discussed clearly show that the noise of the output transistors contribute to the total output noise and therefore the noise model for the high-gain current-

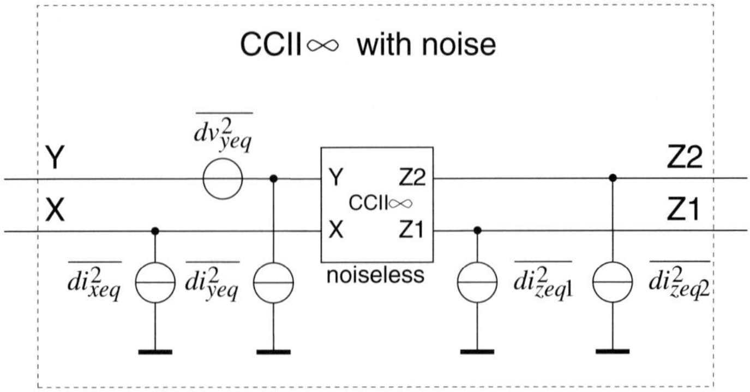

Figure 5.19 The equivalent noise sources of a multi-input high-gain current-conveyor.

conveyor should also include equivalent output current noise sources, as depicted in Figure 5.19. In this model, the equivalent noise sources at Y-terminal are identical to the noise sources of the input CCII-. However, the remaining noise sources are divided between the X-terminal and the two Z-outputs according to the following equations:

$$\overline{di_{xeq}^2} = \overline{di_{x1}^2} + \overline{di_{z1}^2}, \tag{5.23}$$

$$\overline{di_{zeq1}^2} = \overline{di_3^2} + \overline{di_{B14}^2}, \tag{5.24}$$

$$\overline{di_{zeq2}^2} = \overline{di_4^2} + \overline{di_{B15}^2}. \tag{5.25}$$

5.6.4 Other current-mode feedback amplifiers

The current-mode operational amplifier is similar to the high-gain current-conveyor. The input stage is typically a positive second generation current-conveyor. However, a negative current-conveyor can similarly be used as an input stage if the two output terminals are interchanged. Therefore, the only significant difference between the two current amplifier types is the structure of the output stage.

For the same reason, the noise model of the current-mode operational amplifier is also very similar to the noise model of the high-gain current-conveyor. The current-mode operational amplifier has only one input, namely the X-terminal of the input conveyor, while the Y-terminal is always connected to the analogue ground. Therefore, the Y-terminal current noise can be omitted and the Y-terminal voltage noise source moved to the current input, the X-terminal. Similarly, adding noise sources deriving from the output stage is relatively straightforward.

The noise behaviour of current-feedback operational amplifiers is generally modelled by an equivalent voltage noise source at the noninverting input and an equivalent current noise source at the inverting input, as in the simple conveyor noise model of Figure 5.17b. The voltage follower output stage of the current-feedback operational amplifier does not contribute to the total output noise voltage unless there is current noise present in the voltage follower output since this noise slightly adds the input referred current noise.

5.6.5 General notes on current amplifier noise

Most current-mode amplifiers are difficult to optimise for low noise because such amplifiers can be used in wide variety of applications. The best example of this is the high-gain current-conveyor. When the high-gain current-conveyor is used as a replacement for a voltage-mode operational amplifier, the input source follower has a strong impact on the total noise whereas the output transistors do not contribute to the total output voltage noise. However, when a dual output high-gain current-conveyor is connected as a closed-loop current amplifier, the situation is different: the noise due to the input source follower can be neglected while the output transistors exert a significant contribution to the total output current noise.

Nevertheless, it is not possible to design a voltage-mode operational amplifier that would provide optimal noise performance in all applications. Voltage-mode operational amplifiers with a bipolar input differential pair exhibit a low voltage noise but a relatively high current noise, rendering them optimal for low impedance levels. Lower current noise levels can be obtained by using junction or MOS field-effect transistors in the input differential pair. The lowering of the current noise takes place at the expense of the voltage noise and thus FET-input operational amplifiers are better suited to higher impedance levels.

The most straightforward way to compare the noise performances of current-mode and voltage-mode amplifiers is achieved by comparing the commercial bipolar current-feedback operational amplifier to commercial bipolar voltage-mode operational amplifiers. In the case of a typical current-feedback operational amplifier such as AD844 [5], the input referred white noise voltage density is 2 $\mathrm{nV}/\sqrt{\mathrm{Hz}}$, which is relatively low compared to most low-noise voltage-mode operational amplifiers in the market. The input referred white noise current is approximately 10 $\mathrm{pA}/\sqrt{\mathrm{Hz}}$, which is at least ten times higher than the input current noise levels of typical bipolar voltage-mode operational amplifiers. However, most commercial current-feedback operational amplifiers are targeted at video applications, where impedance levels are around 75 Ω and thus current noise has very little effect on such applications.

Similarly, the measurement results of a CMOS current-feedback amplifier [18] presented in Figure 4.6 show relatively comparable input voltage noise levels to typical CMOS voltage-mode amplifiers. This CMOS current-feedback operational amplifier topology does not have current noise present at the noninverting terminal. However, there is still current noise present at the inverting input terminal, rendering CMOS current-feedback amplifiers noisier than CMOS voltage-mode operational amplifiers with high impedance levels.

The dynamic range of current-mode amplifiers can be readily scaled by increasing bias current, which is similarly the case with voltage-mode amplifiers. When signal clipping occurs in the amplifier, both voltages and currents are distorted. Therefore, maximising the dynamic range with low supply voltages does not depend on whether the amplifier uses voltages or currents as a signal. A wide dynamic range is achieved by selecting the most suitable circuit topology for the application and performing the electrical design carefully.

References

[1] S. Franco, "Analytical foundations of current-feedback operational amplifies," in *Proc. IEEE Int. Symposium on Circuits and Systems* (ISCAS-93), Chicago, 1993 pp. 1050-1053.

[2] B. Gilbert, "A precise four-quadrant multiplier with subnanosecond response," *IEEE Solid-State Circuits*, vol. SC-3, pp. 365-373, Dec. 1968.

[3] K. Koli, K. Halonen, E. Tiiliharju, "Current-mode up-converting D/A-interface for mobile communication transmitter applications," *IEEE-CAS Region 8 Workshop on Analog and Mixed IC Design*, Pavia, 1996, pp. 20 -25.

[4] C. Toumazou, F. Lidgey, "Novel current-mode instrumentation amplifier," *Electronic Letters*, vol. 25, pp. 228-230, Feb. 1989.

[5] Analog Devices, Linear Products Databook 1990/91.

[6] Q. Zhu, F. Lidgey, W. Su, "High CMRR, second generation current-mode instrumentation amplifiers," in *Proc. IEEE Int. Symposium on Circuits and Systems* (ISCAS-93), Chicago, 1993 pp. 1326-1328.

[7] F. Lidgey,W. Su, K. Hayatleh, "Novel current-feedback operational amplifier design based on a floating circuit technique ," IEE Colloquium on Analog Signal Processing, Oxford, 1998 , pp. 9/1 -9/4

[8] F. Lidgey, W. Su, "High accuracy, high speed voltage-follower," in *Proc. IEEE Int. Symposium on Circuits and Systems* (ISCAS-98), Monterey, 1998 pp. I460-I463.

[9] W. Su, F. Lidgey, "Common-mode rejection in current-mode instrumentation amplifiers," *Int. J. Analog Integrated Circuit Design and Signal Processing*, vol. 7, pp. 257-260, May 1995.

[10] K. Koli, K. Halonen ,"CMRR Enhancement Techniques for Current-Mode Instrumentation Amplifiers," *IEEE Trans. Circuits and Systems-I*, vol. CAS-47, pp. 622-632, May 2000.

[11] C. Toumazou, A. Payne, F. Lidgey, "Operational Floating Conveyor," *Electronic Letters*, vol. 27, pp. 651-652, April 1991.

[12] A. Fabre, H. Barthelemy, "Composite second-generation current conveyor with reduced parasitic resistance," *Electronic Letters*, vol. 30, pp. 377-378, March 1994.

[13] N. Tan, S. Eriksson, "Low-voltage fully differential class-AB SI circuits with common-mode feedforward,"*Electronic Letters*, vol. 30, pp. 2090-91, Dec. 1994.

[14] E. Bruun, "Analysis of the Noise Characteristics of CMOS Current Conveyors,"*Int. J. Analog Integrated Circuit Design and Signal Processing*, vol. 12, pp. 71-78, Jan. 1997.

[15] G. Palmisano, G. Palumbo, S. Pennisi, *CMOS Current Amplifiers*, Boston/Dorbrecht/London, Kluwer Academic Publishers, 1999, 159 p.

[16] A. Arbel, "Comparison between the noise performance of current-mode and voltage-mode amplifiers," *Int. J. Analog Integrated Circuit Design and Signal Processing*, vol. 7, pp. 221-242, May 1995.

[17] B. Harvey, "Current feedback opamp limitations: a state-of-the-art review," in *Proc. IEEE Int. Symposium on Circuits and Systems* (ISCAS-93), Chicago, 1993 pp. 1066-1069.

[18] K. Manetakis, C. Toumazou, C. Papavassiliou, "A 120 MHz, 12 mW CMOS current feedback opamp," in *Proc. IEEE* Custom Integrated Circuits Conference (CICC-98), Santa Clara, May 1998, pp. 365-368.

Chapter 6

Current-mode continuous-time filters

Since inductors in integrated circuits are feasible components exclusively at the gigahertz range below these frequencies, integrated filters are realised as active continuous-time filters. The inductors can be replaced in the active filters with simulated inductances, using either generalised immittance converters (GICs) or gyrators [1, 2]. However, most active filters are constructed by using lossy and lossless integrators as building blocks [1–3]. Converting a passive filter prototype to a signal-flow-graph and, further, to an integrator-based block diagram is well covered in the literature [1,2,4,5]. In this book, therefore, predominantly the different integrated CMOS realisations of these integrators are concentrated on.

A considerable debate has arisen surrounding the term 'current-mode filter'. In effect, filters can be considered neither voltage- nor current-mode, as in filters both voltages and currents must be taken into account simultaneously. However, active filters can be realised with very different building blocks, which can be considered as voltage-mode or current-mode devices. Therefore, in this book the term 'voltage-mode filter' refers to a filter in which voltage-mode operational amplifiers or operational transconductance amplifiers (OTAs) are used as building blocks. Similarly, when the filter is constructed using current-conveyors or other current amplifiers, it is considered current-mode. Operational transconductance amplifier based filters (OTA-C filters) are also occasionally referred to as current-mode filters, but as OTAs are in most cases used in similar ways to voltage-mode operational amplifiers, the term current-mode is here restricted to building blocks that have at least one low impedance input.

Before moving to current-mode realisations of integrators, first the typical voltage-mode integrator realisations are discussed. However, comparing the performances of different continuous-time active filter building blocks is difficult unless a common figure of merit is used for the different integrator realisations. Quality factor is widely used as a figure of merit in passive filter components and, actually, a quality factor can also be derived for integrators [2].

6.1 Integrator quality factor

Let us first consider an ideal lossless inductor as an integrating element. Thus, the inductor integrates the voltage v_L across the inductor resulting in a frequency dependent current through the inductor $i_L(j\omega) = \frac{v_L}{j\omega L}$. If the inductor is replaced with a lossy inductor with a series resistance R_L as in the lossy element, the resulting integration function is

$$H(j\omega) = \frac{i_L}{v_L} = \frac{1}{j\omega L + R_L}. \tag{6.1}$$

The quality factor of an element is essentially a measure of the energy that is stored compared with the energy that is dissipated in a steady-state sinusoidal excitation. In the case of inductors, the quality factor is derived as [1,2]

$$Q_L(\omega) = \frac{\omega L}{R_L}. \tag{6.2}$$

Comparing Equations (6.1) and (6.2) the quality factor of an integrator can be expressed in a general form as [2]

$$Q_I(\omega) = \frac{\mathrm{Im}\left(\frac{1}{H(j\omega)}\right)}{\mathrm{Re}\left(\frac{1}{H(j\omega)}\right)} = -\frac{\mathrm{Im}\left(H(j\omega)\right)}{\mathrm{Re}\left(H(j\omega)\right)}. \tag{6.3}$$

The quality factor depends strongly on the frequency. Additionally, for active components, the quality factor can be negative. A positive quality factor entails a positive phase error (phase lead) while a negative quality factor entails a negative phase error (phase lag). To demonstrate this effect, the relation between integrator quality factor and phase can be easily derived as [2]

$$\Phi_I = -\frac{\pi}{2} + \arctan\frac{1}{Q_I(\omega)}. \tag{6.4}$$

If integrators with opposite sign quality factors are present in the active filter, it is possible that phase errors in the filter cancel each other out. Therefore, with inventive circuit techniques, the quality factor of an active filter may exceed the quality factors of the components, a situation which is impossible with passive filters.

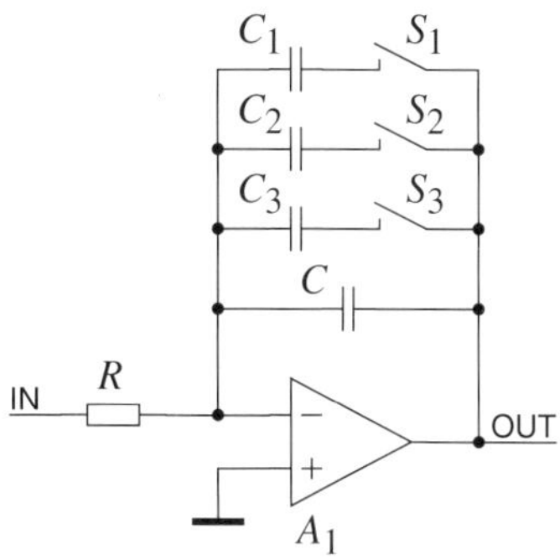

Figure 6.1 Lossless inverting integrator realised with a voltage-mode operational amplifier and integrated passive components.

6.2 Voltage-mode active-RC integrators

Perhaps the most widely known active filter technique uses operational amplifiers, resistors, and capacitors to construct integrators and filters. In most CMOS-processes, there are high quality polysilicon or metal capacitors available, whereas high quality resistors are available in few dedicated analogue CMOS processes or they are available as a costly additional option. Even when both high quality capacitors and resistors are available in the process, the RC time constant may vary by almost $\pm 50\%$ when both process variations and different environment conditions are taken into account [6]. The integrated resistors in particular have large process variations and temperature dependencies. Similarly, the capacitance density of the capacitors may vary from die to die by more than $\pm 10\%$ and this variation does not correlate with the process variations of the resistors.

As the time constant variation is too large for most applications, variable time constants, controlled by an additional calibration or auto-tuning circuit, are needed in integrated RC-active filters. In most cases, a variable integrating time constant is realised as a switched array of parallel capacitors, as shown in Figure 6.1 [6]. The size of the switched capacitors are binary weighted in order to simplify the digital control of time constants. Whether the CMOS-switches should locate at the input or the output of the amplifier depends on the requirements of the application. At the inverting input node, the switch on-resistance has almost no effect on the linearity but the parasitic capacitances of the switch are added to the inverting input and thus more phase lag occurs. Alternatively, at the output node, the switch on-resistance degrades the linearity, particularly at high frequencies, but the parasitic capacitances of the switches have minimal effect on the performance of the integrator.

In the parallel capacitor array, the limited on-resistance of the CMOS-switches used have a minimal effect on the high-frequency performance of the integrator since any resistance in series with the integrating capacitor results in a phase lead at high frequencies. This phase lead reduces the high-frequency phase lag deriving from the limited bandwidth of the operational amplifier. However, the amount of the phase lead deriving from the CMOS-switches is quite insignificant because the main integrating capacitor shunts most of the current past the switches at high frequencies.

If integrated resistors with low temperature dependencies such as thin-film resistors are used, it may be possible to calibrate the filter only once during the testing phase and store the correct switch control data to a small non-volatile memory. Then also the tuning circuit can be included in the testing system rather than the integrated circuit. In most cases, only resistors with large temperature coefficients are available and thus the tuning procedure must be repeated within relatively short intervals and thus an on-chip auto-tuning circuit is required.

There are numerous ways to realise the auto-tuning for the RC-active filter [7,8]. In most systems, there is an accurate clock signal available so that a digital timer circuit can be designed that measures the integration time between two reference voltages and a special control logic increases the active capacitances in the capacitor array, provided that the correct value is found [8]. Alternatively, a reference resistor realised as a switched capacitor circuit is used as a reference slope in an auto-tuning circuit resembling a dual slope integrating A/D-converter widely used in digital multimeters [6, 7]. Similarly, also successive approximation techniques can be used in the tuning procedure [9].

Because of the discrete nature of the time constant control an idle period is usually required in the system so that the switching of the capacitors does not disturb the operation of the system. On the other hand, in the digital control there is no long-term drift in the control circuit itself unlike in a fully analogue continuous-time control loop which cannot hold the found calibrated state for long. Since the digital control can be disabled for relatively long time periods, it can be turned off while unused and thus power is saved and the interference deriving from the auto-tuning circuit is reduced.

Operational amplifiers with high open-loop gain and low distortion are relatively easy to design. Similarly, integrated capacitors have a very low signal dependency and if the CMOS-switches are placed at the inverting input the distortion arising from the switches is similarly low. Therefore, the linearity of active-RC filters using voltage-mode operational amplifiers is usually limited by the signal dependency of the integrated resistors.

The transfer function of the discussed inverting integrator is expressed as

$$H(s) = -\frac{1}{sRC}\frac{A_v(s)}{1+A_v(s)+\frac{1}{sRC}}, \tag{6.5}$$

where $A_v(s)$ is the open-loop gain of the voltage-mode operational amplifier. If one-pole model is assumed with limited DC-gain for the operational amplifier open-loop gain

$$A_v(s) = \frac{\omega_0}{s + \frac{\omega_0}{A_v(0)}}, \tag{6.6}$$

then letting $s = j\omega$ the quality factor of the integrator is expressed using Equation (6.3) as

$$Q_I(\omega) \approx \frac{1}{\frac{1}{\omega RCA_v(0)} - \frac{\omega}{\omega_0}}. \tag{6.7}$$

In most cases the open-loop DC-gain is very large so that the quality factor equation reduces to

$$Q_I(\omega) \approx -\frac{\omega_0}{\omega}. \tag{6.8}$$

This means that for accurate high-Q filters the corner-frequency of the filter should be almost two decades lower than the operational amplifier unity-gain frequency. Furthermore, since the voltage-mode operational amplifier needs to be unity-gain stable even with relatively large capacitive loads, the unity-gain frequency of the operational amplifier is significantly lower than the f_T of the transistors available in the integration process. Therefore this active filter technique is used at relatively low frequencies in low distortion applications.

6.3 OTA-based integrators

Another widely used continuous-time filtering technique uses operational transconductance amplifiers (OTAs) as building blocks. A typical folded cascode OTA is shown in Figure C.3 in Appendix C. Thus, the transfer function of the lossless inverting integrator of Figure 6.2a is

$$H(s) = -\frac{g_m}{(g_o + sC)\left(1 + \frac{s}{p_2}\right)}, \tag{6.9}$$

where g_m is the transconductance of the OTA, g_o is the limited output conductance, and p_2 is the nondominant pole of the OTA, generally deriving from the cascode transistors. Similarly, the quality factor for the lossless integrator is

$$Q_I(\omega) \approx \frac{1}{\frac{g_o}{\omega C} - \frac{\omega}{p_2}}. \tag{6.10}$$

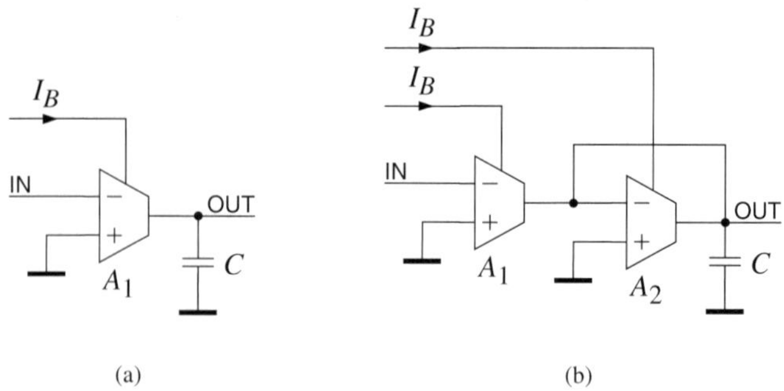

Figure 6.2 Inverting integrators based on operational transconductance amplifiers. a) Lossless OTA-C integrator. b) Lossy OTA-C integrator.

If cascode techniques are used in the OTA output, the output conductance is insignificant enough to be neglected in the quality factor equation, so that it reduces to

$$Q_I(\omega) \approx -\frac{p_2}{\omega}. \tag{6.11}$$

The quality factor of the OTA-C integrator depends on the nondominant pole of the amplifier deriving from parasitic capacitances of an internal node of the amplifier. This pole is at a significantly higher frequency than the unity gain frequency of a typical voltage-mode operational amplifier and thus a better high frequency performance can be obtained with OTA-C filters than with voltage-mode active-RC filters.

Lossy integrators can be realised in active-RC filters simply by adding a resistor in parallel with the integrating capacitor whereas in OTA-C filters, lossy integrators require an additional OTA as an active load resistor, as depicted in Figure 6.2b. On the other hand, in single-ended active-RC filters noninverting integrators require two operational amplifiers whereas an inverting OTA-C integrator can be changed to a noninverting integrator simply by swapping the inputs of the OTA or by feeding signal to both inputs and thus realising the inverting and the noninverting integrators simultaneously with the same OTA. Moreover, the transistor-level realisations of OTAs are commonly quite area and power efficient and thus OTA-C filters require less area and power than active-RC filters.

6.3.1 The effects of process variation and temperature drift

The transconductance g_m of the OTA depends on temperature and other process variations. Therefore, the transconductance is conventionally controlled by the bias current [10–12] although, in some OTA-topologies, the transconductance is controlled by an additional control voltage [13] or by the supply voltage [14]. Because of the temper-

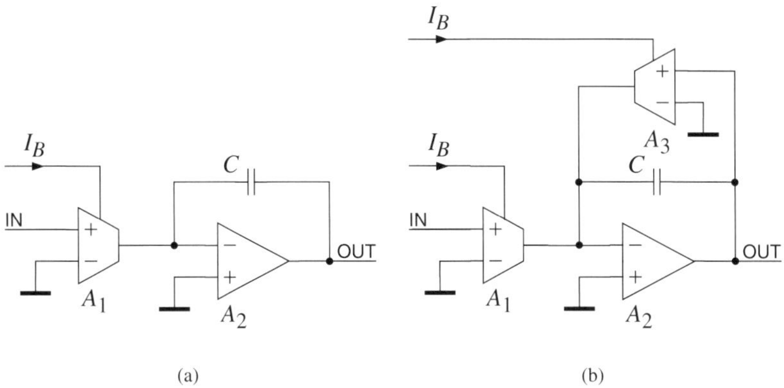

Figure 6.3 Inverting OTA-based integrators with lower sensitivity to parasitic capacitances. a) Lossless OTA-Miller integrator. b) Lossy OTA-Miller integrator.

ature drift of the OTA transconductance, the bias current must be continuously auto-tuned. A typical tuning arrangement [3, 10] includes an OTA-based current-controlled oscillator controlled by a phase-locked loop. There are numerous alternative ways to auto-tune the OTA-C filter. However, almost every method uses a replica OTA-integrator and a reference frequency.

Because the integrating capacitor in OTA-C integrators is grounded, all parasitic capacitances connected to the same node are added to the total integration capacitance. These parasitic capacitances may also affect the shape of the filter transfer function and consequently auto-tuning the filter corner frequency cannot cancel out all effects of this process variation although the same applies to several other nonidealities in the filter. Additionally, these parasitic capacitances are signal dependent, resulting in increased distortion, particularly with high-frequency filters, where relatively small capacitors are required.

Such sensitivity to parasitic capacitances can be reduced by using active Miller-integrators rather than grounded integration capacitors, as described in Figure 6.3 [3, 15]. It is tempting to assume that using an OTA-Miller integrator, rather than a conventional OTA-C integrator, increases the area and the power consumption of the filter. However, the voltage amplifier A_2 does not have to drive a resistive load and the transconductance of input OTA contributes additionally to the total integrator DC-gain and consequently the amplifier A_2 can be as simple as a common-source amplifier stage. Moreover, since the OTA is now driving a low impedance load no cascode transistors are required to increase the OTA output impedance. Therefore, the transistor level realisations of OTA-Miller integrators are almost as simple as conventional OTA-C integrators and thus also competitive area and power efficiency is attained.

When the Miller-integrator is realised with a simple common-source amplifier stage, there is also a right half-plane zero in the frequency response arising from the feed-forward effect of the integrating capacitor. In the voltage-mode active-RC integrators, this effect can generally be neglected because of the large open-loop gain and moderately low output impedance of the operational amplifier. However, this right half-plane zero is easily compensated by adding a resistor in series with the integrating capacitor, as in most Miller-compensated operational amplifiers [3, 5, 15]. This resistance is normally realised with a MOS-transistor operating in triode region. For this transistor, a bias circuit that provides adequate tracking of process and temperature variations is relatively straighforward to realise and thus this nonideality seldom restricts the high-frequency performance of the OTA-Miller filters.

6.3.2 Transconductance linearity

When the transconductance of the OTA is realised with a simple differential pair, described in Figure 6.4a, a significant amount of distortion also results. The large signal equation of the differential pair is derived in Equation (C.21) in Appendix C, resulting in a third order intermodulation distortion equation of

$$IM3 = \frac{3}{32}\left(\frac{\hat{v}_{in}}{V_{GS}-V_t}\right)^2. \tag{6.12}$$

Because of the symmetry in the circuit, even harmonic distortion can be neglected. Theoretically, an intermodulation distortion level of 1% is reached at one third of the maximum signal amplitude and in practice this distortion level is reached at lower signal amplitudes.

Although the OTA transconductance has significant nonlinearities, filter realisations include feedback loops of two or more OTAs, which reduce the low frequency distortion, particularly in low-pass filters. Unfortunately, near the filter corner frequency, distortion reduction is no longer attained, as discussed earlier in the context of current-mirror and current amplifiers. In continuous-time filters, there are normally significant gain peaks near the filter corner frequency reducing the maximum signal amplitude. The distortion in OTA-C filters may be high, even at low frequencies, if the first OTA in the filter is not included in a feedback loop.

The linearity required in most applications results in either low signal levels or large gate voltages when simple differential pairs are used as the transconductance element. Therefore, minimising the distortion and maximising the dynamic range requires a more linear transconductance element, particularly at low supply voltages. A simple method for reducing the distortion is to use two or more differential pairs in series as described in Figure 6.4b [11]. In the case of two differential pairs, the sig-

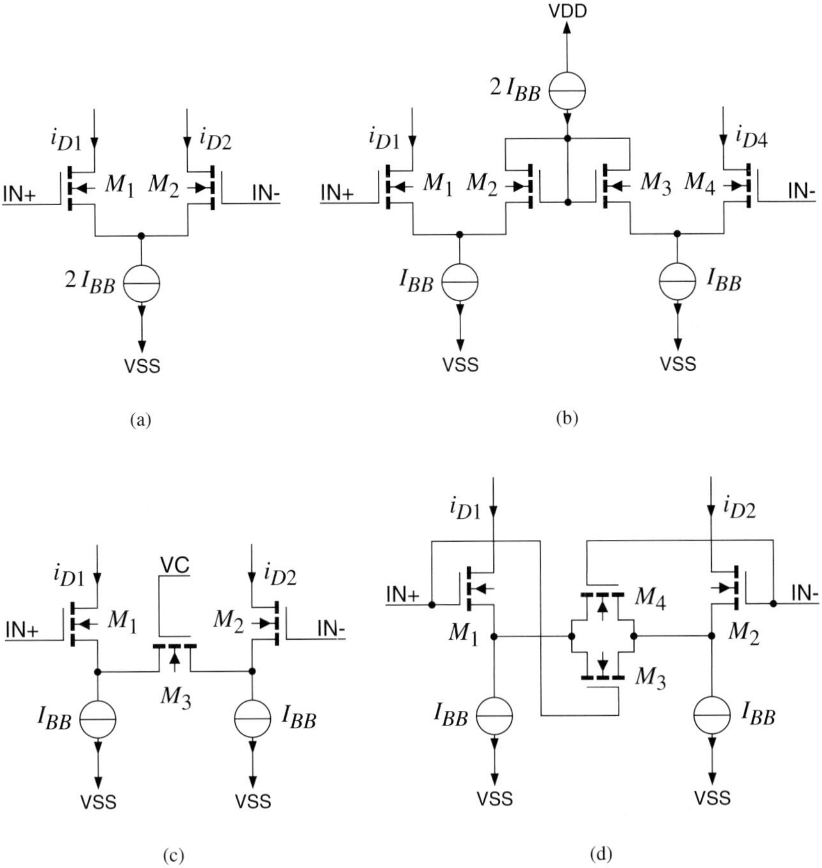

Figure 6.4 Different transconductance realisations. a) A simple differential pair. b) Two differential pairs in series. c) A source degenerated differential pair. d) A modified source degeneration method.

nal amplitude seen by one differential pair is halved, resulting in only a quarter of the distortion that one differential pair would generate. However, adding more differential pairs in series will rapidly decrease the high frequency performance.

A widely used distortion reduction technique is the source degeneration technique. The simplest realisation of a source degenerated differential pair is shown in Figure 6.4c. There, the bias current source is split into two identical current sources and the transistor M_3 is used as a voltage controlled source degeneration resistor. The linear range of the source degenerated differential pair can be extended by the arrangement in Figure 6.4d where two MOS-transistors in parallel are dynamically biased by the input voltage [12]. The linearity of a source degenerated differential pair increases with the degeneration resistance. However, with low supply voltages it may be difficult to realise very large source degeneration resistances without sacrificing too much dynamic and tuning range. Moreover, heavily degenerated transconductances lead additionally to large time constants in the filter and thus this technique is not the optimal choice for high frequency filters.

The transconductance of a MOS-transistor is very linear in the triode region if the drain-source voltage of the transistor is kept constant. In BiCMOS realisations the differential pair transistors can be forced into the triode region with bipolar cascode transistors [15, 16]. Since the transconductance of a bipolar transistor is large compared to MOS-transistors, the drain-source voltages of the differential pair transistors are relatively signal independent and thus low distortion will result. Because the drain-source conductance of a MOS-transistors is low in the triode region the output impedance of the OTA may remain relatively low with simple cascode techniques and as a result OTA-Miller integrators are common with this linearization technique [15].

6.4 Integrators with MOS-resistors

In voltage-mode active-RC filters the resistors can also be realised with MOS-transistors operating in the triode region [17]. Since MOS-resistors have significant nonlinearities, fully differential integrators are widely used in these filters because in this case even order distortion is efficiently rejected. The distortion can be reduced further by using cross-coupled MOS-transistors in the configuration of Figure 6.5. The resistance is controlled by a difference between two control voltages, resulting in an extended tuning range as the resistance increases to infinity, while both control voltages become identical if ideally matched MOS-transistors are assumed. However, very high resistance values cannot be used in practice because of device mismatches and noise.

In the case of both cross-coupled and simple MOS-resistors, rather high control voltages are required to ensure the triode region operation for the MOS-resistors. How-

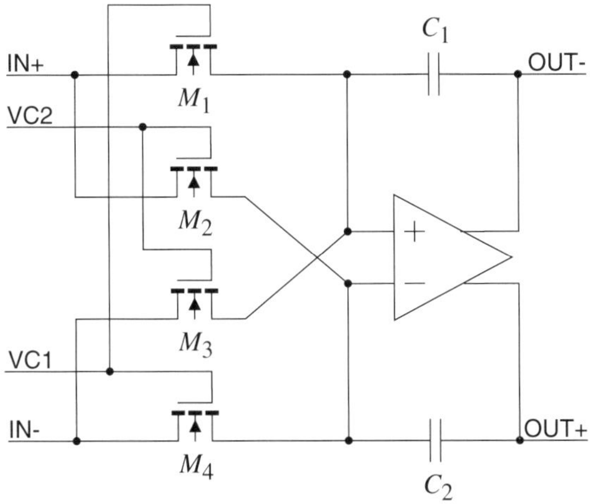

Figure 6.5 Fully differential active-RC integrator with linearised MOS-resistors.

ever, this makes it difficult to achieve a large signal swing with low supply voltages with this filter technique. Since this filtering technique is based on voltage-mode operational amplifiers, only a moderate high-frequency performance is achieved. Moreover, MOS-resistors require continuous tuning. Therefore, this circuit technique is seldom used in the case of modern deep-submicron CMOS-processes. However, in CMOS-processes without a high-quality integrated resistor, a MOS-resistor based active filter may be feasible at low frequency and low distortion applications.

6.5 Current-conveyor based filters

As demonstrated in Chapter 3.1.3, integrators are easily realised with current-conveyors. Since most class-A CCII- implementations avoid output current-mirrors by using instead a current folding, lower distortion can be achieved than with class-A CCII+ implementations. An inverting active-RC integrator based on this negative conveyor is presented in Figure 6.6a. Similarly, MOS-C integrators based on negative conveyors can be realised. In order to reject even order nonlinearities differential integrators are typically used, as depicted in Figure 6.6b. In most cases, this differential integrator requires additionally a common-mode feedback circuit to set the DC-voltages at the outputs.

The transfer function of a lossless current integrator based on a positive second generation current-conveyor was previously derived in Section 3.2.3 as Equation (3.33). For most negative second generation current-conveyors, the frequency dependencies of the forward voltage and current gains can be neglected because the parasitic capacitances at conveyor terminals dominate the frequency behaviour regardless. Conse-

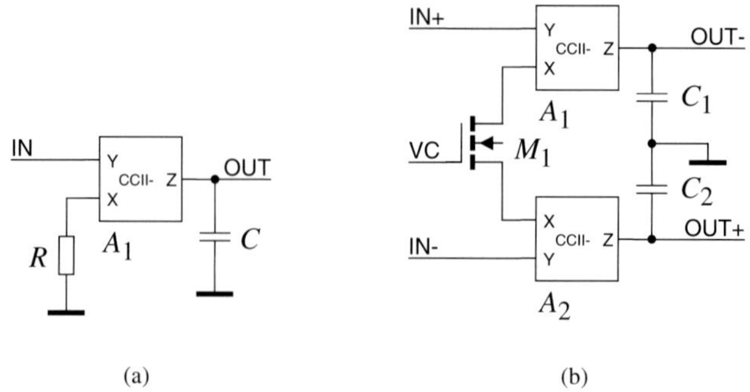

Figure 6.6 Current-conveyor based integrators. a) Inverting lossless active-RC integrator. b) Differential lossless MOS-C integrator.

quently, neglecting the limited DC-gain the transfer function of the inverting voltage integrator can be simplified to

$$A_v(s) = -\frac{1}{s\,(R+Z_x)\,(C+C_z)}\,\frac{1+\frac{s}{z}}{1+\frac{s}{p}}, \tag{6.13}$$

where both the pole and the right half-plane zero are arising from the parasitic capacitance at the X-terminal

$$z = \frac{1}{RC_x}, \tag{6.14}$$

$$p = \frac{1}{(Z_x \parallel R)\,C_x}, \tag{6.15}$$

Based on this transfer function, it is easy to calculate the integrator quality factor as

$$Q_I(\omega) \approx \frac{1}{\omega}\frac{pz}{z-p}. \tag{6.16}$$

The resistance R should be considerably larger than the X-terminal impedance Z_x in order to keep the distortion and time-constant temperature drift low. Therefore, the zero z is at a significantly lower frequency than the pole p. Compared to OTA-C filters, better linearity is obtained with CCII- based filters at the expense of high-frequency performance. However, the high-frequency performance of the conveyor based filters is still good in comparison with voltage-mode operational amplifier based active-RC filters.

If the differential conveyor based MOS-C integrator of Figure 6.6b are compared to the source degenerated differential pair in Figure 6.4c, these two circuits are found very similar. The differential conveyor integrator can be considered a generalised case of the source degenerated differential pair, where the input transistors M_1 and M_2 can

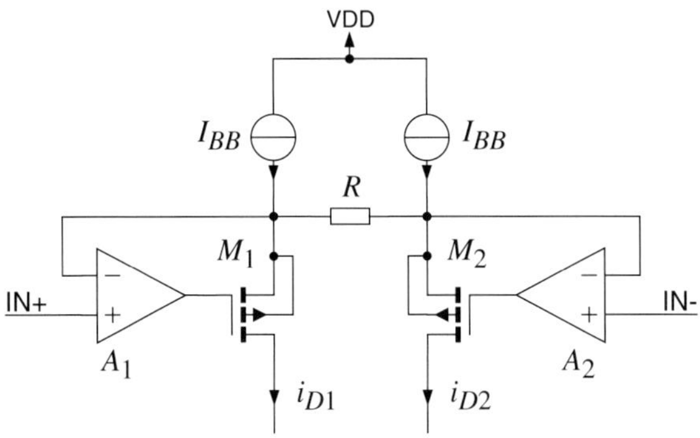

(a)

Figure 6.7 Current-conveyor like source-degenerated transconductance element.

be replaced with more complex circuits. In effect, in most OTA-C filters with source-degenerated transconductance elements, current-conveyor like structures can be found.

As an illustration, the transconductor of Figure 6.7 [20] is presented. In this case additional voltage amplifiers A_1 and A_2 are used to boost the transconductance of transistors M_1 and M_2 so the linearity and accuracy of the transconductor depends almost solely on the source degeneration resistor R. Unfortunately, the voltage amplifiers A_1 and A_2, driving the transistors M_1 and M_2, have limited input and output voltage ranges, reducing the applicable input voltage range of the transconductance element and thus limiting its use in low voltage applications.

There are also transconductors that use feedback current amplifiers in closed-loop to increase the source degeneration accuracy. A high-gain conveyor is found in the core of many linearised transconductors [21,22]. There are other circuit topologies that blur the difference between a high-gain and low-gain conveyor, such as the half-circuit of a differential transconductor displayed in Figure 6.8a [23]. This circuit closely resembles a dual-output high-gain conveyor connected in a closed-loop operating as a CCII+. Since all transistors in the signal path are NMOS-transistors, a level-shift NMOS-transistor M_2 is required for large input voltage range. In most n-well CMOS-processes this means that, because of the bulk-effect, the gate-source voltage of transistor M_2 is high, thus limiting the minimum supply voltage.

A further limitation in this circuit is the signal dependent drain-source voltage of the current-mirror input transistor M_3, which generates distortion as a result of the channel length modulation. However, if the degeneration resistor is realised as a MOS-resistor, the distortion arising from the current-mirror M_3 and M_4 may not add significantly to the total distortion of the circuit. In any event, in an application in which a

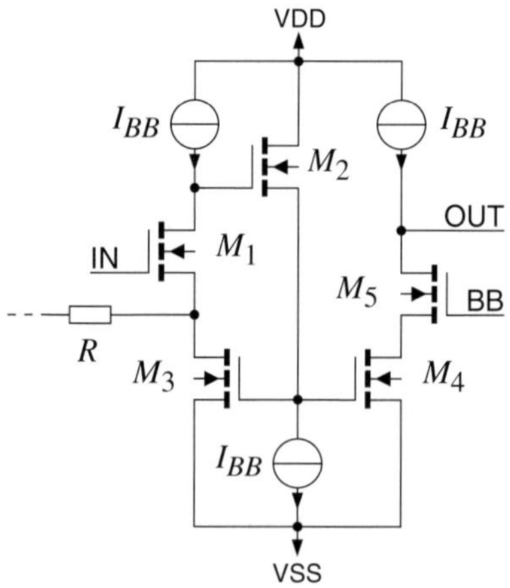

(a)

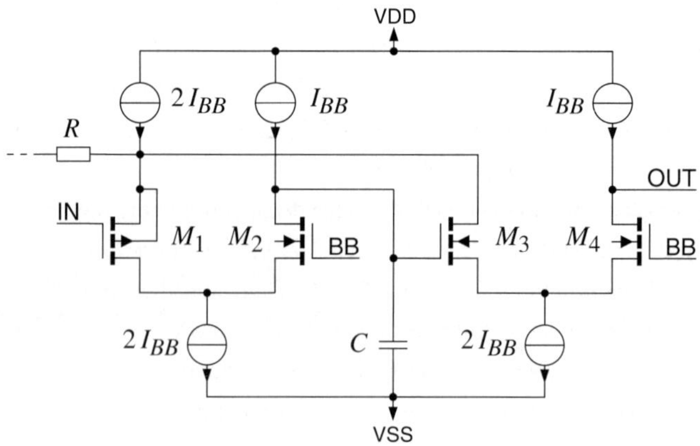

(b)

Figure 6.8 Other current-conveyor based transconductance element. Only the other half of the differential circuit is displayed for simplicity. a) An all-NMOS CCII+ [23]. b) A folded high voltage swing CCII- [7].

wide bandwidth supercedes linearity, this circuit is worth trying out.

Another feedback current amplifier based source-degenerated transconductor is presented in Figure 6.8 [7]. This circuit combines the input structure of a high-gain current-conveyor to a current-mode operational amplifier like output structure. In order to maximise the input voltage swing, the signal path is folded by an additional common-gate NMOS amplifier stage (transistor M_2).

6.6 Current-mirror based filter

Filters can also be constructed with very simple building blocks. Lossy integrators can be realised with current-mirrors and lossless integrators with simple common-source amplifier stages. Because current output signals cannot be shared with other current-inputs, multiple outputs are required to construct filters. This is readily achieved simply by making as many replica outputs as are required, as shown in Figure 6.9 [24]. Similarly, additional current-mirrors are required for constructing noninverting integrators.

The distortion performance of a current-mirror used as a lossy integrator was previously derived in Chapter 2.1.3. Thus, it can be assumed that current-mirror based filters exhibit relatively high nonlinearity near the filter corner frequency. The poor linearity near the filter corner frequency is not, however, critical in all applications. For example, when filtering the output signal of a D/A-converter the signal energy near the corner frequency of the filter is low, as depicted in Figure 6.10 and thus quite large nonlinearity at the corner frequency is tolerated provided that the linearity remains significantly lower above and below the corner frequency.

In addition to the one-transistor lossless integrator realisation of Figure 6.9b, other current-mirror based lossless integrator realisations are also published, such as the single-ended lossless integrator with both inverting and noninverting inputs, shown in Figure 6.11a [25]. In this circuit, a positive feedback path (transistors M_1, M_2 and M_4) is used to cancel the input impedance of the second current-mirror parallel to the integrating capacitor C. The dynamic nonlinearity of this integrator is derived in Appendix D. Those calculations show that, if the integrating capacitor is significantly larger than the parasitic input capacitance of the current-mirrors, the nonlinearity of the dual current-mirror integrator is almost identical to the nonlinearity of the one-transistor integrator of Figure 6.9b.

Similarly, the small-signal transfer function can be derived for both the inverting and the noninverting integrator signal paths as

$$\frac{i_{out}}{i_{in+}} = \frac{\omega_1 \omega_2}{s(s+\omega_1+\omega_2)}, \tag{6.17}$$

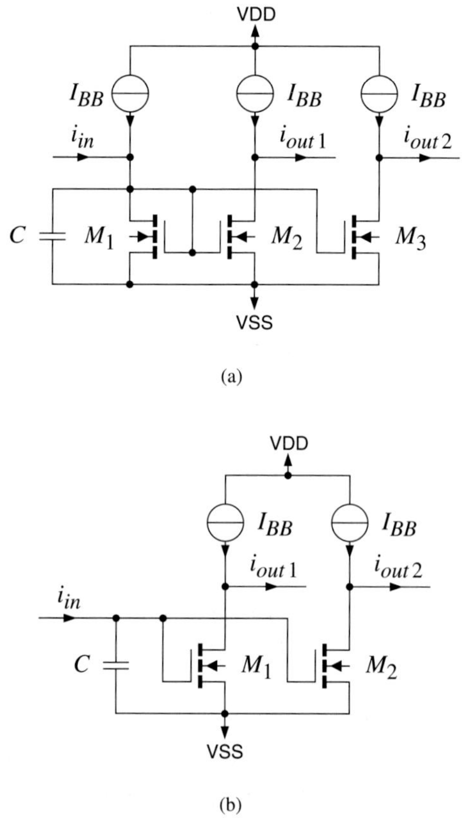

Figure 6.9 Current-mirror as an filter building-block. a) A lossy inverting integrator. b) A lossless inverting integrator.

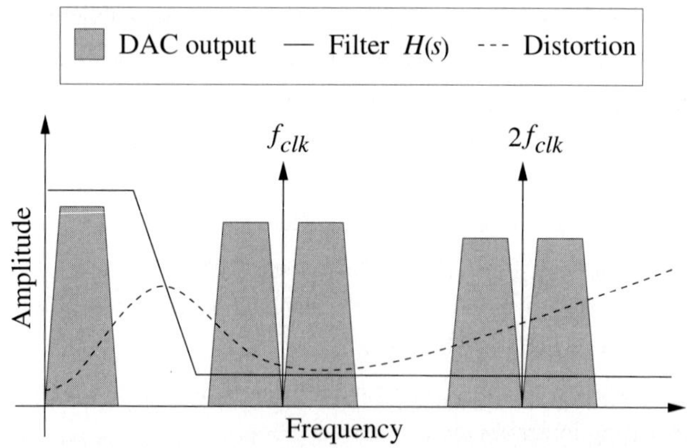

Figure 6.10 The nonlinearity of the current-mirror based filter in a smoothing filter application.

$$\frac{i_{out}}{i_{in-}} = -\frac{\omega_2\left(s+\omega_1\right)}{s\left(s+\omega_1+\omega_2\right)}, \tag{6.18}$$

where $\omega_1 = \frac{g_{m1}}{C_{GS1}+C_{GS2}}$ and $\omega_2 = \frac{g_{m3}}{C+C_{GS3}+C_{GS4}+C_{GS5}}$. Based on these transfer functions, the quality factors can similarly be expressed as

$$Q_{I+}(\omega) = -\frac{\omega_1+\omega_2}{\omega} \approx -\frac{\omega_1}{\omega}, \tag{6.19}$$

$$Q_{I-}(\omega) = \frac{\omega^2+\omega_1\omega_2+\omega_1^2}{\omega\omega_2} \approx \frac{\omega_1}{\omega}\left(1+\frac{\omega_1}{\omega_2}\right). \tag{6.20}$$

Since the zero in the inverting integrator transfer function almost cancels out the pole the quality factor $Q_{I-}(\omega)$ is high. However, this circuit requires cascode current-mirrors in order to minimise the effects of channel length modulation. These cascode transistors add additional poles to the integrator transfer functions and thus a lower quality factor results in practice. Furthermore, this integrator is very sensitive to device mismatches, particularly to the threshold voltage mismatch, because in this circuit positive feedback is used to enhance the integrator DC-gain.

The differential current-mirror based lossless integrator of Figure 6.11b [26] is actually very similar to the single-ended circuit. However, because of the symmetry in the circuit, even order distortion is effectively cancelled out in this circuit as derived in Appendix D. These calculations also show that, if channel length modulation and device mismatches are neglected, this differential integrator performs like an ideal integrator. Therefore, the integrator quality factor solely depends on the parasitic poles deriving from cascode transistors, and the DC-gain is limited by device mismatches and channel length modulation. Furthermore, this circuit exhibits a high input impedance solely for differential input signals, while for common-mode signals, the input impedance is equal to the input impedance of a plain current-mirror.

Because the differential input impedance of the differential integrator is infinite, in the ideal case the circuit can be divided into two parts: two linear capacitors integrating the differential input current and a non-linear fully differential transconductance amplifier. Then, the non-linear output current without the integrating function, as derived in Appendix D, is

$$i_{OUT}(v_{in}) = i_{OUT+}\left(\frac{v_{in}}{2}\right) - i_{OUT-}\left(-\frac{v_{in}}{2}\right) = b_1 v_{in} - \frac{1}{2}\frac{b_2^2}{b_1+j\omega C}v_{in}^3, \tag{6.21}$$

where $b_1 = g_m = \sqrt{2\beta I_{BB}}$ and $b_2 = \frac{1}{2}\beta$. Below the unity gain frequency, this nonlinearity is almost identical to the nonlinearity of a simple MOS differential pair (Figure 6.4a) derived in Equation (C.22) in Appendix C. At higher frequencies, the third order component of the differential current-mirror based integrator decreases with frequency leading in lower distortion than in the differential pair. However, in practical circuits,

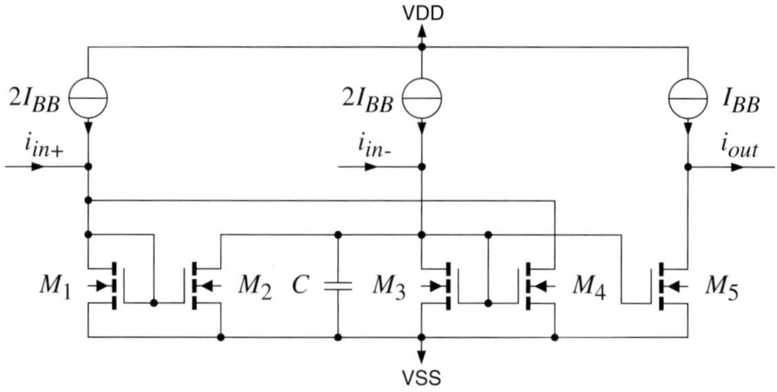

(a)

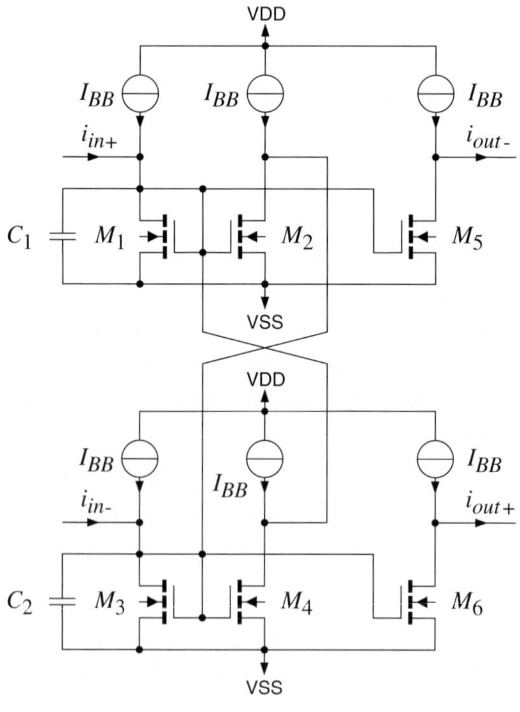

(b)

Figure 6.11 a) A single-ended dual current-mirror lossless integrator [25]. b) A differential current-mirror based lossless integrator [26].

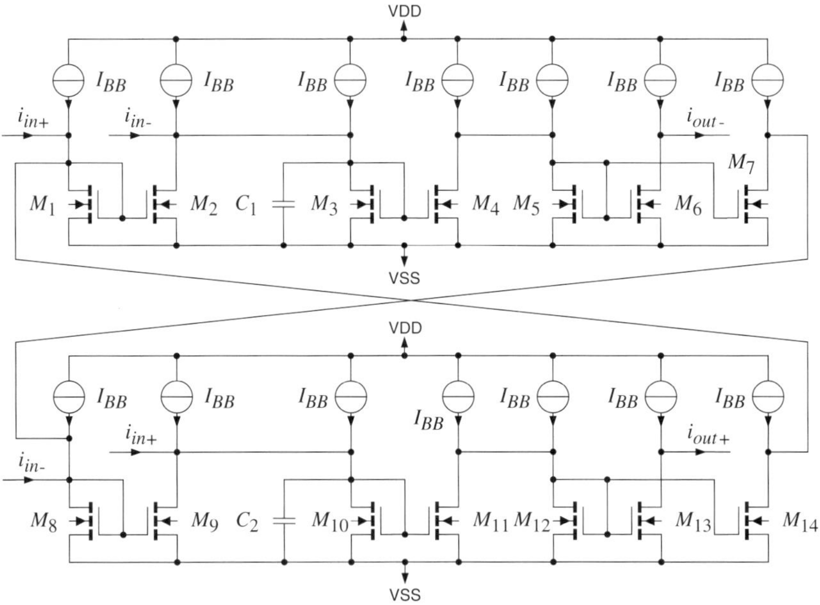

Figure 6.12 An enhanced differential current-mirror based integrator [28].

there are other sources of high-frequency nonlinearity such as cascode devices that make this advantage over the MOS differential pair relatively insignificant.

The differential integrator outperforms the single-ended integrator both in linearity and in quality factor. Furthermore, the single-ended circuit uses almost as much current and die area as the differential version. In low voltage high frequency applications, the differential integrator may also perform more effectively than an OTA-C filter, using simple differential pairs as the transconductance element, because the voltage needed to ensure saturation region operation for the differential pair tail-current source is not required in the current-mirror based integrator. Furthermore, in the current-mirror based integrator, all transistors in the signal path are NMOS-transistors. However, relatively large saturation voltages are required for the mirror transistors to minimise the sensitivity to device mismatches and thus the usable filter frequency range is limited.

The sensitivity to mismatches can be reduced by such special circuit techniques as that presented in Figure 6.12 [28]. The differential DC-gain of this integrator is derived in [28] as

$$A_{dm}(0) \approx \frac{A_{mi}A_{mo} + A_{mo}}{\frac{6g_{ds}}{g_m} + 1 - \varepsilon - A_{mi}A_{mo}}, \tag{6.22}$$

where ε represents the total mismatch between the two current-mirror chains and A_{mi} and A_{mo} represent the current-mirroring ratios of the additional input and output current-mirrors

$$A_{mi} = \frac{W_2}{L_2}\frac{L_1}{W_1} = \frac{W_9}{L_9}\frac{L_8}{W_8}, \tag{6.23}$$

$$A_{mo} = \frac{W_7}{L_7}\frac{L_5}{W_5} = \frac{W_{14}}{L_{14}}\frac{L_{13}}{W_{13}}. \tag{6.24}$$

If both A_{mi} and A_{mo} are equal to one, the sensitivity to the mismatch ε is the same as that in the original differential integrator. However, if it is assumed that $A_{mi}A_{mo} < 1$ and $A_{mo} > 1$, then choosing, for example, $A_{mi} = \frac{1}{10}$ and $A_{mo} = 9$ the sensitivity to mismatches is significantly reduced.

The additional current-mirrors add at least two poles to the integrator transfer function. Furthermore, the pole deriving from the output mirrors is at a relatively low frequency since the required current gain leads to large gate areas. Therefore, this integrator is not suitable for high-frequency applications. However, this circuit structure additionally improves the common-mode rejection of the integrator. Therefore, input stages in a low voltage current-mode system may prove a suitable application for this circuit.

6.7 High-gain current-conveyor based filters

As described in Chapter 4.4, the high-gain current-conveyor can be used as a direct replacement for a voltage-mode operational amplifier and thus active-RC integrators can also be realised, as seen in Figure 6.13a. Because of the relatively simple circuit structure of the high-gain conveyor, higher filter corner frequencies or lower power consumption are more feasible than in the case of conventional voltage-mode amplifiers. Furthermore, in most single-ended filter topologies, additional inverting amplifiers are required that can be realised with a dual-output CCII∞ without resistors leading to enhanced high-frequency performance and reduced area.

Similarly, low-gain current-conveyors can be constructed with this amplifier. Therefore, source-degenerated transconductance elements can be constructed with a dual-output CCII∞, additionally making the OTA-C filter possible, as shown in Figure 6.13b. Consequently, the input voltage swing in the Y-terminal will limit the maximum signal swing of the filter whereas in the active-RC realisation almost a rail-to-rail Z-output swing is possible. Normally, in the active-RC approach, the input structure of the CCII∞ can be straightforward since the Y-terminal is used exclusively for setting the DC-voltage level and thus a part of the input structure can be moved into the bias circuit and shared with several amplifiers. Moreover, because no replica outputs are required for the CCII∞ based active-RC integrators, a higher power and area efficiency is reached with the active-RC realization than with the source-degenerated

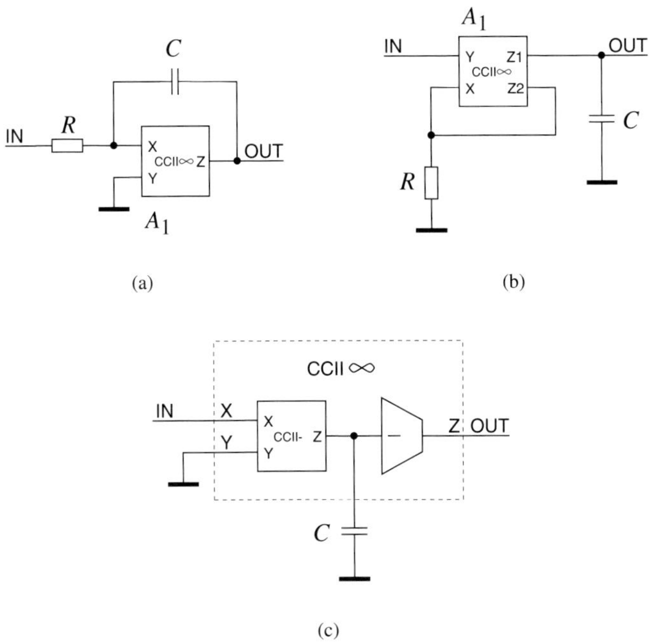

Figure 6.13 High-gain conveyor based integrators. a) A lossless inverting active-RC integrator. b) An OTA-C type lossless noninverting integrator. c) Alternative OTA-C type integrator implementation.

OTA-C approach, particularly when high-quality integrated resistors are used in both filter realizations.

In addition to these two high-gain current-conveyor based integrator realisations, a third way to realise an integrator is possible, in which the integrating capacitor is connected to the internal high-impedance node, resulting in a current-input OTA-C type integrator. In this integrator, the integrating time constant is set by the transconductance of the output stage while the input CCII- is used exclusively as a current-buffer ensuring a high impedance in the intermediate node.

The output stage is, however, a common-source amplifier stage and thus similar distortion performance to that in the simple current-mirror based filters will result. Similarly, if a current-mode operational amplifier rather than a high-gain current-conveyor is used, a differential amplifier is used as the output stage and similar distortion performance results to that of OTA-C filters with simple differential pairs as transconductance elements.

In this third type of a CCII∞ based integrator, large distortion occurs, particularly near the corner frequency. Therfore, even a slight reduction in distortion would

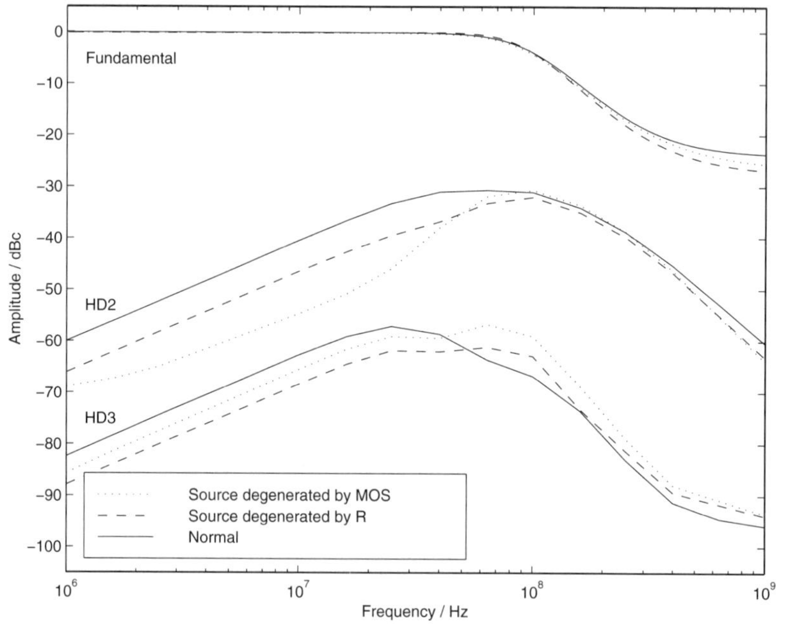

Figure 6.14 The simulated distortion of the CCII∞ implementation of Figure 4.16 operating as a CCII+ with and without source degeneration in the output transistors M_3 and M_4.

be welcome. For example, the source degeneration is easy to realise in the output common-source amplifier stage. The effect of source degeneration is easy to verify with the CCII∞ implementation of Figure 4.16. Two different source degeneration methods are used: the output transistors M_3 and M_4 are resized from 100/5 to 100/2.5 and either a linear 2136 Ω resistor or a NMOS-transistor with aspect ratio of 41/5 and a gate bias of 1.75 V is added to the output transistor sources. In both cases, the same effective transconductance as in the original CCII∞ implementation is targeted so that the corner frequency remains unaltered. The results of the distortion simulations of these two source degeneration methods is compared with the simulated distortion of the original circuit in Figure 6.14. In all simulations, the input signal is 20 μA, which is 20% of the bias current.

The results show that the distortion is quite effectively reduced below the corner frequency. The MOS-degeneration method in particular effectively reduces the second-order distortion. Since the source degeneration MOS-transistors operate in the triode region, these transistors generate second order distortion, which seems to cancel out part of the second order distortion of the main output transistors M_3 and M_4. This kind of cancellation normally depends on the signal amplitude and process parameter variation. Moreover, the rejection of the third order distortion is equally effective in

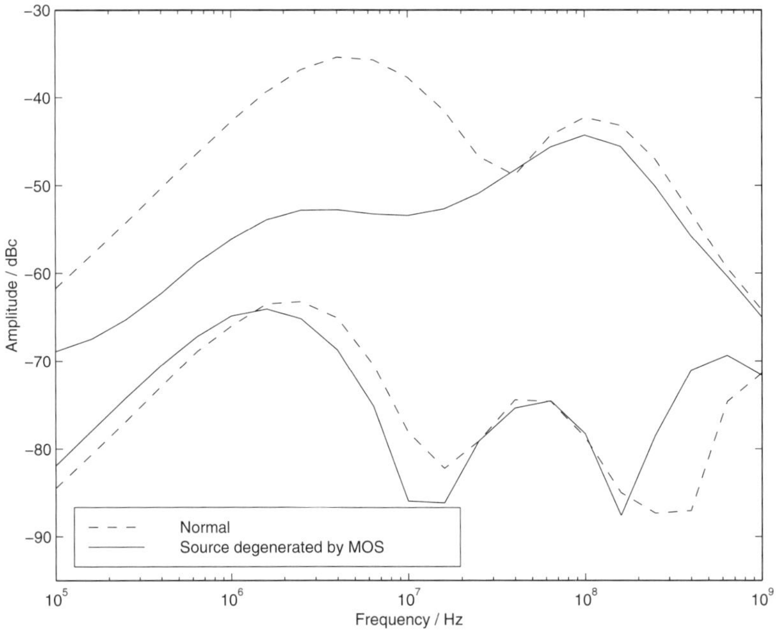

Figure 6.15 The simulated distortion of the CCII+ connected CCII∞ with and without source degeneration and an additional 10 pF capacitor in parallel with the gate-source capacitances of the output transistors M_3 and M_4 resulting in a corner frequency $\omega_0 = 2\pi 6.4$ MHz.

both degeneration methods and thus the MOS-resistors retain no advantage over the linear resistors in differential circuits, where the second-order distortion is effectively reduced even without the help of source degeneration.

According to the simulation results, it seems that close to the corner frequency, source degeneration no longer reduces the distortion. In this region, the non-linear input impedance of the X-terminal, in conjunction with the parasitic capacitances at this terminal, add to the total distortion. When constructing a lossy integrator with this circuit, by placing an additional capacitor in parallel with the gate-source capacitances of the output transistors M_3 and M_4, the integrator pole and the distortion peak arising from the output transistor are moved to lower frequencies. At the same time, the distortion peak deriving from the nonlinearity at the X-terminal remains unchanged. However, since this distortion appears significantly above the lossy integrator corner frequency, it is effectively attenuated in the output, as seen in the simulation results of Figure 6.15.

In this CCII∞ implementation, NMOS output transistors and a PMOS input current buffer were selected for better comparison with an NMOS current-mirror. Therefore, the nondominant pole because of the PMOS input stage and the dominant pole because of the NMOS output stage are exceptionally close to each other. If opposite type of transistors are used in both the input and output stages, these poles become wider apart and thus lower high frequency distortion results. Because the nonlinearity at the X-terminal generates largely second order distortion, differential structures help to mitigate the effects of this secondary source of distortion.

6.8 Multi-output current integrator with a linearised transconductor

In order to maximise the high frequency performance with low supply voltages, simple current-mirror based integrators would be satisfactory if nonlinearity were not a problem. However, even in the case of differential current-mirror based filters, the nonlinearity of the transconductance element is only comparable to the nonlinearity of a simple differential pair.

The nonlinearity of the CCII∞ in the CCII+ configuration is very similar to the nonlinearity of a cascode current-mirror and therefore source degeneration can be successfully used to lower the distortion of current-mirror based filters as well. Unfortunately, since source degeneration decreases the effective transconductance of the mirror transistors, as high filter corner frequencies cannot be reached as with current-mirrors without a source degeneration.

When comparing the lossy OTA-C integrator of Figure 6.16a to a lossy integrator based on a dual output CCII∞ of Figure 6.16b, one practical difference emerges, even if both circuits use identical transconductance elements. In the dual output CCII∞ integrator, the transconductance elements are inside the same amplifier cell so that a minimum distance and optimal matching conditions are easily achieved in the circuit layout and consequently as low distortion as possible is maintained at low frequencies, particularly in low-pass filter applications. Careful layout techniques can be used in OTA-C filters but in most cases still greater distances between circuit elements critical for accurate matching will result.

Based on such arguments, an optimal continuous-time filter building block for high-speed low-voltage and low distortion applications should be a current-input device with a linearised transconductance element as an output stage. Furthermore, this linearised transconductance element should be readily scalable for more outputs with a minimal number of additional circuit elements. In addition, as low distortion applications are targeted, it would be illogical to neglect differential signals in this filter building block.

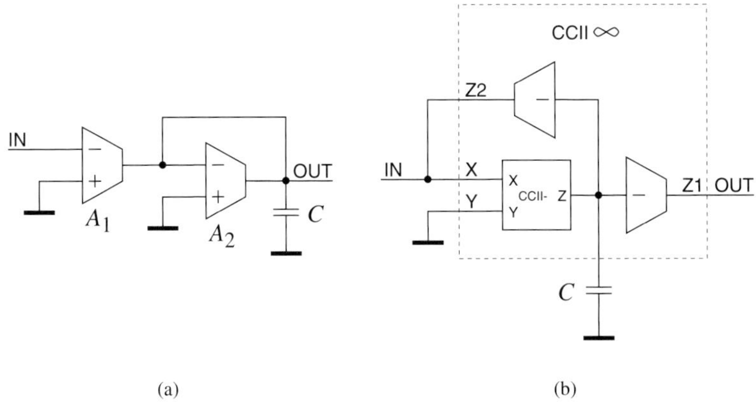

Figure 6.16 Lossy integrator implementations a) based on an OTA b) based on a CCII∞.

6.8.1 Linearization by drain current difference

For optimal accuracy in simulations, a linearisation method using MOS transistors operating in one operation region only is preferred. This operation region is preferably the saturation region deriving from the faster operation and lower noise compared to the other operation regions of the device. For maximising the voltage swing in the integrating node, the linearisation principle should use only transistors with their sources connected to the supply rails. The transconductance element should be easily multiplied and scaled in order to realise current-mode ladder filters.

A linearisation principle which meets these requirements is presented in Figure 6.17 [11]. A pair of matched MOS-transistors with sources connected to ground are driven with a differential signal with a well defined common-mode voltage level, resulting in MOS-transistor drain currents

$$i_{D1} = \frac{\beta}{2}\left(\frac{v_{in}}{2} + V_{CM} - V_T\right)^2, \tag{6.25}$$

$$i_{D2} = \frac{\beta}{2}\left(-\frac{v_{in}}{2} + V_{CM} - V_T\right)^2. \tag{6.26}$$

When the difference of these currents is calculated, a linear output current results as

$$i_{D1} - i_{D2} = \beta v_{in}\left(V_{CM} - V_T\right). \tag{6.27}$$

An interesting feature of this linearisation principle is that the theoretical maximum current amplitude for linear operation is four times larger than the quiescent current

$$I_Q = \frac{\beta}{2}\left(V_{CM} - V_T\right)^2. \tag{6.28}$$

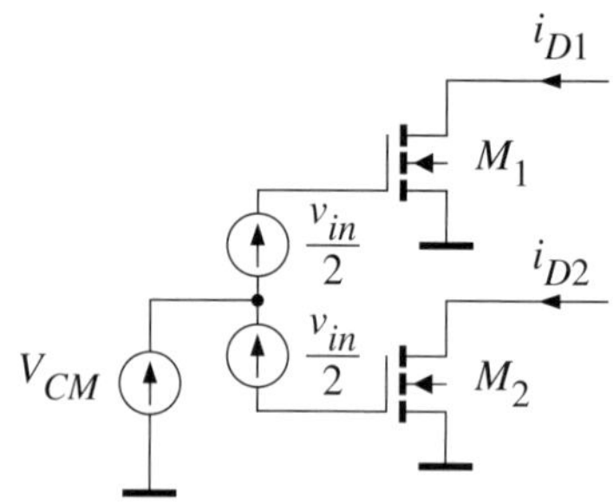

Figure 6.17 Two MOS transistors with a balanced input voltage

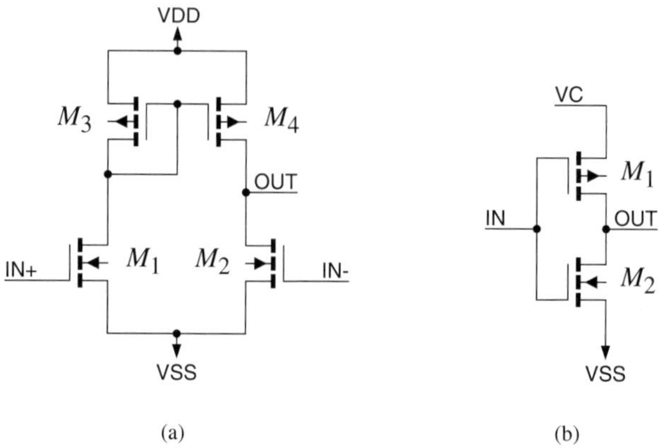

Figure 6.18 Linearised transconductors based on the linearisation principle of Figure 6.17.

Clearly, MOS-transistors fall into the weak inversion operating range before turning off entirely and consequently the practical linear range is commonly around $3I_Q$. The drawback of this linearisation technique is that normal current sources cannot be used to bias the transistor pair since the two output currents have a signal dependent DC-component.

A straightforward transistor level realisation of this linearisation principle would be to use a current-mirror to invert one of the drain currents, resulting in the circuit shown in Figure 6.18a [11]. However, this results in adding a new source of distortion, particularly at high frequencies. This becomes yet worse if large signal swing is required, because the distortion of the current-mirror is very large when the signal amplitude reaches the bias current level. Similarly, the PMOS current-mirror in conjunction with the NMOS input transistors is detrimental to the overall distortion performance.

One of the drain currents can also be inverted by using a PMOS-transistor rather than an NMOS-transistor. Consequently, the input signal is similarly turned into a

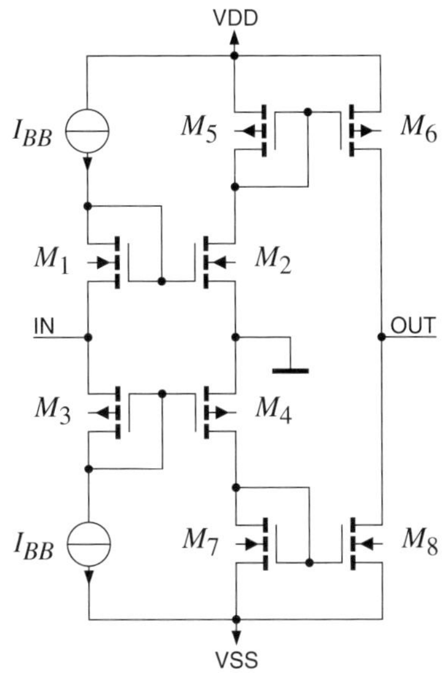

(a)

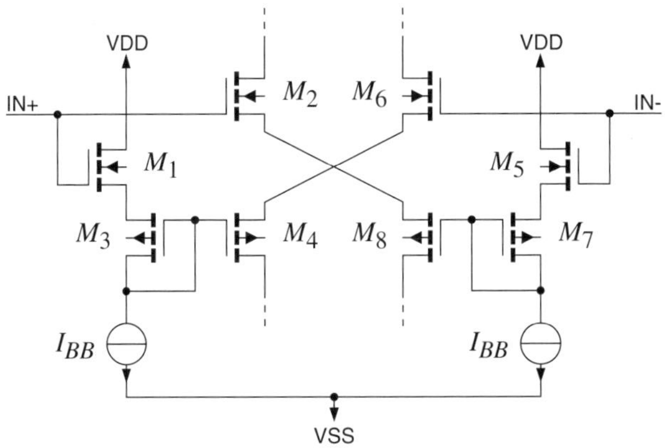

(b)

Figure 6.19 a) A CCII+ as a linearised transconductor. b) A differential class-AB transconductor based on the same principle.

simple single-ended signal, resulting in a plain CMOS-inverter [14]. In this case, the linearisation accuracy depends on the matching of the process parameters between NMOS- and PMOS-transistors. However, in modern CMOS-processes, there is only a weak correlation, if any, between these transistor types.

Tuning the transconductance of an inverter involves tuning its supply voltage, which normally results in that an additional servo amplifier is used to supply the current into the filter while regulating the supply voltage level. This circuitry needs headroom to operate, thereby increasing the minimum operating voltage requirement. In large high frequency filters, the limited output impedance of this control amplifier may lead to crosstalk between different integrator stages, resulting in erroneous transfer function or even in oscillation.

The X-terminal impedance of the push-pull second generation current-conveyor is linearised by the same principle, as the earlier derived Equation (3.44) shows. Therefore, this circuit can readily be converted into a transconductance amplifier merely by grounding the X-terminal and using the Y-terminal as the voltage input and the Z-terminals as the current-output, as depicted in Figure 6.19a. However, the limited input impedance at the voltage input may limit the DC-gain of lossless integrators realised with this transconductor.

A higher input impedance can be reached with the differential input structure of Figure 6.19b [30], where the input signal is fed to the gates of the input NMOS-transistors. Cross-connecting the sources of the four input transistors M_2, M_4, M_6, and M_8 results in a similar effect to that of the grounding the X-terminal of the push-pull CCII+. In the schematic, the output structures are omitted. However, the drain currents of transistors M_2 and M_4 are mirrored, as in the conveyor, to form a single-ended output and, if a differential output is required, identical output structure can be repeated for the transistors M_6 and M_8.

Both complementary circuits require a considerable amount of supply voltage in order to operate. Furthermore, in the differential circuit of 6.19b, the input voltage range is shifted towards the positive supply rail, so that both the input and output can operate together in a very limited voltage range even with relatively large supply voltages. Since the output structures in both circuits use current-mirrors, the high frequency performance of these two circuits is limited. However, because of the push-pull operation, even order distortion components are at least partially canceled. Additionally, in the differential circuit even order distortion components arising from the input transistors are effectively cancelled, leading to better linearity than in the push-pull conveyor case.

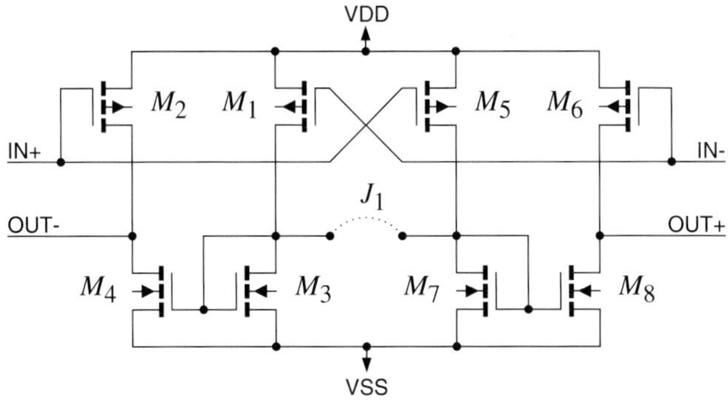

Figure 6.20 A circuit realising both class-AB and dynamic biased differential transconductors just by opening or closing the jumper J_1.

6.8.2 Linearisation by dynamic biasing

The large signal high frequency linearity can be improved by using a dynamic biasing technique. Thus, a dynamic bias current can be generated by taking the average of the two drain currents i_{D1} and i_{D2}

$$i_{DB} = \frac{i_{D1} + i_{D2}}{2} = \frac{\beta}{2}(V_{CM} - V_T)^2 + \frac{\beta}{8}v_{in}^2. \tag{6.29}$$

The equation shows that the quiescent bias current depends on the common-mode input voltage V_{CM} and that the bias current depends on the square of the differential input voltage v_{in}. When this current is subtracted from both drain currents, the result is two linear output currents:

$$i_{OUT+} = i_{D1} - i_{DB} = \frac{\beta}{2}v_{in}(V_{CM} - V_T), \tag{6.30}$$

$$i_{OUT-} = i_{D2} - i_{DB} = -\frac{\beta}{2}v_{in}(V_{CM} - V_T). \tag{6.31}$$

Ideally, all nonlinearities are cancelled already in the single-ended output. The linearisation accuracy is degraded by the transistor mismatches and phase errors deriving from the bias circuitry but because of the differential nature of this structure, even order distortion terms are further reduced.

The nonidealities of the dynamically biased transconductor can be compared with the class-AB transconductor of Figure 6.18a by referring to the example circuit presented in Figure 6.20. When the jumper is left open, a differential output version of the class-AB transconductor with NMOS current-mirrors and PMOS input transistors results. Similarly, when the NMOS current-mirror inputs are short-circuited together with the jumper J_1, the dynamically biased transconductor results. All four NMOS-transistors are assumed identical as in the case of the PMOS input transistors. NMOS current-mirrors are selected in order to minimise the errors in the linearisation, particularly at high frequencies.

One might consider a case whereby a sinusoidal input signal with an amplitude corresponding to the theoretical limit for linear operation is fed to the transconductor

$$v_{in} = 2(V_{GS} - V_T)\cos\omega t. \tag{6.32}$$

Then the signal at the NMOS-mirror inputs is

$$i_{D1} = \frac{I_Q}{4}(3 + 4\cos\omega t + \cos 2\omega t), \tag{6.33}$$

$$i_{D5} = \frac{I_Q}{4}(3 - 4\cos\omega t + \cos 2\omega t), \tag{6.34}$$

when the jumper is open. Similarly, when the jumper is closed a dynamic bias current is fed into the current-mirrors

$$i_{DB} = \frac{I_Q}{4}(3 + \cos 2\omega t). \tag{6.35}$$

Therefore, in the dynamically biased case, the current-mirrors must carry a signal with only a 33% modulation index with the same input amplitude that leads to a 100% modulation index in the class-AB case. Thus, significantly lower distortion is generated in the dynamically biased current-mirrors and a superior high-frequency accuracy is obtained to that of the class-AB version.

The dynamically biased transconductor additionally rejects common-mode signal by a common-mode feed-forward mechanism [31] provided that all input transistors remain in the saturation region. However, the transconductance of the circuit depends on the input common-mode voltage and consequently the circuitry driving this transconductor should also reject a common-mode signal. Moreover, transistor mismatches may also lead to leaking of the squared signal into the signal path.

The input transistors of the transconductor can also be biased in the triode region rather than the saturation region [32]. Thus, the bias current-mirrors carry primarily DC-signal since the input transistors operate quite linearly, which may lead to improved common-mode rejection. However, for the same reason, the output current swing of this transconductor does not exceed the quiescent current.

6.9 Design case: A 1 MHz current-mode low-pass filter

6.9.1 Filter building blocks

In order to construct filters with the dynamically biased transconductor previously discussed, certain additional circuitry is required to ensure a proper operation, as can be seen in the lossy integrator realisation presented in Figure 6.21. Since the dynamically biased transconductor requires a well balanced differential input signal, an additional driver amplifier providing these input conditions for the transconductor is placed at the input of the integrator. This driver amplifier controls the quiescent current and the transconductance of the transconductor with a common-mode feedback loop and increases the differential open-loop gain of the integrator by providing a high output impedance with cascode current-sources.

The transconductor uses PMOS transistors as main elements (MP1A-2B) in order to maximise the distance between the dominant pole caused by these PMOS-transistors and the nondominant pole caused by the NMOS-transistors MN1 and MN2 at the input of the transimpedance driver. The dynamic biasing current is generated by two additional PMOS-transistors (MP3 and MP4) of equal size. The bias current generated is then subtracted from the output currents by an NMOS-mirror with a mirroring ratio of 1/2. From the transconductor circuit realisation, it is clear that by adding two NMOS and two PMOS transistors, an additional differential current output is formed and the

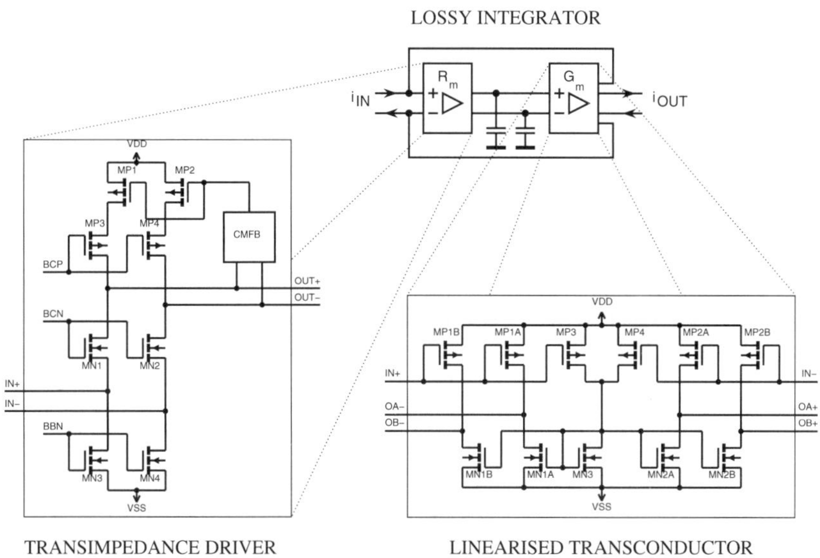

Figure 6.21 The principle of a lossy current-mode integrator using dynamically biased linearised transconductors.

output signal is easily scaled simply by tuning the transistor aspect ratios.

Although in this integrator the driver amplifier is referred to as a transimpedance amplifier, this integrator is a differential version of the high-gain conveyor based lossy integrator of Figure 6.16b and consequently the driver amplifier can also be considered a negative second generation current-conveyor with differential X- and Z-terminals. However, in earlier publications [33–35] the circuit principle was easier to describe without explaining the entire current-conveyor tradition. Since the function of the driver amplifier is to convert a small differential current into a large balanced voltage referring to this amplifier as a transimpedance amplifier is logical in this case.

The transimpedance driver amplifier

The detailed schematic of the driver amplifier is presented in Figure 6.22 and the aspect ratios of the transistors are listed in Table 6.1. A high output impedance for differential signals is achieved with the cascode current source MP1-4, which is controlled by the common-mode sensing double differential pair MNC1-4 [33–35]. Similarly, the common-mode feedback loop lowers the output impedance for common-mode signals.

The effect of supply voltage variation is minimised by mirroring the drain currents of the common-mode sensing transistors symmetrically to the main amplifier with transistors MPC1-4 and MNC7-8. The common-mode sensing circuit limits the minimum supply voltage to approximately 2.5 V in the case of the 1.2 μm CMOS-process used if the sensing circuit operates in strong inversion. The common-mode sensing circuit operates correctly regardless of the MOS operation region but it may be difficult to obtain a sufficient differential voltage swing in weak inversion.

The aspect ratios of the common-mode sensing transistors MNC1-4 are relatively small in order to limit the bandwidth of the common-mode feedback loop thus preventing common-mode oscillations. The nondominant pole of the common-mode feedback loop is caused by the PMOS current-mirrors (MPC1, MPC1 and MP1-3). However, this pole is at a significantly lower frequency than the nondominant pole of the differential signal path since this pole is caused by the input NMOS transistor MN1 and MN2.

The integrator capacitance area can be reduced to one fourth of the original area by using a floating capacitor between positive and negative signal paths rather than two separate grounded capacitors [3]. However, with this driver amplifier, the common-mode feedback remains uncompensated and thus common-mode oscillations may occur. Therefore there must be a significant amount of grounded capacitance at the output of the driver amplifier to ensure stability.

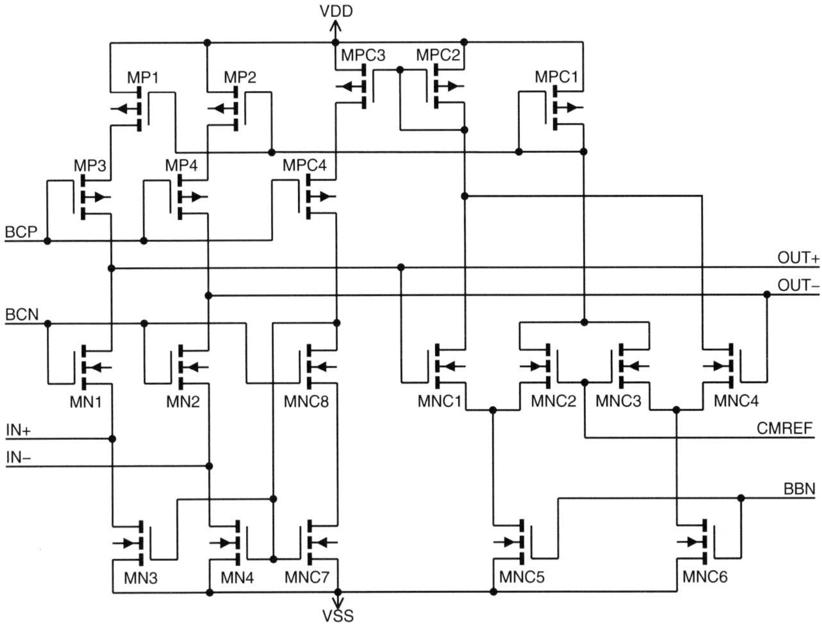

Figure 6.22 The driver in detail.

Transistors	Aspect ratio / μm/μm
MN1, MN2 and MNC8	200/1.2
MN3, MN4 and MNC5-7	50/3
MNC1-4	5/10
MP3, MP4 and MPC4	50/1.2
MP1, MP2 and MPC1-3	50/3

Table 6.1 Transistor dimensions of the transimpedance driver amplifier of Figure 6.22.

Transistors	Aspect ratio / μm/μm Figure 6.23a	Figure 6.23b
MN1A-2B	60/3	
MN3	120/3	
MN4	25/3	
MN5	25/3	-
MN6	100/1.2	
MN7	100/1.2	-
MP1A-2B, MP3 and MP4	60/6	
MP5	50/3	25/3
MP6	100/1.2	

Table 6.2 Transistor dimensions of the transconductor of Figure 6.23.

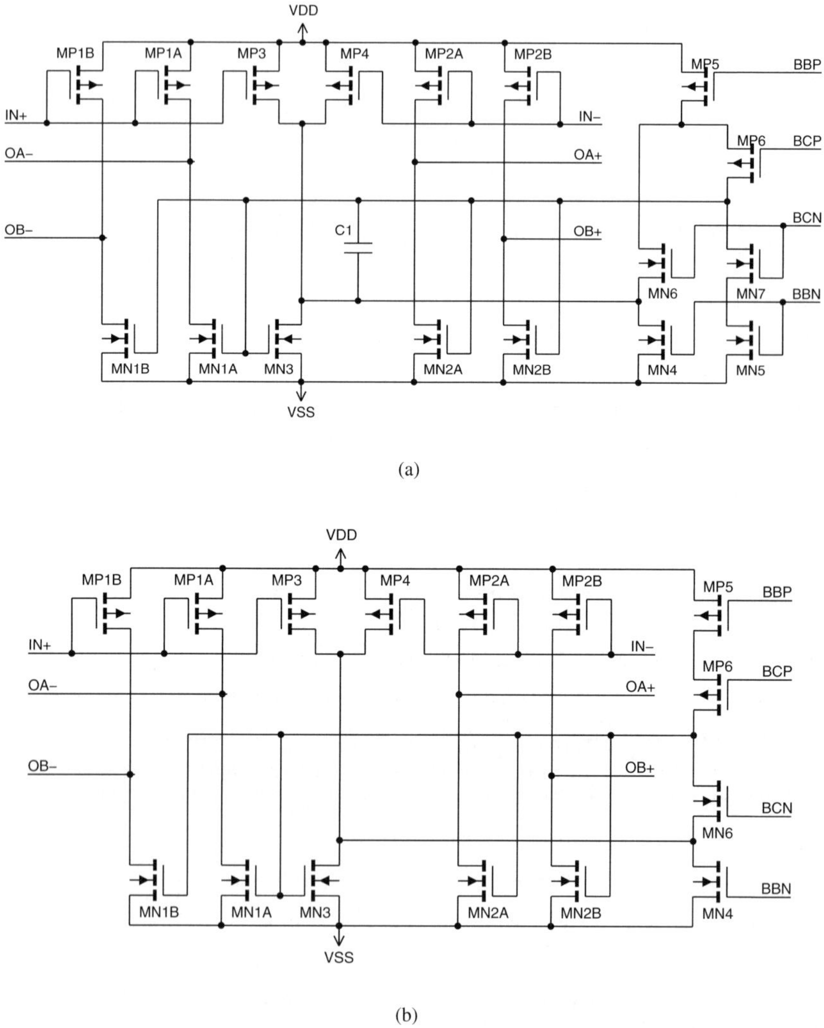

Figure 6.23 The transconductor in detail. a) The first prototype. b) The simplified implementation of later designs.

Multiple-output linearised transconductance element

The detailed schematic of the dynamically biased transconductor is presented in Figure 6.23a. The transconductance of the output MOS-transistors MP1A-MP2B is linearised by the dynamic biasing current generated by the transistors MP3 and MP4. In order to minimise the effect of the channel length modulation the common drain node of transistors MP3, MP4 and MN3 is set to the same potential as the driver amplifier inputs. This is achieved by means of a double-folded cascode structure of transistors MN4-7 and MP5-6. The feedback loop of the current-mirror is compensated with a feed-forward 5 pF compensation capacitance C1.

After the fabrication of the first prototype, it became apparent that a simpler way exists to drive the gates of the NMOS bias mirror, as shown in Figure 6.23b. At the same time, the compensation capacitor C1 was eliminated, leading to significant savings in the layout area. Because the variation of the current in the NMOS bias mirror is low, biasing the NMOS cascode transistor MN6 is not critical. The aspect ratios of the transistors in the modified bias mirror structure are identical to the earlier version, excluding the PMOS current source transistor MP5, which must supply only half of the previous current.

Temperature drift compensation of the integrator time constant

The transconductances in this filter depend on temperature and process variations. As discussed earlier, there are various automatic tuning schemes to lock the filter time constants to an external reference, resistor or reference frequency [3]. When using a frequency reference, feed-through from the tuning circuit may cause problems and consequently a separate tuning time slot is usually needed in the system level. Furthermore, this tuning must be performed periodically deriving from the temperature drift. In this filter, however, a different way to adjust continuously the transconductances of the filter is used. This is achieved by means of a CMOS current reference circuit using an off-chip reference resistor.

The detailed schematic of the CMOS current reference is presented in Figure 6.24. The main reference circuit uses a first generation current-conveyor constructed of two low-voltage cascode current mirrors. The NMOS current mirror has a mirroring ratio of one but the aspect ratio of the PMOS-transistor MP1 is significantly larger than that of MP2. Therefore, an off-chip resistor R_{EXT} is used as a source degeneration resistor so that the mirroring ratio of the PMOS mirror is equal to one at one specific current value, thus generating a stable reference current. This type of current reference conventionally uses transistors in the weak inversion to produce a PTAT current

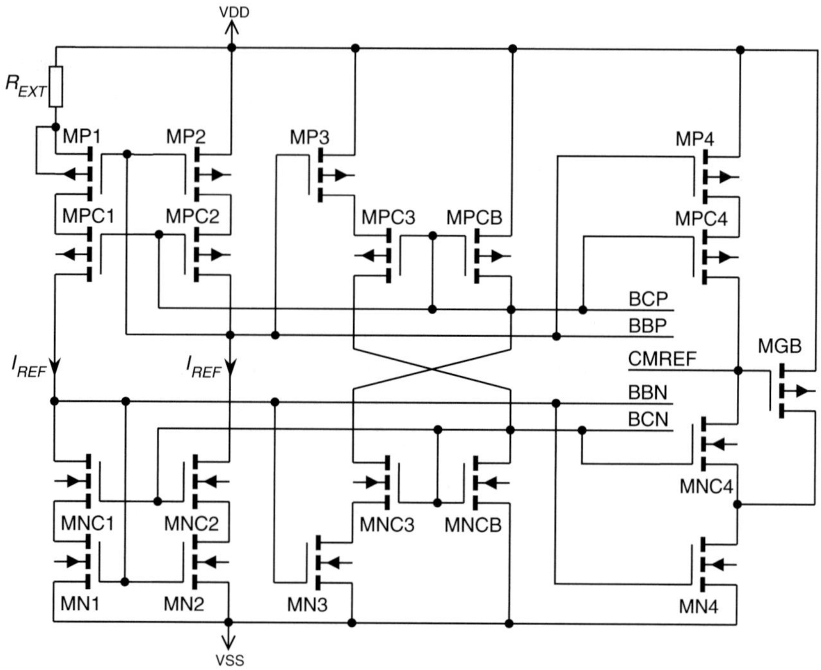

Figure 6.24 The current-reference circuit without start-up circuitry

Transistors	Aspect ratio / μm/μm
MN1-4	50/3
MNC1-3	50/1.2
MN4	25/3
MNC4	50/1.2
MNCB	8/3
MP1	275/3
MP2 and MP3	50/3
MPC1-3	50/1.2
MP4	25/3
MPC4	25/1.2
MPCB	10/3
MGB	60/6

Table 6.3 Transistor dimensions of the current-reference of Figure 6.24.

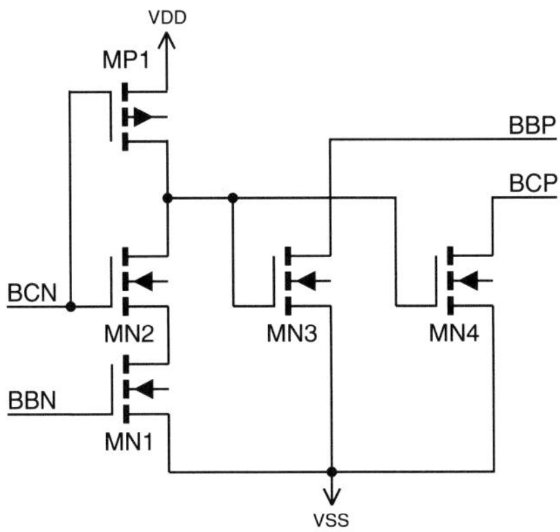

Figure 6.25 The startup circuit of the current reference of Figure 6.24.

Transistors	Aspect ratio / μm/μm
MN1 and MN2	50/1.2
MN3 and MN4	5/4
MP1	5/20

Table 6.4 Transistor dimensions of the startup circuit of Figure 6.25.

reference [36]

$$I_{REF} = \frac{kT}{qR_{EXT}} \ln A \tag{6.36}$$

if

$$\frac{W_{MP1}}{L_{MP1}} = A\frac{W_{MP2}}{L_{MP2}}. \tag{6.37}$$

The PTAT-current may still not be the most effective choice for the MOS-transistor transconductance temperature drift compensation. Therefore, a case where all the transistors operate in the strong inversion is considered, resulting in a different reference current [21,37]:

$$I_{REF} = \frac{2\left(\sqrt{A}-1\right)^2}{A\beta_{MP1}R_{REF}^2}. \tag{6.38}$$

With the 1.2 μm CMOS-process employed, the resulting reference current is approximately 20 μA with a 10 kΩ resistor R_{EXT}.

The effect of the channel length modulation is minimized by the use of cascode transistors in the reference mirrors and by the symmetrical cascode bias voltage generation. For this reason, the reference voltage for the common-mode feedback circuits of the integrators are generated by an additional five transistor circuit (MGB, MN4, MNC4, MP4 and MPC4) which replicates the DC-voltages of the transconductor.

The CCI-based current reference core has an additional stable operation point, when all of the transistors are in the off state. Therefore, the startup circuit shown in

Figure 6.25 is designed to ensure nonzero reference current. In this circuit, transistors MN1, MN2, and MP1 form a logic function from the NMOS bias voltages BBN and BCN. If any of these voltages are too low, transistors MN3 and MN4 increase the current through the PMOS current-mirror and the PMOS cascode bias transistor MPCB. This increases the current flowing into the NMOS-current mirror and the NMOS cascode bias transistor MNCB until the NMOS bias voltages BBN and BCN are sufficiently increased, which is when the startup circuit shuts itself down.

The biasing technique described reduces the temperature dependency of the filter time constants to approximately -70 ppm/K in simulations and -100 ppm/K in measurements. Although the CMOS current reference can reduce temperature and process variation dependencies of the filter, it cannot generally reduce the effect of the capacitance variation when metal or polysilicon capacitors are used. However, if the capacitors are realised with PMOS transistors, both the reference current and the capacitance depend at least partially on correlated process parameters such as oxide thickness and thus lower time constant variation results. In any event, the capacitance variation can be tuned out in the testing phase by adjusting the resistance of the external resistor R_{EXT}. The absolute accuracy without trimming is still more than adequate for many applications, such as anti-aliasing and smoothing, even with metal or polysilicon capacitors.

6.9.2 The first filter realisation

With the differential dynamically biased integrator, a third order elliptical low pass filter suitable for smoothing and anti-alias applications is implemented. The passive prototype of the filter is presented in Figure 6.26a and the component values of the filter are presented in Table 6.5. For this circuit, three state variable equations can be derived as, for example, in [4] as

$$-v_1 = -\frac{1}{s(C_1+C_2)}\left(i_s - \frac{v_1}{R_s} + v_3 sC_3 - i_2\right), \tag{6.39}$$

$$-i_2 = -\frac{1}{sL_1}(v_1 - v_3), \tag{6.40}$$

$$v_3 = -\frac{1}{s(C_2+C_3)}\left(-sC_2 v_1 - i_2 + \frac{v_3}{R_l}\right). \tag{6.41}$$

These equations are easily represented as a block diagram using inverting integrators. Based on this block diagram, the differential current integrator ladder filter implementation is derived, as presented in Figure 6.27, which also shows the off-chip components used.

The test chip uses differential inputs and a single-ended output. The input voltage-

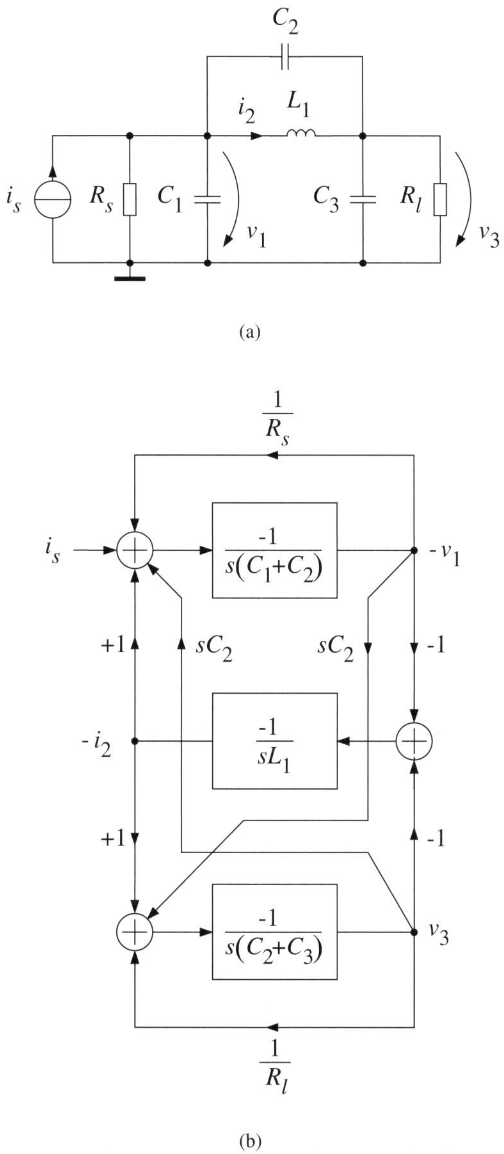

Figure 6.26 a) The used ladder filter prototype. b) The block diagram of the filter.

to-current conversion is achieved using an on-chip polysilicon resistor and an optional off-chip resistor. An on-chip resistance at the input is needed in order to prevent the input stray capacitances degrading the phase margin of the first integrator. In the output current-to-voltage conversion, the same combination of on- and off-chip resistance is used solely to increase the DC-gain accuracy. The filter capacitances are realised with polysilicon capacitors and therefore transmission zeroes are easily realised with floating capacitors.

The transconductances and capacitors in the filter are often scaled so that signal maxima at all filter intermediate nodes are as closely matched as possible so that maximum dynamic range can be reached. However, the signal maxima occur near the filter corner frequency and, since this filter is targeted at smoothing or antialias applications, the signal energy is not generally very high at this frequency range. Additionally, in this third order filter the peaks in the frequency responses are relatively low and thus the optimised dynamic range is merely a few decibels better than original. For these reasons, all outputs in the multi-output transconductors are unscaled.

Integrator Q-enhancement

The high-frequency behaviour of the differential integrator used is rather similar to simple single-ended high-gain current-conveyors. Therefore, the integrator quality factor can be approximated as

$$Q_I(\omega) \approx -\frac{p_{in}}{\omega}, \tag{6.42}$$

where p_{in} is the pole cause by the parasitic capacitances at the input of the driver amplifier (Figure 6.22) and on the transconductances of the input transistors MN1 and MN2:

$$p_{in} = \frac{g_{mMN1}}{C_{INP}} = \frac{g_{mMN2}}{C_{INM}}. \tag{6.43}$$

It is relatively easy to increase the transconductance of such NMOS input transistors is merely by increasing the bias current and aspect ratio. However, in order to minimise power consumption, it is preferable to use as low bias currents as possible.

When the filter is realised with relatively low bias currents, the filter transfer function will differ from the transfer function of the prototype filter. This is normally compensated by using predistorted filter coefficients that take into account the non-dominant pole of the integrator. Alternatively, optimisation can be used to find component values that match more closely the prototype filter. However, neither technique can accomodate the effects of process and temperature variation on the filter transfer function.

In this filter, the capacitance values listed in Table 6.6 were calculated using the ideal component values, taking into account the parasitic capacitances. The effect of

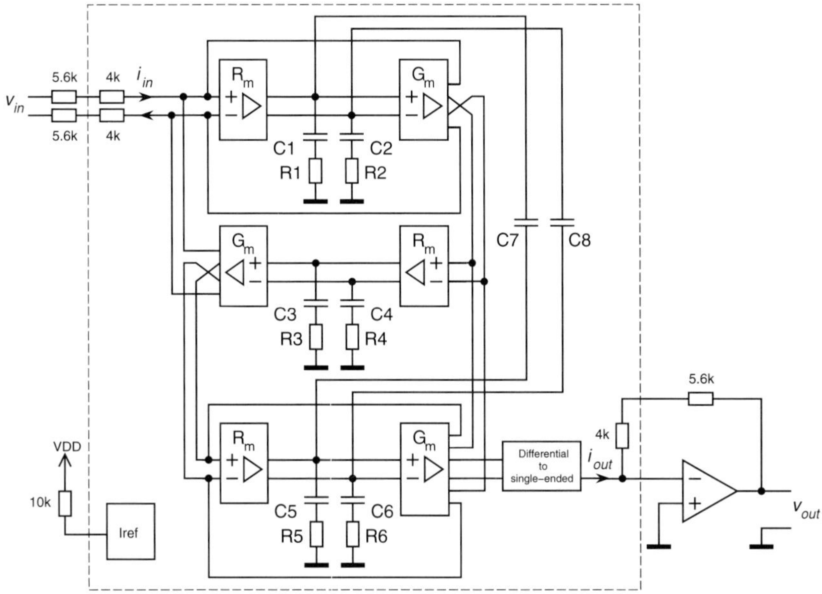

(a)

Figure 6.27 The realised filter including off-chip interfaces.

Component	Value
R_s and R_l	1 Ω
C_1 and C_3	1.0938 F
C_2	0.399 F
L_1	0.8255 H

Table 6.5 The normalised component values of the passive filter of Figure 6.26a. The passband ripple of this elliptical filter is 0.3 dB.

Capacitor	Capacitance / pF
C1 and C2	8.282
C3 and C4	6.018
C5 and C6	7.909
C7 and C8	4.000

Table 6.6 The final filter capacitance values after taking into account parasitic capacitances.

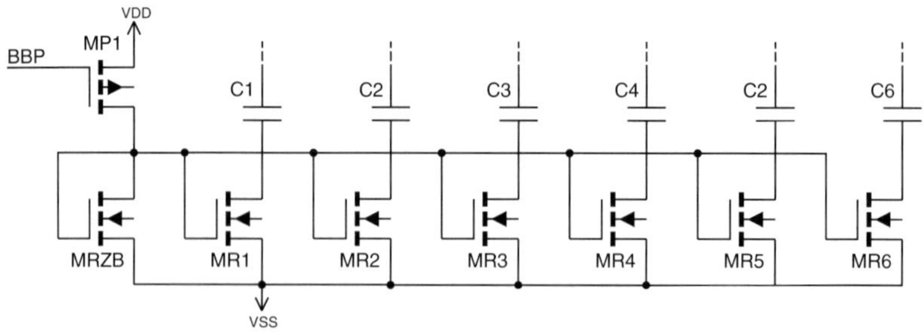

Figure 6.28 The realisation of the Q-enhancement resistors.

Transistors	Aspect ratio / μm/μm
MRZB	8/8
MR1 and MR2	126.8/2
MR3 and MR4	90/2
MR5 and MR6	121.2/2
MP1	50/3

Table 6.7 Transistor dimensions of the Q-enhancement circuit of Figure 6.28.

the integrator non-dominant poles is minimised by Q-enhancement techniques. This is achieved by adding a resistor R_z in series with the integrating capacitor [3, 38]. Thus, a left half-plane zero $z_{comp} = \frac{1}{R_z C}$ is added to the transfer function, resulting in an integrator quality factor of

$$Q_I(\omega) \approx \frac{1}{\omega} \frac{z_{comp} p_{in}}{p_{in} - z_{comp}}. \tag{6.44}$$

There are two alternative methods of realising Q-enhancement that involve by adding zeroes to the integrator transfer functions. By letting the resistance

$$R_z = \frac{1}{g_{mMN1}} \frac{C_{INP}}{C}, \tag{6.45}$$

the integrator quality factor becomes infinite. Thus, the compensation resistor is added in series with every integrating capacitor in the filter, as shown in Figure 6.27. As a side-effect of this method, attenuation is decreased at frequencies well above the filter corner frequency.

Alternatively, by letting the resistance

$$R_z = \frac{2}{g_{mMN1}} \frac{C_{INP}}{C}, \tag{6.46}$$

the resulting quality factor is $+\frac{p_{in}}{\omega}$ and consequently, in this integrator, the same amount of phase lead occurs as that of the phase lag in an uncompensated integrator. For example, in this third order ladder filter, the second integrator forms a feedback loop with both the first and the third integrator and, as a result, adding the resistor to the second integrator cancels out the phase errors simultaneously in both integrator loops, as in the revised filter implementation of Figure 6.35.

The resistance R_Z is realised with an NMOS transistor biased in the triode region. The control voltage of the NMOS resistors is generated from the internal reference current with a diode-connected NMOS transistor as shown in Figure 6.28. Because of the tracking nature of the MOS-resistor biasing, the sensitivity of the filter to temperature and process variation is reduced. With this technique, a deeper notch is achieved than with optimised component values, without the phase lag compensation. The optimal transistor aspect ratios listed in Table 6.7 were found in simulations by sweeping the width of the transistors MR1-6 until best matching with the prototype filter was accomplished.

Experimental results

The filter is fabricated with a 1.2 μm CMOS-process. The micro-photograph of the filter is presented in Figure 6.29. While the area of the whole chip is 2.2 mm^2, the filter area is only 0.15 mm^2/pole. The filter operates with down to a 3 V single supply and the current consumption of the whole filter with interfaces is 850 μA at room temperature (only the filter 230 μA/pole). The quiescent currents of the output transistors of the integrators are nominally 10 μA and the driver amplifier main branch currents are 20 μA. The measured dynamic range of the filter is 65.5 dB when referenced to output signal with 1% total harmonic distortion.

Nine of the ten fabricated filter chips operated well and their measured frequency responses are presented in Figure 6.30. All responses are very similar, so the circuit principle seems to work. However, the pass band ripple of the filter is higher than in simulations. Similarly, the stop band notch is not as deep as in simulations. Therefore, the Q-enhancement does not work as well as it should.

Only Level 2 SPICE-models were available for the CMOS-process that was used. Unfortunately, these simple, old models cannot predict the high-frequency behaviour of the MOS-transistor with sufficient accuracy to predict the correct values for the Q-enhancement resistors. Since in the filter chip the gate node of the Q-enhancement resistors MR1-6 was accessible as an output pin, one filter chip was manually tuned with an external voltage bias to the correct frequency response. The manually tuned chip was selected so that the reference current of the filter remained as close as possible to the nominal value. The optimal value for this voltage bias was found to be 1.0 V. The original and tuned frequency responses are shown in Figure 6.31 along with a simulation with the same voltage bias.

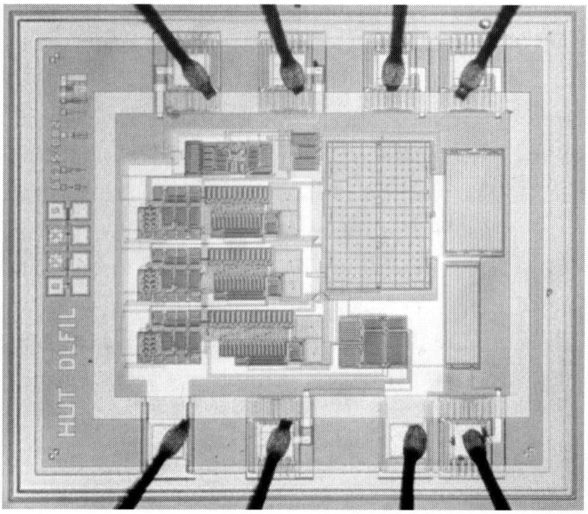

Figure 6.29 The micro-photograph of the filter.

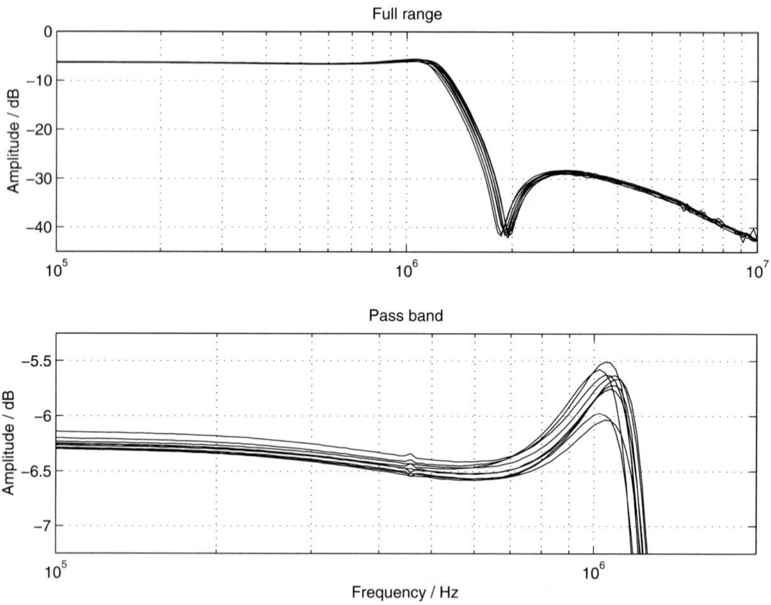

Figure 6.30 The measured frequency responses of the nine working filter chips.

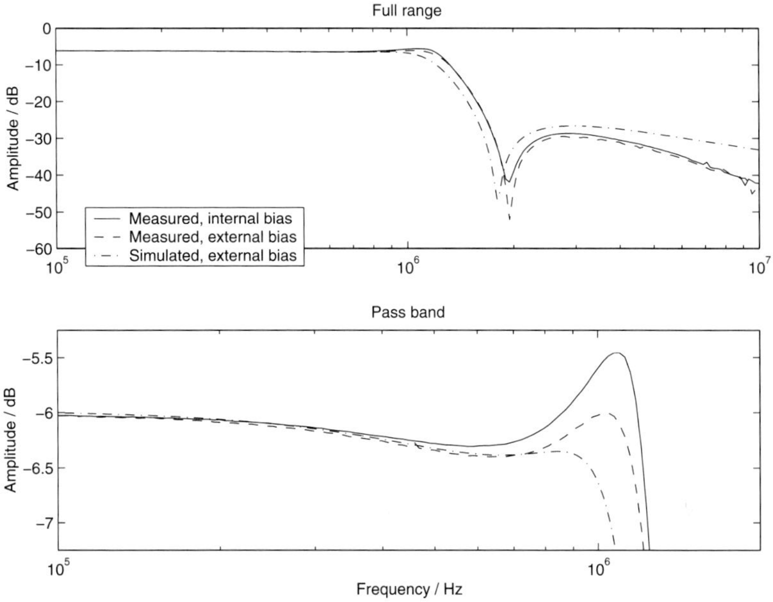

Figure 6.31 The effect of inaccurate MOS-models to the operation of the Q-enhancement.

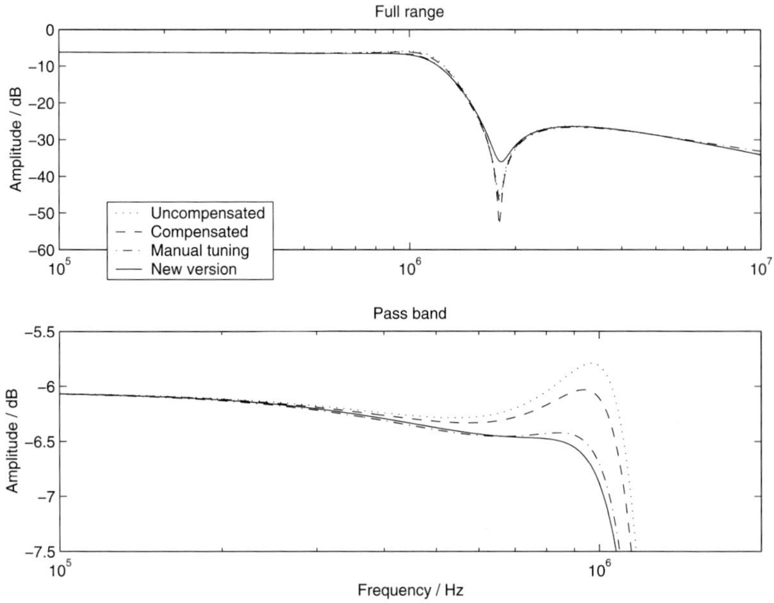

Figure 6.32 The effect of Q-enhancement in simulations.

According to these results, it seems that the simulation models used are too optimistic in predicting the high-frequency performance of the transistors. Therefore, in the second test chip, the Q-enhancement circuitry was re-designed so that the targeted frequency response was the simulation result of the old filter with the 1.0 V bias, as shown in Figure 6.32.

The measured total harmonic distortion of the first filter is below -40 dB up to the signal peak level of 60 μA, which is 50 % more than the theoretical maximum of the linearisation principle. The second order distortion (Figure 6.33a) dominates the total harmonic distortion and is virtually independent of frequency, while the third order distortion (Figure 6.33b) rapidly decreases with frequency.

The measured second order distortion is significantly larger than expected, based on the simulations. The used Level 2 SPICE-models assume only square-law nonlinearity for transistors in saturation. Since the linearization technique used cancels out this nonlinearity entirely, overly optimistic results in distortion simulations will result. Similarly, the differential structure of the circuit will also reject even order distortion too radically in simulations. Therefore, no distortion simulation results were displayed for this filter.

The previously mentioned shortcomings in the simulation of the filter do not explain the relatively large and frequency independent second-order harmonic distortion. However, there is a design error in the differential to single-ended converter realised with a high-swing cascode PMOS current-mirror (Figure 6.34 and Table 6.9). The cascode transistors should have a larger bias voltage, since the mirror transistors fall into the triode region at high signal peaks. This unsymmetrical clipping generates the observed second order distortion. Furthermore, this operation on the edge of the saturation and triode region also increases the sensitivity to threshold voltage mismatch, thus increasing distortion. The Level 2 MOS-models used do not predict the transition region between saturation and triode region and consequently an error like this is easy to make.

In any event, to use a current-mirror as a differential to single-ended converter in low distortion applications is unwise, as is frequently mentioned earlier in this book. However, those chapters discussing current-mirror noise were written after the filter was designed. Therefore in later designs based on this dynamically biased transconductor, differential outputs are used.

6.9.3 The second test chip

As mentioned earlier, a revised version of the filter shown in Figure 6.35 was designed and fabricated. This filter uses differential outputs so that differential to single-ended conversions are realised with external circuitry if required. Additional 4 kΩ polysilicon

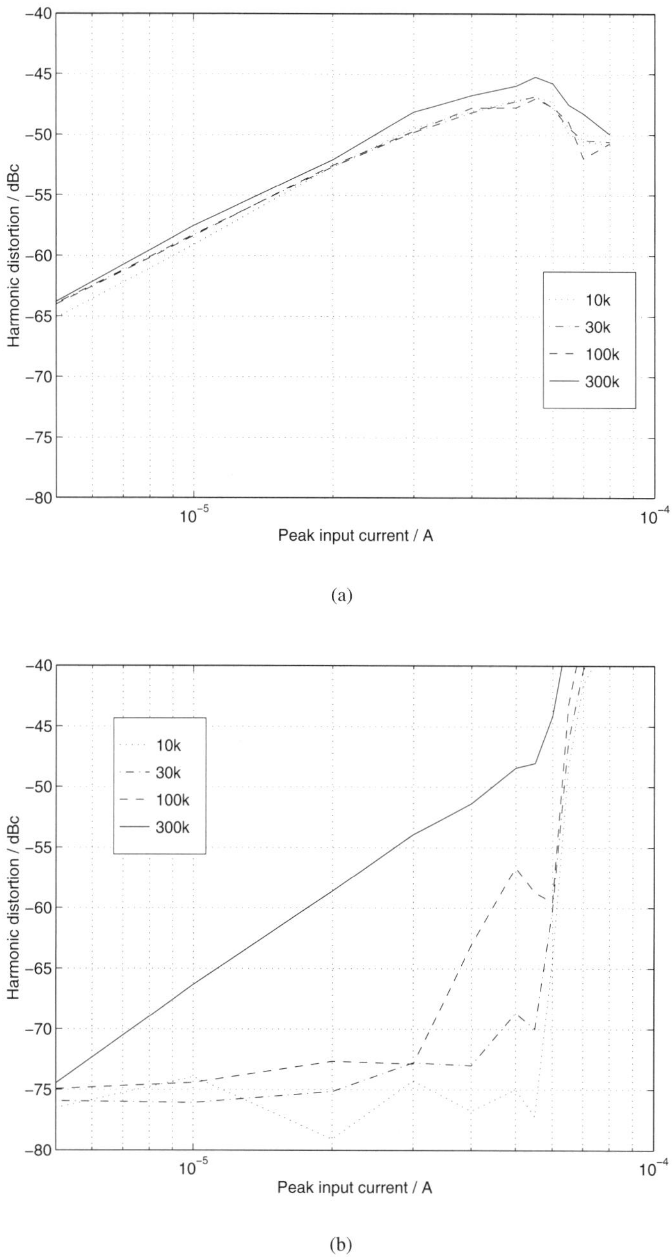

Figure 6.33 The measured distortion vs. frequency. a) The second order harmonic distortion b) The third order harmonic distortion.

Chip area	all	2.2 mm^2
	per pole	0.15 mm^2
Current consumption	whole chip	850 μA
	per pole	230 μA
THD	100 kHz 120 μA$_{pp}$	-47 dBc
IM3	250 kHz 40 μA$_{pp}$	-65 dBc
	400 kHz 40 μA$_{pp}$	
Output noise	BW 10 MHz	22.6 nA
Dynamic range	THD 1%	65.5 dB

Table 6.8 The performance characteristics of the first filter chip.

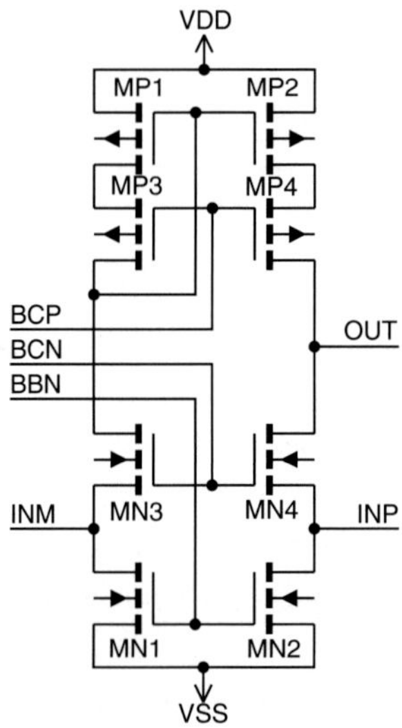

Figure 6.34 The differential to single-ended conversion circuit.

Transistors	Aspect ratio / μm/μm
MN1 and MN2	200/3
MN3 and MN4	800/1.2
MP1 and MP2	200/3
MP3 and MP4	200/1.2

Table 6.9 Transistor dimensions of the differential to single-ended conversion circuit of Figure 6.34.

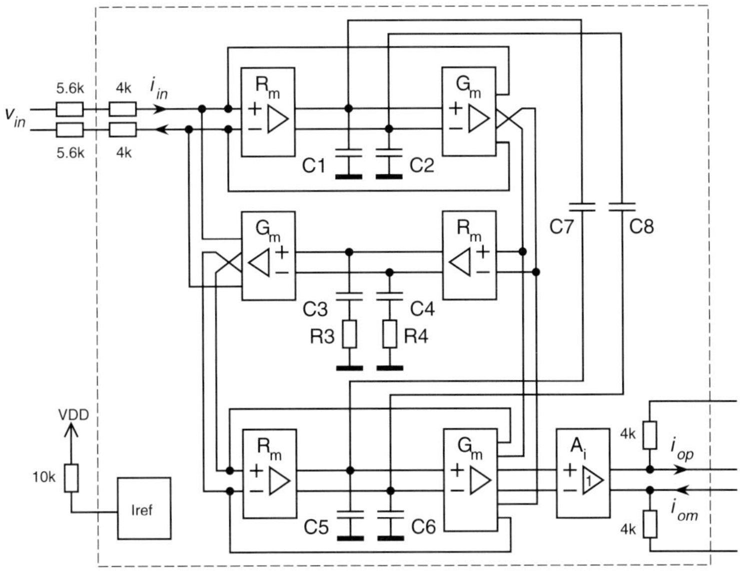

Figure 6.35 The revised filter with differential outputs and simplified Q-enhancement.

resistors are added to the output in order to minimise potential pass band gain errors. Similarly, the two resistors can be used to calibrate the frequency response of the external voltage to current converter.

The transconductors were implemented in this filter with the simpler circuit topology of Figure 6.23b. This filter uses the simpler Q-enhancement method, whereby compensation resistors are added solely to the middle integrator. The new aspect ratios for these MOS-resistors is set to 14/2 and the bias circuit is identical to the first filter implementation. Additionally, two transistors with an external gate bias are added in parallel to the two MOS-transistors in case the Q-enhancement circuit should malfunction. Similarly, because the bias voltage of the PMOS cascode transistors was found to be slightly too low, a degree of fine tuning was made to the aspect ratios of the current reference.

Alternate driver implementation

Because it was not certain that the measured second order distortion was due to the differential to single-ended converter, a second filter with an enhanced transimpedance driver topology was designed. Mismatches in transconductor transistors may generate common-mode currents that are not entirely rejected by the common-mode feedback of the driver amplifiers. Consequently, the remaining common-mode signal may modulate the transconductance of the transconductors, thus increasing second-order distortion.

These common-mode errors can be further reduced by using a common-mode feedforward structure in the transimpedance driver, as shown in Figure 6.36 [31]. Because of the increased input capacitance of this circuit topology, three times larger bias currents are used than in the original driver. The input impedance of this driver is significantly lower than in the first implementation and thus low frequency accuracy can be enhanced with this topology.

Although the driver amplifier is more complicated than the original implementation, the transconductor can be simplified. In fact, the NMOS-current mirror in the transconductor does not need cascode transistors because the drain-source voltages of all transistors in the transconductor can be set to the same potential simply by using the same DC input voltage level in the driver and transconductor NMOS-mirrors. Thus, the transconductors can be realised as in the Figure 6.21 depicting the principle of the integrator.

This current-mirror based driver can be used to extend the time constant range of the filter to lower frequencies. Because the transconductance of the transconductor topology used is identical to that of a single MOS transistor biased with the quiescent current of the transconductor, realising large time constants may lead to too low bias currents or too large gate voltages. However, because the output conductance of a MOS-transistor is proportional to drain current, halving the current-mirroring ratio by halving the aspect ratios of the output branch transistors MN2A, MN2B, MN6A, MN6B, MP2A, MP2B, MP6A, and MP6B, for example, will double the output impedance of the driver while halving the effective transconductance. Thus, scaling the driver output transistors will scale the integrator time constant while the open-loop DC-gain remains unchanged.

The differential current-buffers used at the output of the filters are shown in Figure 6.37 and the transistor dimensions are listed in Table 6.11. For the sake of simplicity, only one half of the drivers is shown. The output buffer for the filter using the original driver topology is based on the output current-mirror of the first filter chip. The transistor aspect ratios are the same and only the gates of the PMOS-mirror are now rewired to implement two cascode current-sources.

Because the DC-voltage levels at the transconductor outputs are different in the second filter, using the alternate driver topology, the output current buffer must also be different. Therefore, in the circuit of Figure 6.37b, regulated cascode stages are used. This minimises the DC-voltage differences at the transconductor transistor drains in the last integrator. Similarly, regulated cascodes ensure minimal distortion for the output stage.

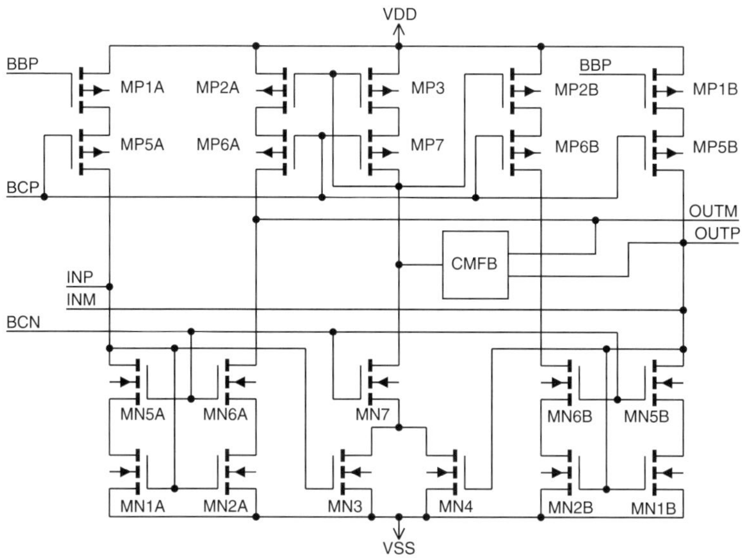

Figure 6.36 The alternate transimpedance driver implementation.

Transistors	Aspect ratio / μm/μm
MN1A-2B	240/2
MN3 and MN4	120/2
MN5A-6B and MN7	240/1.2
MP1A-2B and MP3	150/3
MP5A-6B and MP7	150/1.2

Table 6.10 Transistor dimensions of the alternate transimpedance driver of Figure 6.36.

Transistors	Aspect ratio / μm/μm Figure 6.37a	Figure 6.37b
MN1	800/1.2	400/1.2
MN2	200/3	
MP1	200/3	
MP2	200/1.2	
MN3	-	200/1.2
MN4	-	60/3
MP3	-	25/3

Table 6.11 Transistor dimensions of the output current buffers of Figure 6.37.

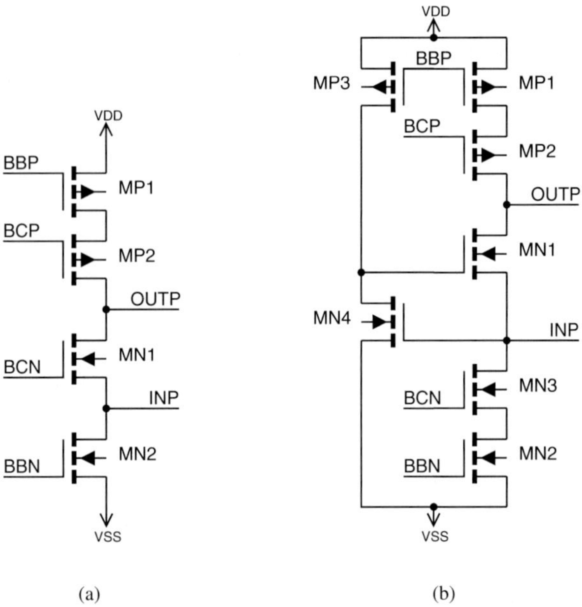

Figure 6.37 a) A half-circuit of the differential output buffer based on first chip with a single-ended output. b) A half-circuit of the differential output buffer used in the filter using alternate transimpedance drivers.

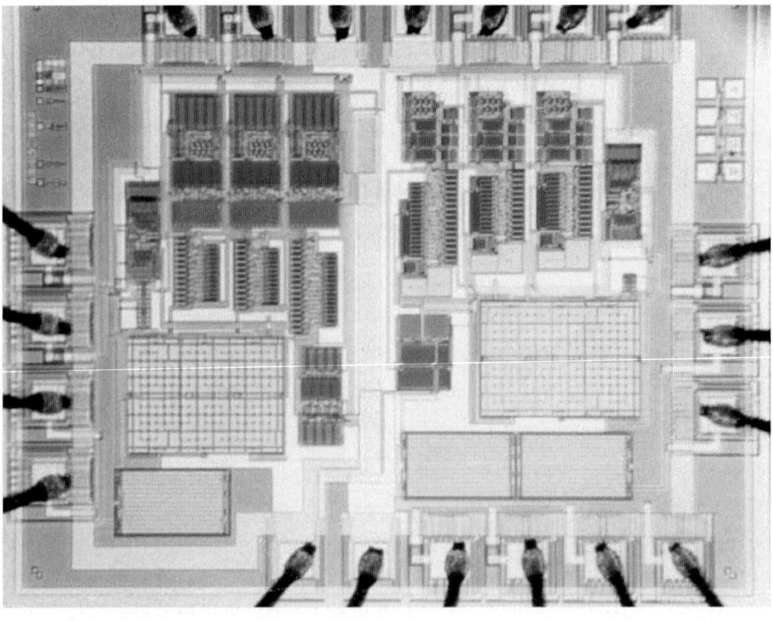

Figure 6.38 The microphotograph of the second filter chip.

Experimental results

None of the received ten chips malfunctioned and thus the frequency responses of all chips could be measured. The measurement results for both filters are shown in Figure 6.39. The measured pass-band ripple of the filter using the original driver implementation (Figure 6.39a) is now very close to the ideal response. Similarly, the stop band notch is as deep as expected. The corner frequency variation is also moderately low when the fact that all chips use the same external 10 kΩ resistor to set the bias current is taken into account. However, one of the frequency responses is still significantly apart from the other, perhaps caused by larger offset voltages in the current reference.

The frequency responses of the filter using the alternate driver topology of Figure 6.39b show a lower variation between chips. The low frequency gain variation in particular is extremely low compared to the filter using simpler drivers. This is accounted for by the lower input impedance of the alternate driver. The pass-band ripple is higher in this filter than in the simpler implementation. However, the capacitor and Q-enhancement resistors were identical in both filters, so equally accurate frequency responses were not expected. The effect of the common-mode feedforward on the filter frequency response variation cannot be evaluated because the number of samples is insufficient to permit statistical analyses.

The harmonic distortion was first measured from only one output current branch. In addition, differential outputs were combined into one single-ende signal by using the separate current-mode instrumentation amplifier described in Figure 5.8. The distortion measurement results for both new filters are presented in Figure 6.40. The second order harmonic distortion of the filter using the simple driver topology (Figure 6.40a) is significantly better than the second order distortion of the first filter chip, proving that the linearity of the first filter chip was indeed limited by the output differential to single-ended converter of Figure 6.34. Even the single-ended second order distortion of the revised filter is lower than the distortion of the first chip with the differential to single-ended converter.

The second-order distortion of the filter using common-mode feedforward (Figure 6.40b) is surprisingly large in the single-ended output configuration. The transconductor used in conjunction with the common-mode feedforward driver includes no cascode devices. This transconductor is more sensitive to channel length modulation, which may account for the higher single-ended second order distortion. Furthermore, there are no common-mode feedforward circuitry at the output to further reject the second order distortion.

The measurement results show that the common-mode feedforward does not improve the linearity of the filter. However, slightly less variation in the filter frequency response can be expected with this circuit topology. Additionally, with this driver,

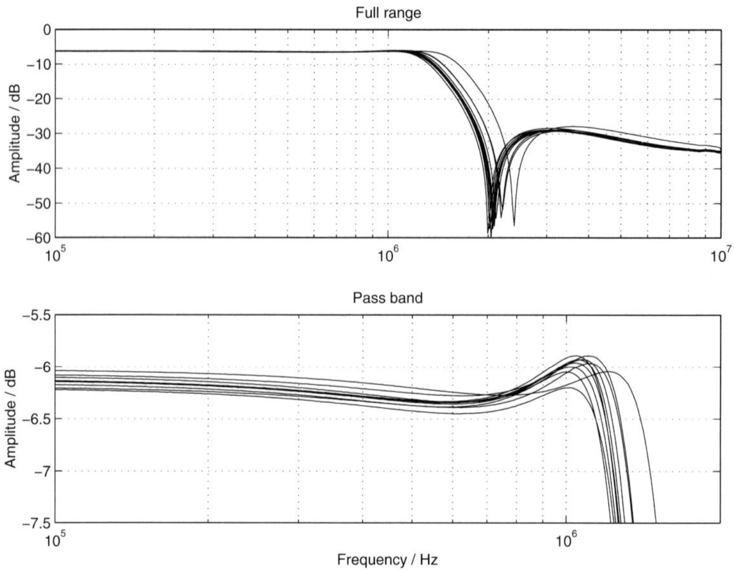

(a)

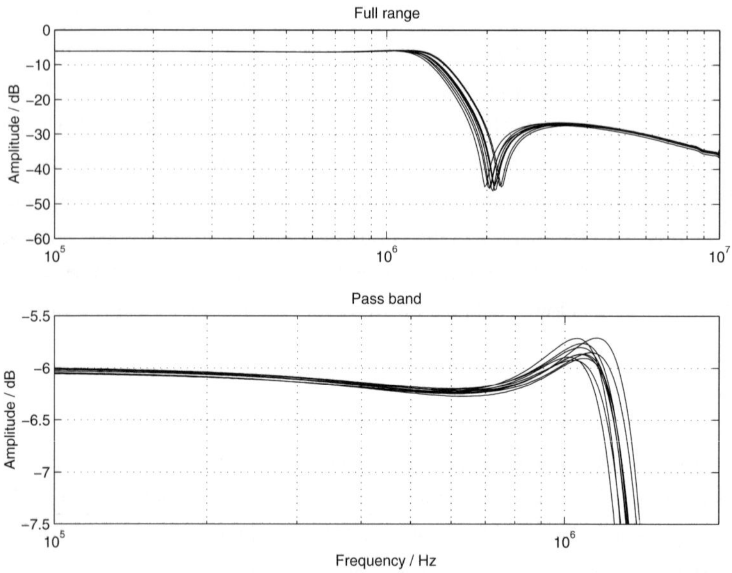

(b)

Figure 6.39 The measured frequency responses of the second test chip using differential outputs. a) Simple driver topology. b) Current-mirror based driver using common-mode feedforward.

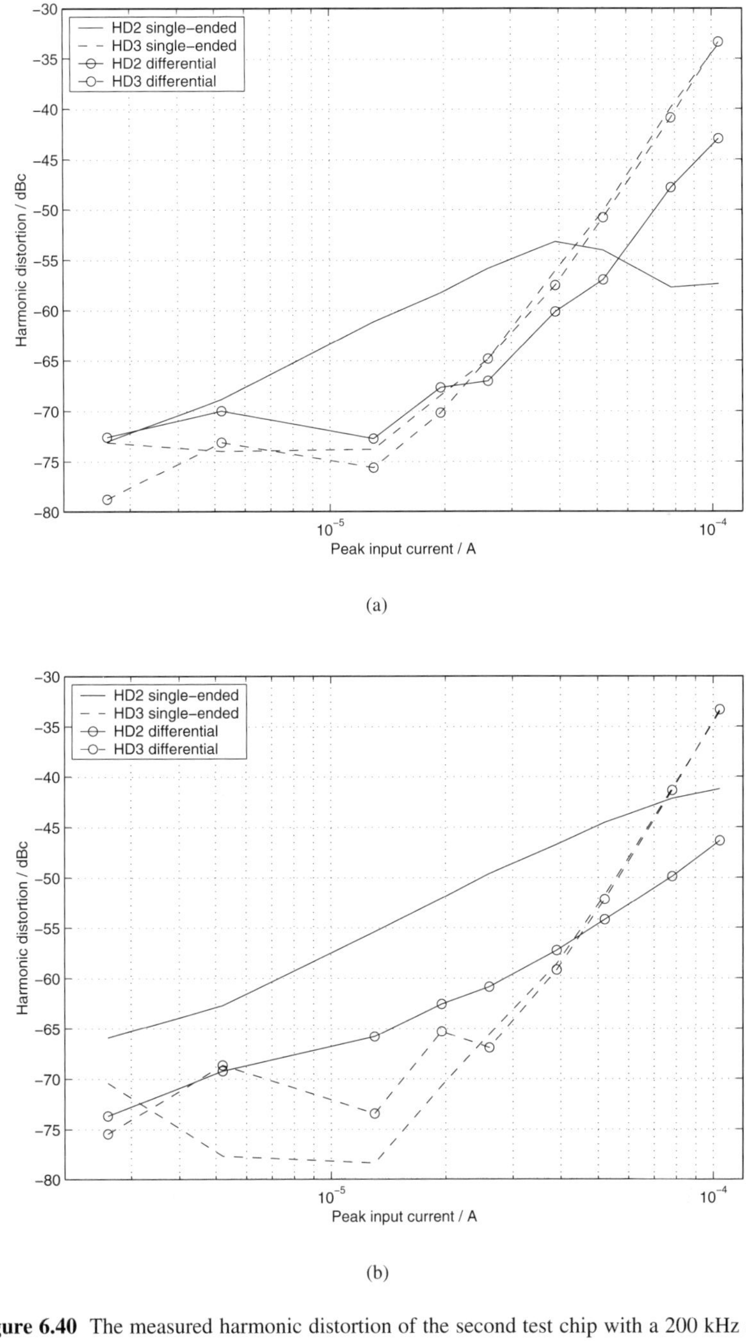

Figure 6.40 The measured harmonic distortion of the second test chip with a 200 kHz sinusoidal input signal. a) Simple driver topology. b) Current-mirror based driver using common-mode feedforward.

larger filter time constants can be realised than with the simple driver. Therefore, there are applications where this common-mode feedforward technique may have advantages over other solutions.

6.10 Final remarks

The current-mode filter implementation described is similar to many OTA-C filters. However, the described filter implementation has advantages over voltage-mode OTA-C filters. The linearisation method employed, using dynamic biasing, exhibits excellent linearity at high frequencies compared to most linearisation techniques using MOS-transistors in saturation. If better linearity is required, MOS-transistors operating in the triode region must instead be used [15, 16, 32, 39].

The linearization method used works well with current-integrators since the transconductance tuning that involves setting the common-mode voltage level is achieved locally inside the integrator. Similarly, the scalability of the dynamically biased transconductor is beneficial particularly in current-mode filters. Because the filters designed were implemented with a relatively old 1.2 μm CMOS-process, the full potential for high-frequency applications was not demonstrated. However, with this filter topology, a fifth order 50 MHz smoothing filter for a 10 bit D/A-converter is additionally successfully implemented [40, 41].

Unlike OTA-C filters, the filter building block used can also be used to implement differential active-RC filters. Then, very low distortion results by virtue of the linearised output stage. Similarly, this circuit can be used since a general purpose differential low distortion high-gain current amplifier. Therefore, based on these circuit topologies, it is possible to design a set of analogue standard cells for wide application range.

References

[1] L. Bruton, *RC-Active Circuits: Theory and Design*, New Jersey, Prentice Hall, 1980, 523 p.

[2] R. Schaumann, M. Ghausi, K. Laker, *Design of Analog Filters - Passive, Active RC, and Switched Capacitor*, New Jersey, Prentice Hall, 1990, 608 p.

[3] Y. Tsividis, "Integrated continuous-time filter design - an overview," *IEEE J. Solid-State Circuits*, vol. 29, March 1994, pp. 166-176.

[4] R. Gregorian, G. Temes, *Analog MOS Integrated Circuits for Signal Processing*, New York, John Wiley & Sons, 1986, 598 p.

[5] K. Laker, W. Sansen, *Design of Analog Integrated Circuits and Systems*, New York, McGraw-Hill, Inc, 1994, 898 p.

[6] A. Durham, W. Redman-White, J. Hughes," Circuit architectures for high linearity monolithic continuous-time filtering," *IEEE Trans. Circuits and Systems-II*, vol. CAS-39, Sep. 1992, pp. 651-657.

[7] J. Hughes, N. Bird, R. Soin, "Self-tuned RC-active filters for VLSI," *Electronics Letters*, vol. 22, Sep. 1986, pp. 993-994.

[8] S. Lindfors, J. Jussila, K. Halonen, L. Siren, "A 3V continuous-time filter with on-chip tuning for IS-95," *IEEE J. Solid State Circuits*, vol. SC-34, Aug. 1999, pp. 1150-1154.

[9] J. Franca, F. Nunes, "Successive approximation tuning of monolithic continuous-time filters," *Electronics Letters*, vol. 28, Aug. 1992, pp. 1696-1697.

[10] H. Khorramabadi, P. Gray, "High-frequency CMOS continuous-time filter," *IEEE J. Solid State Circuits*, vol. SC-19, Dec. 1984, pp. 939-948.

[11] R. Torrance, T. Viswanathan, J. Hanson, "CMOS voltage to current transducers," *IEEE Trans. Circuits and Systems*, CAS-32, Nov. 1985, pp. 1097-1104.

[12] F. Krummenacher, N. Joehl, "A 4-MHz CMOS continuous-time filter with on-chip automatic tuning," *IEEE J. Solid State Circuits*, vol. SC-23, June 1988, pp. 750-758.

[13] V. Gopinathan, Y. Tsividis, K. Tan, R. Hester, "Design considerations for high-frequency continuous-time filters and implementation of an antialiasing filter for digital video," *IEEE J. Solid State Circuits*, vol. SC-25, Dec. 1990, pp. 1368-1378.

[14] B. Nauta, "A CMOS transconductance-C filter technique for very high frequencies," *IEEE J. Solid-State Circuits*, vol. 27 , Feb. 1992, pp. 142-153.

[15] C. Laber, P. Gray, "A 20-MHz sixth-order BiCMOS parasitic-insensitive continuous-time and second-order equalizer optimized for disk-drive read channels," *IEEE J. Solid-State Circuits*, vol. 27 , Feb. 1992, pp. 142-153.

[16] R. Alini, A. Bascirotto, R. Castello, "Tunable BiCMOS continuous-time filter for high-frequency applications," *IEEE J. Solid-State Circuits*, vol. 27 , Dec. 1992, pp. 1905-1915.

[17] M. Banu, Y. Tsividis, "Fully integrated active RC filters in MOS technology," *IEEE J. Solid-State Circuits*, vol. 18, Dec. 1983, pp. 644-651.

[18] Z. Czarnul, "Modification of Banu-Tsividis continuous-time integrator structure," *IEEE Trans. Circuits and Systems*, CAS-33, July 1986, pp. 714-716.

[19] B. Song, "CMOS RF circuits for data communications applications," *IEEE J. Solid-State Circuits*, vol. 21, April 1986, pp. 310-317.

[20] Z. Chang, D. Haspeslagh, J. Verfaillie, "A highly linear CMOS G_m-C bandpass filter with on-chip frequency tuning,"*IEEE J. Solid-State Circuits*, vol. SC-32, March 1997, pp. 388-397.

[21] S. Willingham, K. Martin, A. Ganesan, "A BiCMOS low-distortion 8MHz low-pass filter," *IEEE J. Solid-State Circuits*, vol. SC-28, Dec. 1993, pp. 1234-45.

[22] J. Rijns, "CMOS low-distortion high-frequency variable-gain amplifier," *IEEE J. Solid-State Circuits*, vol. SC-31, July 1996, pp. 1029-1034.

[23] I. Mehr, D. Welland, "A CMOS continuous-time G_m-C filter for PRML read channel applications at 150 Mb/s and beyond," *IEEE J. Solid-State Circuits*, vol. 32, April 1997, pp. 499-513.

[24] J. Ramírez-Angulo, M. Róbinson, E. Sánchez-Sinencio, "Current-mode continuos-time filters: two design approaches," *IEEE Trans. Circuits and Systems-II*, CAS-39, June 1992, pp.337-341.

[25] S. Lee, R. Zele, D. Allstot, G. Liang, "CMOS continuous-time current-mode filters for high-frequency applications," *IEEE J. Solid-State Circuits*, vol. 28, March 1993, pp.323-329.

[26] S. Smith, E. Sánchez-Sinencio, "Low voltage integrators for high-frequency CMOS filters using current mode techniques," *IEEE Trans. Circuits and Systems-II*, CAS-43, Jan. 1996, pp.39-48.

[27] R. Zele, D. Allstot, "Low-power CMOS continuous-time filters," *IEEE J. of Solid-State Circuits*, vol. SC-31, pp. 157-168, Feb. 1996.

[28] X. Quan, S. Embadi, E. Sánchez-Sinencio, "Improved fully-balanced current-mode integrator," *Electronics Letters*, Vol. 34, pp. 1-3, Jan. 1998.

[29] S. Dupuie, M. Ismail, "High frequency CMOS transconductors," In: C. Toumazou, F. Lidgey, D. Haigh (ed.), *Analogue IC design: the current-mode approach*, Peter Peregnirus Ltd on the behalf of IEE: London, UK, 1990, pp. 181-238 (Chapter 5).

[30] E. Seevinck, R. Wassenaar, "A versatile CMOS linear transconductor/square-law function circuit," *IEEE J. of Solid-State Circuits*, vol. SC-22, pp. 366-377, June 1987.

[31] N. Tan, S. Eriksson, "Low-voltage fully differential class-AB SI circuits with common-mode feedforward," *Electronics Letters*, vol. 30, Dec. 1994, pp. 2090-91.

[32] F. Rezzi, A. Baschirotto, R. Castello, "A 3V 12-55MHz BiCMOS pseudo-differential continuous-time filter," *IEEE Trans. Circuits and Systems-II*, CAS-42, Nov. 1995, pp. 896-903.

[33] K. Koli, K. Halonen, "A fully differential class-AB switched-current integrator for signal processing," *IEEE J. Solid-State Circuits*, vol. SC-32, Feb. 1997, pp. 238-244.

[34] K. Koli, K. Halonen, E. Tiiliharju, "A 3V 1MHz dynamically biased 100 ppm/K temperature-stabilized CMOS continuous-time elliptical low-pass filter," *Proc. 1996 European Solid-State Circuit Conference (ESSCIRC'96)*, Neuchâtel, Sep. 1996, pp. 92-95.

[35] Koli, K., Halonen, K., "Current-mode temperature compensated continuous-time CMOS transconductance-C filter," *Analog Integrated Circuits and Signal Processing*, vol. 15, Jan. 1998, pp. 59-70.

[36] E. Vittoz, "The design of high-performance analog circuits on digital CMOS chips," *IEEE J. Solid-State Circuits*, vol. SC-20, June 1985, pp. 657-665.

[37] J. Steininger, "Understanding wide-band MOS transistors," *IEEE Circuits and Devices*, May 1990, pp. 26-31.

[38] A. Nedungadi, R.Geiger, "High-frequency voltage-controlled continuous-time lowpass filter using linearized CMOS integrators," *Electronics Letters*, vol. 22, July 1986, pp. 729-731.

[39] S. Lindfors, K. Halonen, M. Ismail, "A 2.7-V elliptical MOSFET-only g_mC-OTA filter," *IEEE Trans. Circuits and Systems-II*, CAS-47, Feb. 2000, pp. 89-95.

[40] M. Kosunen, K. Koli, K. Halonen, "A 50 MHz 5th order elliptic LP-filter using current mode Gm-C topology," *Proceedings of the IEEE International Symposium on Circuits and Systems,* 31. May-3. June 1998, Monterey, California, USA, vol. I, pp. 512-515.

[41] M.Kosunen, J. Vankka, M. Waltari, L. Sumanen, K. Koli, K. Halonen, "A CMOS quadrature baseband frequency synthesizer/modulator," *Proceedings of the European Solid-State Circuits Conference*, 22 - 24 September 1998, The Hague, The Netherlands, pp. 340-343.

Chapter 7

Current-mode logarithmic amplifiers

Most non-linear functions are in effect current-mode circuits based on translinear principles, either using the exponential behaviour of the bipolar transistors [1–3] or the square law behaviour of MOS-transistors in saturation [4]. By using current signals throughout the system, these circuits can be further simplified by omitting unnecessary voltage-to-current and current-to-voltage conversions, as is shown in the balanced frequency mixer example in Chapter 5.3.

The accuracy of bipolar transistor based translinear circuits is normally degraded deriving from the non-zero base current and Early-effect. Typically, the effects of these nonidealities can be reduced by placing the critical components inside a feedback loop of an operational amplifier. However, placing an active component inside a feedback loop can boost the loop gain and add additional poles to the transfer function thus making additional compensation techniques necessary. Alternatively, we could use current amplifiers rather than voltage-mode operational amplifiers in these applications and perhaps overcome some of the problems.

In this chapter, logarithmic amplifiers are used as an example of non-linear current-mode applications. Based on this illustrative application, various ways to enhance circuit performance by using current-amplifiers can be tested. Logarithmic amplifiers are widely used in radio receivers as signal strength meters controlling variable gain amplifiers. Such applications additionally require rectifiers or peak detectors to detect the amplitude of the signal. Therefore one can discuss a wide variety of non-linear functions with different logarithmic amplifier implementations.

7.1 Diode-feedback logarithmic amplifiers

A logarithmic function can be realised with a diode or a bipolar transistor as they have an almost exact logarithmic relationship between voltage and current. The v_{BE} of a bipolar transistor follows logarithmically the collector current i_C for at least six decades as

$$v_{BE} = \frac{kT}{nq} \ln \frac{i_C}{I_S}, \tag{7.1}$$

where k is the Bolzman constant, T is the absolute temperature in Kelvins, n is a process dependent constant between one and two, q is the electron charge and finally I_S is the saturation current. The series resistance at the emitter will limit the logarithmic operation at high current levels but increasing the emitter area will extend the logarithmic operation a little further.

7.1.1 Voltage-mode operational amplifier based realizations

This principle can be utilised with a simple circuit realisation of Figure 7.1a, where the bipolar transistor is placed as a feedback element in an operational amplifier circuit in inverting voltage amplifier configuration, resulting in an output voltage

$$v_{OUT} = -\frac{kT}{nq} \ln \frac{v_{IN}}{I_S R_1}. \tag{7.2}$$

There are two temperature dependent terms in this equation, namely the temperature T itself and the saturation current I_S so that a degree of temperature compensation is conventionally required. Similarly, the logarithmic function should be referenced to a well known current level rather than the temperature and process variation dependent saturation current I_S.

Both requirements are satisfied with the circuit topology of Figure 7.1b, where a matched pair of bipolar transistor is used, so that the difference between the two base-emitter voltages is

$$v_{BE1} - v_{BE2} = \frac{kT}{nq} \ln \frac{i_{C1}}{i_{C2}} = \frac{kT}{nq} \ln \frac{v_{IN}}{I_{REF} R_1}. \tag{7.3}$$

Then, the resulting output voltage can be expressed as

$$v_{OUT} = -\frac{kT}{nq} \left(1 + \frac{R_2}{R_{TC}}\right) \ln \frac{v_{IN}}{I_{REF} R_1}. \tag{7.4}$$

Since this output voltage equation remains temperature dependent, the resistor R_{TC} is conventionally a special resistor with a positive 3300...3500 ppm/K temperature coefficient. However, these resistors are quite rare as discrete components. Similarly,

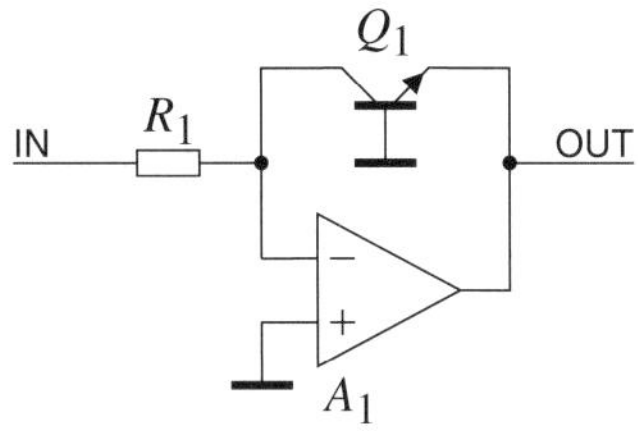

(a)

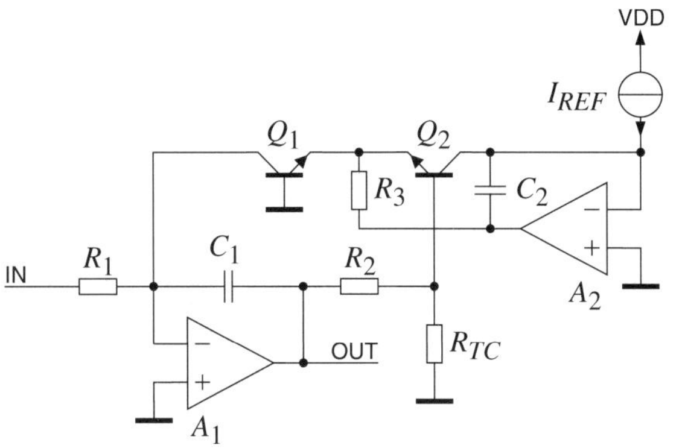

(b)

Figure 7.1 a) The diode-feedback logarithmic amplifier principle. b) A temperature compensated logarithmic amplifier.

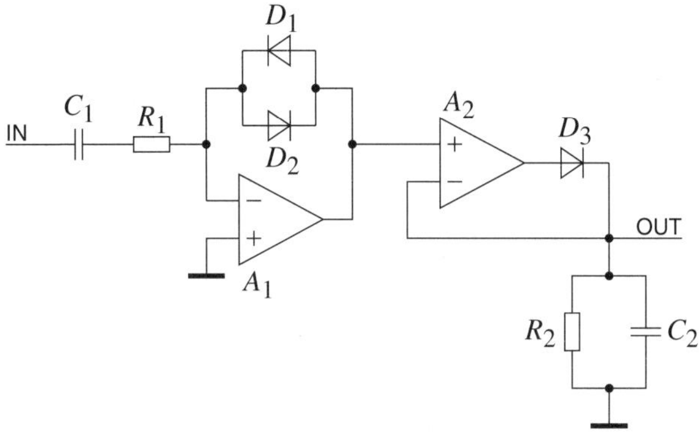

Figure 7.2 A logarithmic peak detector realised with voltage-mode operational amplifiers.

in discrete circuits the temperature tracking between the resistor R_{TC} and the bipolar transistor pair is difficult to maintain. It may similarly be possible to realise this temperature dependent resistor ratio in integrated circuits by using two different resistor materials, but in this case the output voltage will be sensitive to process variation.

Any parasitic capacitances in the two feedback loops of the logarithmic amplifier lead to phase lag at high frequencies and thus the internal compensation of the operational amplifiers cannot ensure stability in this application. Therefore, additional compensation capacitors C_1 and C_2 are added to the circuit. These two capacitors shunt off the active transistor circuit at high frequencies thus ensuring stability.

A very common application for logarithmic amplifiers is the detection of signal amplitude. Since the logarithmic amplifiers discussed require unipolar current signals, they cannot be used to process AC-signals with a large dynamic range. However, in Figure 7.2 a logarithmic peak detector is shown that can be used in such applications. In this circuit, the feedback network uses two diodes rather than one so that both positive and negative signal peaks are compressed by a logarithmic function. The positive peaks are then detected by the peak detector circuit constructed of the amplifier A_2, the diode D_3 and a discharging time constant set by R_2 and C_2. In order to minimise the effects of offset voltages, a DC-decoupling capacitor C_1 is added to the input.

The output voltage of this logarithmic signal amplitude detector is still strongly temperature dependent and without a reference level the same way as with the circuit in Figure 7.1a. Therefore, an additional temperature compensation circuit is needed after the peak detector.

The feedback-loop of the logarithmic amplifier is always closed whereas in the peak detector the loop is closed only when the input signal is higher than the last detected peak. Therefore, the settling time is not so crucial for the operation in the logarithmic amplifier, and a higher dynamic range and a wider bandwidth can be realised if the logarithmic amplifier precedes the peak detector.

7.1.2 Design case: CCII∞ based logarithmic peak detector

All of the logarithmic amplifiers discussed have an identical problem with high frequency performance. The dynamic impedance of the feedback network is inversely proportional to the input signal level and thus, in a logarithmic amplifier with a 60 dB dynamic range for example, the feedback impedance varies three decades. As a consequence of the gain-bandwidth product limitation of the voltage-mode operational amplifier, reaching a 1 MHz minimum bandwidth with a 60 dB dynamic range, for example, would require an operational amplifier with a gain-bandwidth product in the range of 1 GHz.

Most current-mode feedback amplifiers can overcome this gain-bandwidth product limitation as was demonstrated in Chapter 4. A current-feedback operational amplifier, for instance, can maintain constant closed loop bandwidths up to 20...40 dB of closed loop gain. However, this is achieved by keeping the feedback impedance constant while varying the input resistor, as Equation 4.3 shows and consequently the current-feedback operational amplifier has a gain-bandwidth product limited operation in such applications.

The closed loop corner frequency of a high-gain current conveyor in in effect relatively independent of the feedback impedance, as revealed by derived Equation 4.37 in Chapter 4.4, where the condition for a constant closed-loop bandwidth operation is that the output impedance of the conveyor should be higher than the feedback impedance. However, as the simulation results of the design example CCII∞ in Figure 4.19 show, the Miller-effect in the output stage may also limit the bandwidth at high closed loop gains.

BiCMOS implementation of a CCII∞

Because bipolar transistors are necessary in this application, the logarithmic peak detector is implemented with a BiCMOS-process. The 1.2 μm BiCMOS-process used has vertical npn-transistors with an f_t of 7 GHz and slow lateral pnp-transistors. Therefore, in the designed BiCMOS high-gain current-conveyor presented in Figure 7.3 [5, 6], only npn-transistors are used to enhance the circuit performance.

The required high output resistance is achieved with cascode transistors. The output current-source supplying 1.6 mA of bias current to the output common-source stage is a standard PMOS cascode current-source. In the output common-source stage, a bipolar transistor Q1 is used as the cascode device to reduce the Miller effect in MN3 and to push the pole deriving from the cascode transistor to as high frequencies as possible.

The input stage of the conveyor is a voltage-follower realised with a PMOS differential pair driving a PMOS source-follower. Consequently, the nondominant pole cause by this stage is at relatively low frequencies. For this reason, adequate compensation is achieved with an off-chip 330 pF compensation capacitor C_C.

Logarithmic peak detector implementation

The schematic of the logarithmic amplifier and the peak detector is shown in Figure 7.4 [5,6]. The feedback network includes two diode connected transistors (Q1 and Q2) for symmetrical output. Before the peak detector stage, there is an additional low pass filter realised with a voltage-follower, A3, and an internal RC-network with a -3 dB corner frequency of approximately 3 MHz (R_1 and C_1). The purpose of this stage is to

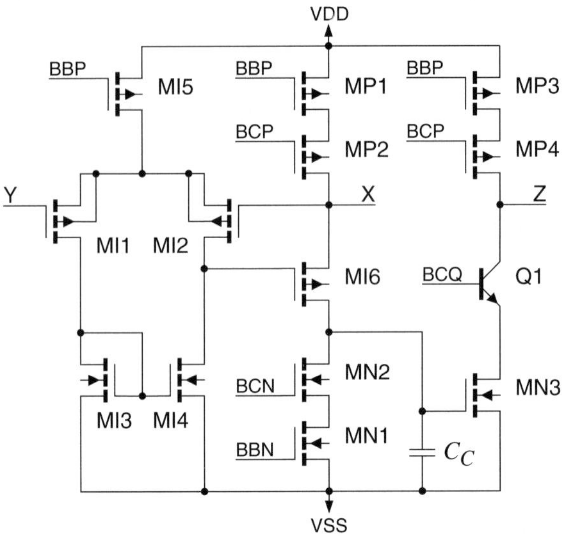

Figure 7.3 The detailed schematic of the designed BiCMOS CCII∞.

Transistors	Aspect ratio / μm/μm
MP3, MP4 and MN3	400/1.2
MP1, MP2, MI5, MN1 and MN2	100/1.2
MI1 and MI2	300/1.2
MI3 and MI4	50 1.2
Q1	12/1.2 (emitter)

Table 7.1 The transistor dimensions of the amplifier in Figure 7.3. The compensation capacitor C_C is an off-chip 330 pF capacitor.

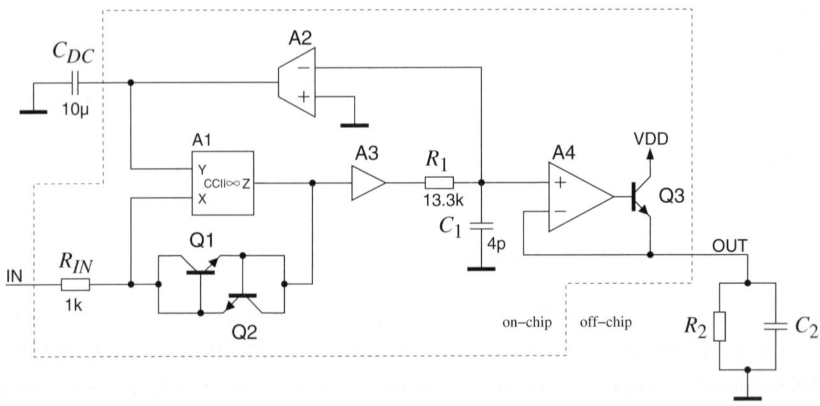

Figure 7.4 The schematic of the designed logarithmic peak detector.

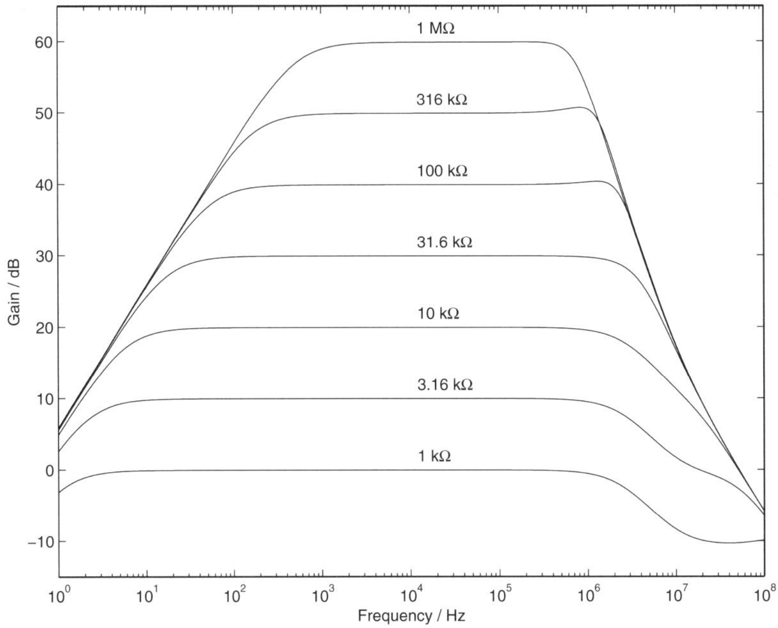

Figure 7.5 The simulated frequency response at the output of the BiCMOS CCII∞ with a feedback resistance varied from 1 kΩ to 1 MΩ.

filter out the possible overshooting in the detected waveforms. This enables accurate peak detection for a wider range of input signals.

Because the input offset at the X-terminal of the high-gain current-conveyor is mainly current, the offset is not cancelled with a capacitor in series with the input resistor R1. Therefore, the offset cancellation is realised with an additional feedback loop consisting of an operational transconductance amplifier A2 and an external 10 μF capacitor C_{DC}. This circuit additionally cancels the offsets of the following buffer stage but leaves the voltage offset of the OTA. Because the offset cancellation feedback loop includes the logarithmic amplifier, the high pass corner frequency is proportional to the feedback impedance, thus leading to the relative large capacitor value. Because the OTA is basically a DC-amplifier, its offset can be minimised by using large area input transistors and cross-coupled centroid-symmetrical layout techniques. An offset compensation SC-network could be added to the input of the OTA for additional reduction of the offset voltage.

In order to achieve fast settling, the peak detection is realised with an emitter-follower rather than a diode. The emitter-follower is driven by an NMOS differential amplifier stage A4 and, because the voltage swing at the output of the differential stage is small, the peak detector response time is shorter than with a complete operational amplifier. Furthermore, because the dynamic range of the signal has been reduced by the logarithmic amplifier, such a simple circuit is practical. The discharge circuit of R_2 and C_2 is an external RC-network for various application requirements.

In order to demonstrate the constant closed-loop bandwidth operation of the BiCMOS CCII∞ designed, the logarithmic amplifier is simulated using a feedback resistor rather than the diode-connected transistors Q1 and Q2. In this simulation (Figure 7.5), the feedback resistance is varied from 1 kΩ to 1 MΩ while the input resistor R1 is kept at a constant resistance value of 1 kΩ, showing almost constant bandwidth operation up to a 60 dB voltage gain. By using smaller input resistances, yet higher gains could be reached but in this case the maximum amplifier output current of 1.6 mA of the BiCMOS CCII∞ limits the maximum signal handling capability.

A test of the fabricated chips revealed that the DC-feedback network failed to work so that the output was stuck close to the supply rails. However, disabling the DC-feedback network by connecting the conveyor Y-terminal to the analogue ground enabled the circuit work to adequately. Therefore, the measurements were made without the offset compensation network.

The measurement results with a 500 kHz sinusoidal input signal show logarithmic operation with approximately 55 dB dynamic range, as seen in Figures 7.6 and 7.7. The positive peak output voltage (solid line in Figure 7.7) is limited by the maximum source current of the amplifier whereas the emitter resistance of Q1 limits the logarithmic behaviour of the negative peak voltages (dashed line in Figure 7.7) at high signal levels.

There is a significant deviation from the ideal logarithmic function in the measured peak output voltage also in the middle range, partly as a result of the drift of offset voltage, both in the measurement set-up and the logarithmic amplifier itself. Furthermore, the slight peaking on the frequency response at some feedback impedance levels demonstrated in Figure 7.5 may cause a degree of error. This error mechanism can be reduced at the expense of bandwidth by increasing the compensation capacitor C_C of the high-gain conveyor. However, there is yet another source of error in the behaviour of the logarithmic amplifier as the output voltage depends on the emitter current rather than the collector current, thus leading to a different equation for the base-emitter voltage:

$$v_{BE} = \frac{kT}{nq} \ln\left(\frac{i_E}{I_S}\frac{\beta}{\beta+1}\right). \tag{7.5}$$

However, the current gain factor β is relatively low in most integration processes, a factor which similarly varies with the bias current level. As a result, less accurate logarithmic behaviour can be achieved with this circuit topology than with topologies controlling the collector current.

At signal levels below -50 dBm the accuracy is limited not only by the bandwidth

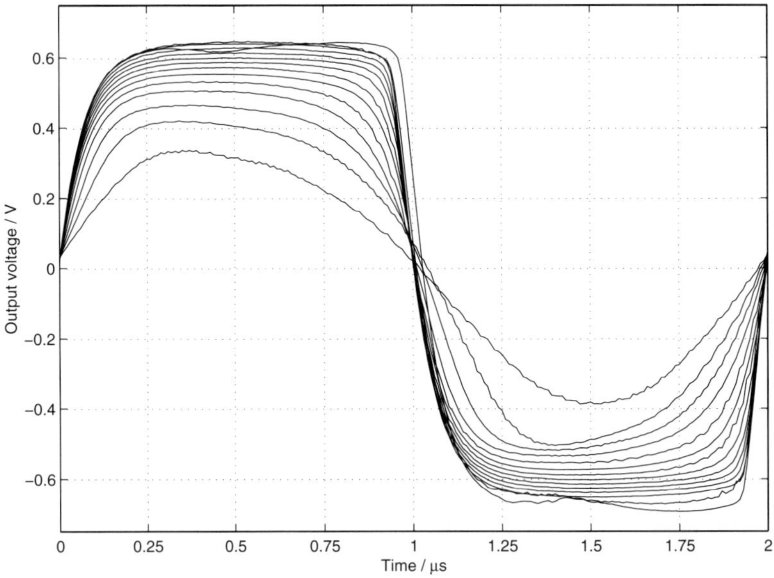

Figure 7.6 The measured output waveforms of the logarithmic amplifier when input signal amplitude is varied from -55 dBm to +10 dBm in 5 dBm steps.

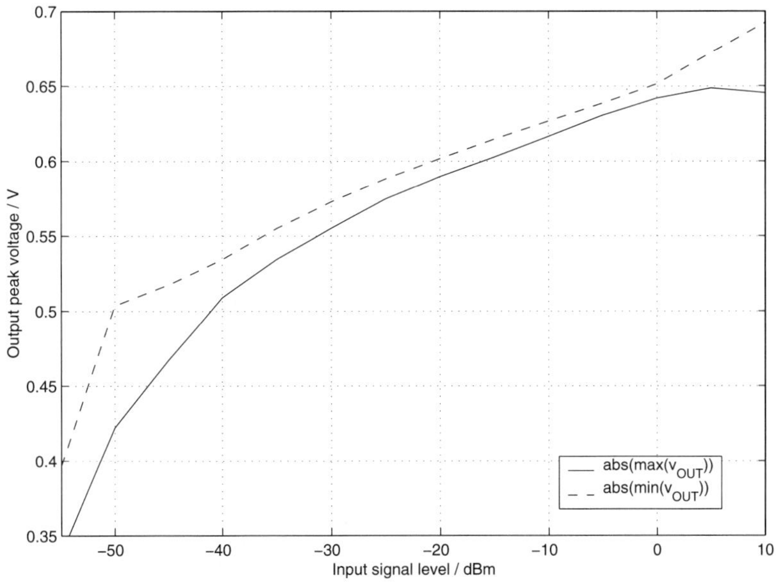

Figure 7.7 The minimum and maximum values of the waveforms of Figure 7.6 plotted against the input signal level.

and the input offset of the high-gain current-conveyor, but also by the time required to discharge the base-emitter junction charges of Q1 and Q2. Therefore, the used high-gain conveyor succeeded in eliciting the full performance from the feedback diodes. If a higher dynamic range is required, this could be achieved solely by increasing the current drive capabilities of the conveyor so that the maximum input signal level can be increased. In this case, the emitter resistances will limit the logarithmic accuracy with large signals unless larger emitters areas are used. This increases the base-emitter junction charging time, limiting the high-frequency operation at small signals levels even further. Therefore, the dynamic range can be increased only by few decibels without decreasing the logarithmic amplifier bandwidth with this circuit principle.

Post processing of the logarithmic output voltage

The output voltage of the logarithmic peak detector remains referenced to the temperature and process dependent saturation current I_S. Additionally, there are other temperature and process dependencies present in the output voltage that should be compensated for. Therefore, additional post processing is required for the detected logarithm of signal amplitude. Because of the peak detector the processed signal is virtually a DC-signal, thus relaxing design specifications.

Temperature compensation principle There are three temperature dependent terms in the detected peak signal, namely the temperature T itself, the saturation current I_S, and the current gain factor β. The saturation current can be cancelled out merely by subtracting another base-emitter voltage from the output voltage of the logarithmic amplifier, resulting in an output equation similar to Equation 7.3. In order to prevent the variation of β from affecting the logarithmic accuracy, this reference voltage should similarly be generated by using a diode connected transistor. Although an accurate reference level can be provided with this method, certain temperature dependencies remain uncompensated.

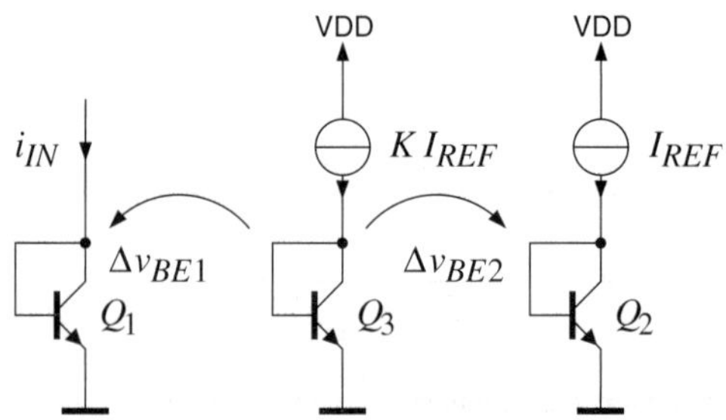

Figure 7.8 The temperature compensation principle.

A more temperature independent logarithm can be produced by using two reference base-emitter voltages rather than one, as seen in Figure 7.8 [6]. The v_{BE} of the transistor Q_1 represents the output voltage of the logarithmic amplifier. Additionally, transistors Q_2 and Q_3 are biased with two reference currents I_{REF} and KI_{REF}. If the voltage differences $\Delta v_{BE1} = v_{BE1} - v_{BE3}$ and $\Delta v_{BE2} = v_{BE2} - v_{BE3}$ are divided, both I_S and T are cancelled out:

$$\frac{\Delta v_{BE1}}{\Delta v_{BE2}} = \frac{\frac{kT}{nq}\ln\frac{i_{IN}}{KI_{REF}}}{\frac{kT}{nq}\ln\frac{I_{REF}}{KI_{REF}}} = 1 - \frac{1}{\ln K}\ln\frac{i_{IN}}{I_{REF}}. \tag{7.6}$$

The Δv_{BE1} goes to zero when the input current i_{IN} reaches the current KI_{REF} and thus the reference current KI_{REF} sets the full scale of the logarithmic output, while the reference current I_{REF} sets the reference level when the sign of the output changes.

Divider principle Although the voltage differences Δv_{BE1} and Δv_{BE2} are easily produced, the division of these voltages may lead into complicated circuits. One of the simplest implementations of a division function is the current controlled variable current mirror shown in Figure 7.9 [3, 7]. The output current of this translinear circuit depends on the ratio of the emitter currents i_A and i_B of the transistors Q_1 and Q_2:

$$i_{OUT} = \frac{i_A}{i_B} i_{IN}. \tag{7.7}$$

Therefore, this circuit can be used both as a one-quadrant (i.e. all signals positive) multiplier and divider.

The current mirroring accuracy is effected by the base currents of the transistors Q_1 and Q_2 and by the Early-voltages of the transistors Q_3 and Q_4. The base current induced errors can be avoided using an extra MOS-transistor buffer in the input. The Early-voltage error can be reduced by adding cascode transistors to the collectors of transistors Q_3 and Q_4 or otherwise simply by keeping the collector-emitter voltages of these transistors as equal as possible.

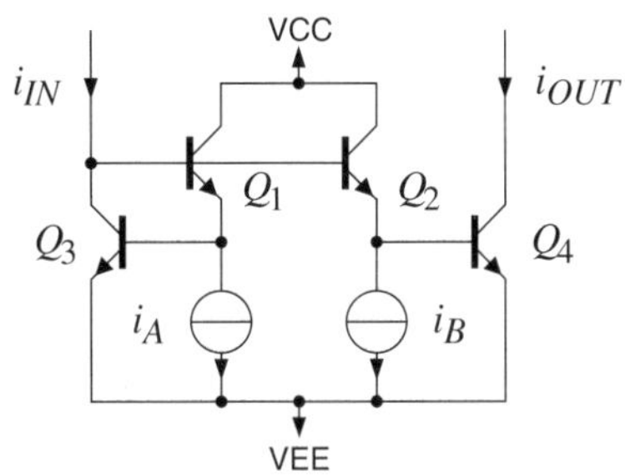

Figure 7.9 The current-controlled variable current-mirror.

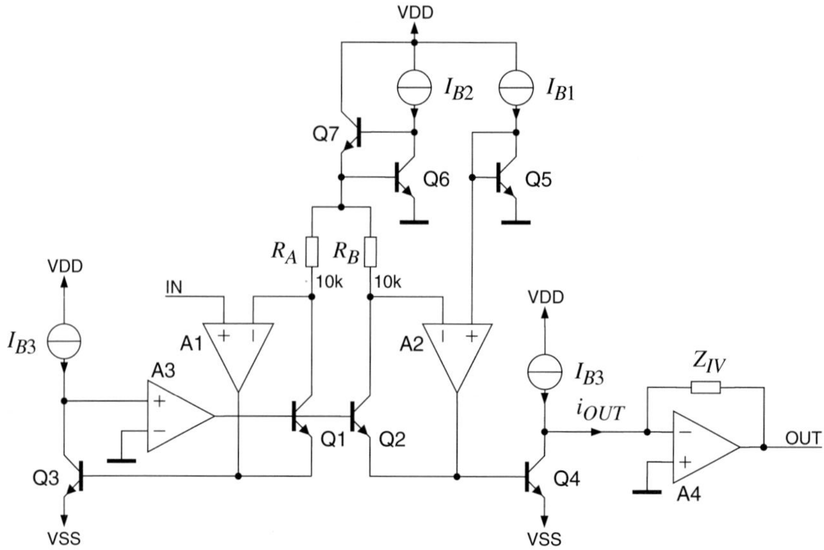

Figure 7.10 The temperature compensation circuit.

Temperature compensation circuit The simplified schematic of the temperature compensation circuit designed is presented in Figure 7.10. The variable current mirror consists of transistors Q1-4 and amplifiers A1-4 and resistors R_A and R_B. The base currents of npn-transistors Q1 and Q2 are buffered with a MOS differential amplifier A3 in order to enhance the current mirror accuracy. The Early-voltages of the transistors Q3 and Q4 do not affect the mirroring accuracy because both transistors have equal collector-emitter voltages. The collector potential of Q3 is forced to virtual ground with the amplifier A3 and the collector potential of Q4 is forced to virtual ground with the operational amplifier A4. The amplifier A4 and the feedback impedance Z_{IV} work either as a current to voltage converter (Z_{IV} is resistive) or as an integrator for AGC-loop control (Z_{IV} is capacitive).

The diode voltage differences are converted into the bias currents of the transistors Q1 and Q2 in the resistances R_A and R_B. The full scale reference is produced with transistors Q6 and Q7. If the input voltage v_{IN} exceeds the base-emitter voltage of Q6, the collector current of Q1 reduces to zero and thus the circuit ceases to operate correctly. This problem can be avoided by choosing the full scale reference larger than the maximum output voltage of the logarithmic amplifier, which is limited by the current sourcing capability of the high-gain current-conveyor.

If the input voltage v_{IN} of the temperature compensation circuit is the detected peak value of the logarithmic amplifier output voltage

$$v_{IN} = \frac{kT}{q} \ln \frac{\hat{v}_{IN}}{R_{IN} I_S}, \tag{7.8}$$

then the equation for the output current is

$$i_{OUT} = \frac{R_B \left(v_{IN} - v_{BE6}\right)}{R_A \left(v_{BE5} - v_{BE6}\right))} I_{B3} - I_{B3}. \tag{7.9}$$

By using Equation (7.6) and letting $R_A = R_B$ the output current equation is reduced to

$$i_{OUT} = -\frac{I_{B3}}{\ln K} \ln \frac{\hat{v}_{IN}}{R_{IN} I_{B1}}, \tag{7.10}$$

where the constant K in the equation is

$$K = \frac{I_{B2}}{I_{B1}} \frac{A_{E6}}{A_{E5}}, \tag{7.11}$$

where A_{E5} and A_{E6} are the emitter areas of the transistors Q5 and Q6. The bias currents I_{B1}, I_{B2} and I_{B3} are converted from an external reference voltage V_{BG} to a current with a resistor R_{BG} and are

$$I_{B1} = \frac{I_{B2}}{2} = 10 I_{B3} = \frac{V_{BG}}{R_{BG}}. \tag{7.12}$$

The output current i_{OUT} is converted to a voltage v_{OUT} with the operational amplifier A4 and the impedance Z_{IV}:

$$v_{OUT} = \frac{V_{BG}}{10 \ln K} \frac{Z_{IV}}{R_{BG}} \ln \frac{R_{BG} \hat{v}_{IN}}{R_{IN} V_{BG}}. \tag{7.13}$$

The output voltage is now a logarithm of the ratio of the peak input voltage and the bandgap voltage.

Accuracy considerations There are three resistor ratios, $\frac{Z_{IV}}{R_{BG}}$ $\frac{R_{BG}}{R_{IN}}$ and $\frac{R_B}{R_A}$, and a current mirroring ratio of 10 (the currents I_{B1} and I_{B3}), which can all be realised quite accurately in an integrated circuit. In addition, the constant K which involves a current ratio of two and an emitter area ratio of two, can be realised accurately with a careful layout design. There is, however, a systematic error mechanism present in the constant K: the transistor Q6 is not diode connected as in the principle (Figure 7.8), but instead uses an extra transistor Q7 for sourcing the collector current for transistors Q1 and Q2. If the transistor current gain factor β is low and the collector current of the Q7 differs substantially from the collector current of Q6, there is an input voltage dependent term in the constant K. This error can not be completely cancelled, because the collector current of the Q7 is input voltage dependent.

The collector-emitter voltage of Q1 similarly depends on the input voltage. The v_{CE} of Q1 is in the range of two diode voltages and the input voltage variation is a few hundreds of millivolts and consequently the variation of v_{CE} is less than 15 % and the error arising from the Early-voltage is small.

The amplifiers A1, A2 and A3 contribute also to the total error. The amplifier A3 buffers only the base currents of the transistors Q1 and Q2 and only large input offsets may lead to current mirroring errors arising from finite Early-voltages of the transistors Q3 and Q4. Therefore, the A1 is realised with a simple differential amplifier stage. The amplifiers A1 and A2 require a voltage gain of at least 60 dB in order not to contribute to the total error and are thus realised as a two-stage Miller-compensated CMOS-operational amplifier. Because the generated Δv_{BE}:s are small, the offsets of the amplifiers A1 and A2 and the peak detector circuit affect the absolute accuracy. The offset voltages do not change substantially over the temperature range, so the temperature compensation is not degraded by them.

Feedback loop compensation Because both low power CMOS operational amplifiers A1 and A2 drive one bipolar transistor emitter (Q1 or Q2) and one base (Q3 or Q4), the normal Miller frequency compensation does not work because of the low impedance load. In the feedback loop consisting of the amplifier A1, the transistor Q1, and the resistor R_A, the bipolar transistor acts more like a cascode transistor than as a feedback element. Therefore, the 4 pF Miller-capacitor C_{C1} is connected to the collector of the transistor Q1, rather than the output node of A1, as shown in Figure 7.11.

The mirror input feedback loop (A3, Q1 and Q3) could be compensated in the same manner, but because the transistor Q1 may turn off in the case of large signal levels, it was compensated with an external 10 nF capacitor C_{C3} to the ground. For the same reason, the feedback loop of the amplifier A1 is protected by a 1.5 pF shunt capacitor C_{C2} to resist the turning off of transistor Q1.

Simulation results Figure 7.12a depicts the simulated output current sweeps of the temperature compensation circuit. Because of the complexity of the entire logarithmic circuit, the input voltage of the temperature compensation circuit is generated by means of a single diode-connected transistor and a current source, in preference to the logarithmic amplifier and peak detector. The simulations show accurate operation near the intended operating amplitude. The accuracy decreases further from the reference level (354 μA). In most extreme case, at a 1 μA current level the curves are within 4.5 % in the temperature range of -40...+120 °C.

The error of the logarithmic amplifier is generally referred to the input signal level

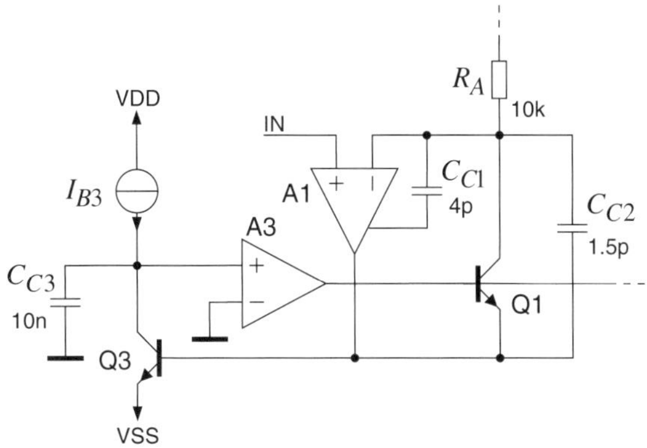

Figure 7.11 The feedback loop compensation principle.

[8]. In this case,the logarithmic function is numerically fitted into a function

$$i_{OUT} = p_1 20\log_{10}\frac{i_{IN}}{I_{REF}} + p_0. \tag{7.14}$$

After evaluating the first order polynomial coefficients p_0 and p_1, the input referred error can be calculated in decibels

$$\varepsilon_{log} = 20\log_{10}\frac{i_{IN}}{I_{REF}} - \frac{i_{OUT} - p_0}{p_1}. \tag{7.15}$$

Based on this principle the simulated input referred error of this temperature compensated logarithmic amplifier is displayed in Figure 7.12b.

This error plot shows that the output current deviates from the ideal output, which is most probably caused by the emitter resistances of transistors Q5 and Q6. The emitter areas of these two transistor are chosen relatively small as the diode connected feedback transistors of the high-gain conveyor based logarithmic peak detector should also be small in order to minimise the base-emitter junction capacitance limiting high-frequency operation at low signal levels. The temperature dependency of the output signal is caused by the temperature dependency of the bipolar transistor current gain. Clearly, in the simulation, random variations that increase the output errors are excluded. Nevertheless, the accuracy is adequate for controlling receiver AGC-loops. Furthermore, in these applications, the accuracy near the reference level is the most important requirement as the AGC-circuitry tries to adjust the signal amplitude precisely to this reference level.

The temperature compensation circuit was included in the same chip with the high-gain current-conveyor logarithmic peak detector. However, because of an error in the offset compensation circuit, the total accuracy of the logarithm could not be evaluated.

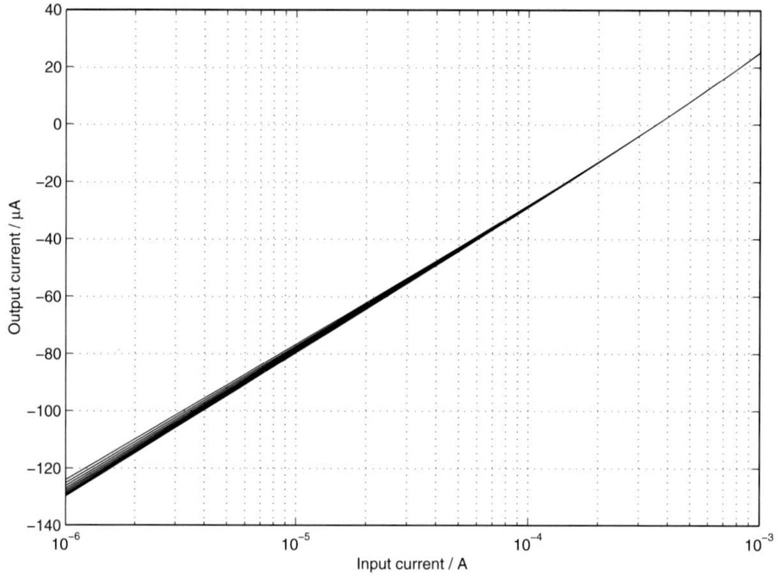

(a)

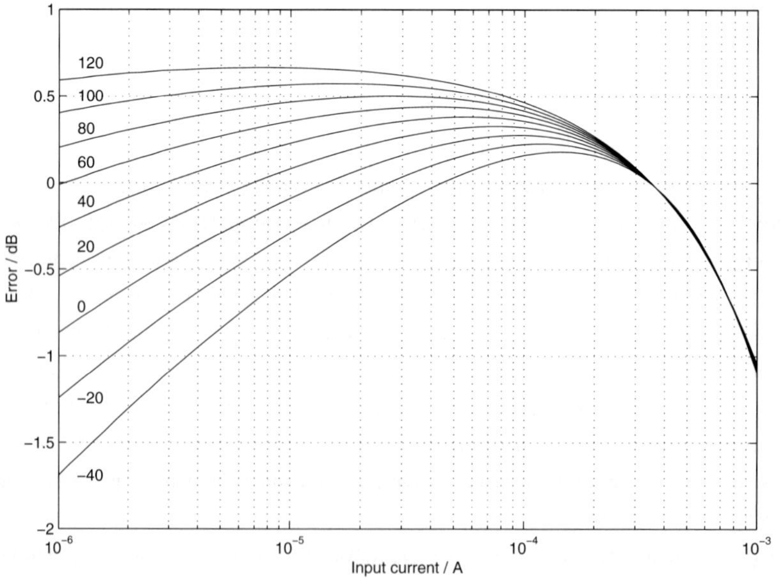

(b)

Figure 7.12 a) The simulated output current vs. input current with the temperature varied from -40°C to +120 °C in 20 °C steps. b) The input referred error in decibels of the same simulation.

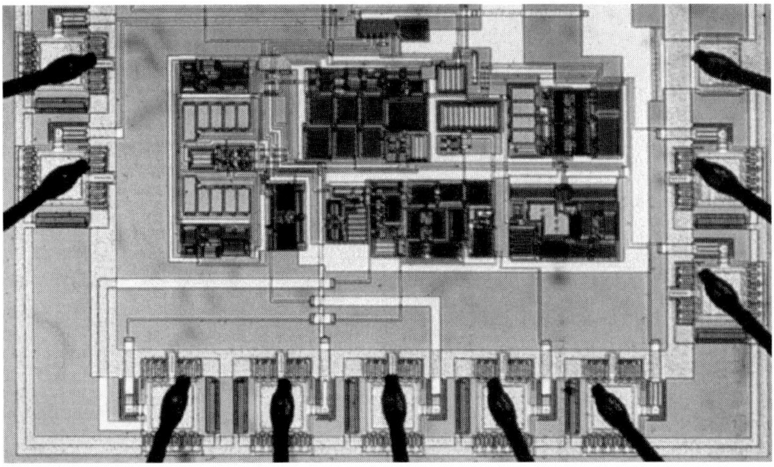

Figure 7.13 The microphotograph of the logarithmic peak detector and the temperature compensation circuit.

Nevertheless, the temperature compensation circuit operated functionally as predicted in simulations. The processed chip shown in Figure 7.13, which includes other circuitry, has a die area of 3.5 mm^2, of which the logarithm circuit discussed takes up one half. The power consumption of the circuit is approximately 25 mW with a 4.5 V supply voltage.

Final remarks on the design

Because of the error in the offset compensation network the performance of the logarithmic could not be fully evaluated. This design additionally demonstrates the importance of additional test pins in a prototype chip as there was not enough of them to discover the problem. In any case, a minor error in generating the bias voltages for example could easily have generated so much current offset at the conveyor X-terminal that the offset compensation network was no longer capable of correcting it. Furthermore, the current driving capabilities of the offset compensation OTA is also very limited as a very low transconductance was required and then quite a minor amount of leakage currents at the OTA output could disturb the operation of the offset compensation network.

A new process run was not made because the specifications changed so that the supply voltage fell to 3 volts, making this circuit topology inapplicable. Nevertheless, the circuit demonstrated that high-gain current-conveyors can extend the frequency range of many non-linear analogue functions. For example, the multiplier/divider circuit used as the core of the temperature compensation circuit could easily be realised

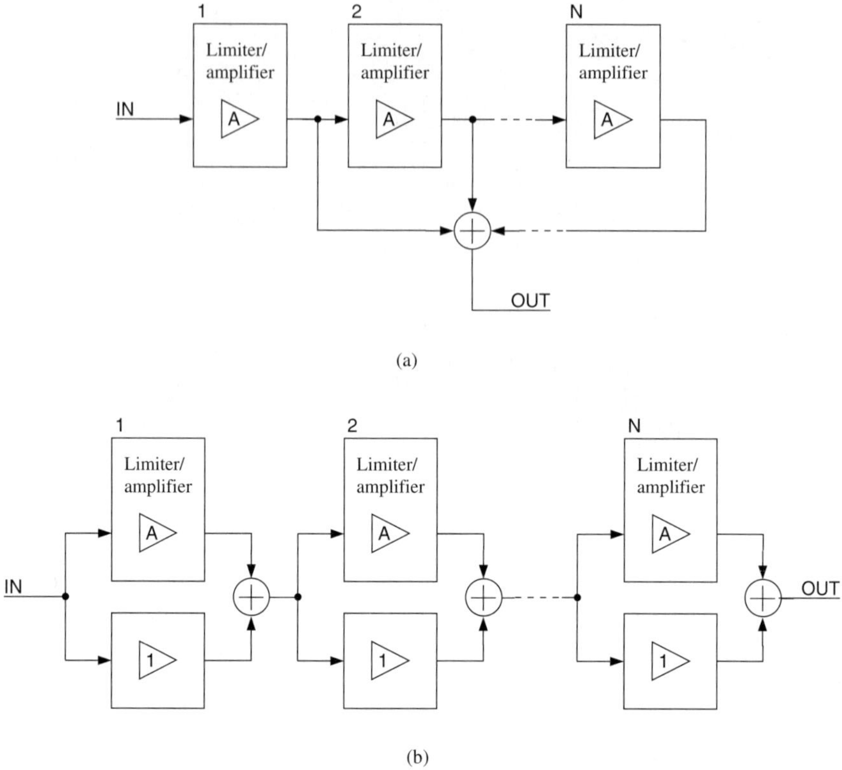

Figure 7.14 Piece-wise linear approximation of the logarithm function. a) Cascaded limiting amplifiers stages. b) A cascade of parallel connected limiting amplifiers and unity gain buffers.

using high-gain conveyors, thus enabling a constant wide bandwidth over a relatively large signal amplitude range. In this circuit, operational amplifiers were used because the wide bandwidth was not required and an unnecessarily wide bandwidth would increase the total output noise.

7.2 Pseudologarithmic amplifiers

Logarithmic amplifiers can also be realised without resorting to the logarithmic behaviour of the pn-junction. A piece-wise linear approximation of the logarithm function can be realised using a cascade of limiting amplifiers [9–12]. The most common way to use these limiting amplifiers is shown in Figure 7.14a where all the outputs of N cascaded limiting amplifiers are combined to form the logarithmic output signal.

The limiting amplifier is in most pseudologarithmic amplifiers a simple differential amplifier stage as in the bipolar implementation in Figure 7.15a. In this case, neglecting the base currents and letting $R_1 = R_2 = R$, the large signal equation for the output voltage is [13]

$$v_{OUT+} - v_{OUT-} = I_{BB} R \tanh\left(\frac{nq}{kT}\frac{v_{IN-} - v_{IN+}}{2}\right). \tag{7.16}$$

A a consequence of the tanh-function the limiting action is relatively smooth thus reducing the error of the piece-wise linear approximation.

Because every amplifier stage adds a degree of delay to the signal, the output signal waveform becomes distorted at very high frequencies. In applications where only the signal amplitude is important, this does not pose a problem. Moreover, in most logarithmic amplifiers aimed for signal strength application, each of the limiting amplifier outputs are either half- of full-wave rectified before summing [10–12]. This output signal is then integrated, thus forming an output signal solely dependent on the signal energy.

In radar applications, for example, logarithmic amplifiers are used to compress the dynamic range of the signal without distorting the phase information of the received signal. In this case, the circuit principle of Figure 7.14a does not perform adequately anymore. Less error in the phase information is generated by forming the amplifier cascade from a limiting amplifier and a linear unity-gain amplifier connected in parallel as depicted in Figure 7.14b. The two parallel connected amplifiers are relatively straightforward to realise, for example by using two differential pairs where one of the pairs is linearised by emitter-degeneration resistors, as shown in Figure 7.15b [9].

7.2.1 Limiting CMOS voltage amplifiers

The CMOS implementations of limiting voltage amplifies are conventionally based on a simple differential amplifier stage, as demonstrated in Figure 7.16. The other differential pair (M_3 and M_4) forms a summing output which is tied together with corresponding outputs of other limiting amplifiers in the pseudologarithmic amplifier, thus forming the logarithmic output.

If saturation region operation for the differential pair transistors is assumed, and $\beta_1 = \beta_2 = \beta$ and $R_1 = R_2 = R$, the large signal differential output voltage equation can be derived as in Equation C.21 in Appendix C:

$$v_{OUT} = v_{OUT+} - v_{OUT-} = v_{IN}\frac{R}{2}\sqrt{\beta\left(4I_{BB} - \beta v_{in}^2\right)}, \tag{7.17}$$

if $|v_{IN}| = |v_{IN+} - v_{IN-}| \leq \sqrt{\frac{2I_{BB}}{\beta}}$. Similarly, the small-signal voltage gain of the am-

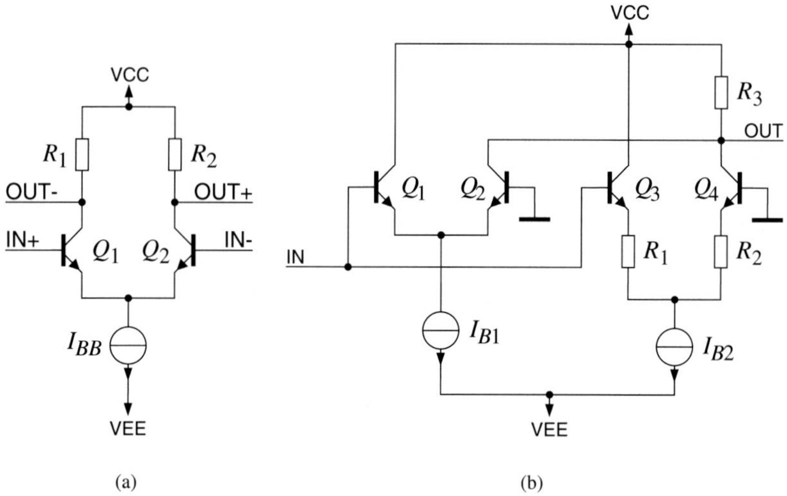

Figure 7.15 a) A bipolar differential limiting voltage amplifier b) A bipolar single-ended implementation of a parallel connected limiting amplifier and unity gain buffer [9].

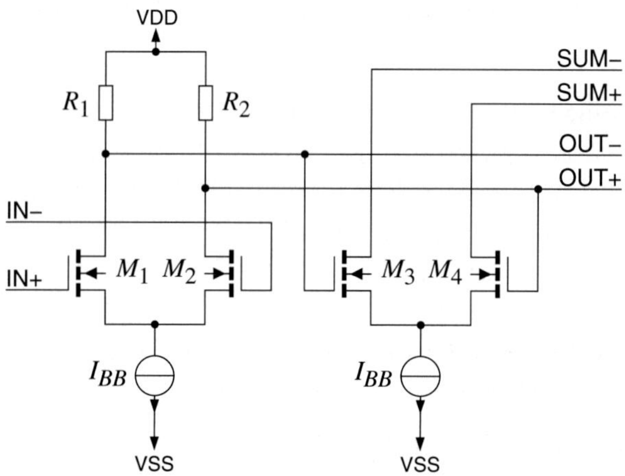

Figure 7.16 CMOS implementation of a differential limiting voltage amplifier providing the summing outputs for the logarithmic output.

plifier is

$$A_v = R\sqrt{\beta I_{BB}}, \tag{7.18}$$

and the limiting voltage level at the output is

$$\max(v_{OUT}) = RI_{BB}. \tag{7.19}$$

Therefore, both the gain and the limiting level are process and temperature dependent. However, in a cascade of several limiting amplifiers the maximum output voltage of the limiting amplifier does not control the limiting level to the same extent as the maximum input voltage of the next amplifier. Therefore, the limiting voltage level caused by the input saturation is based on the Equation (7.17) as

$$v_{lim} = \sqrt{\frac{2I_{BB}}{\beta}}. \tag{7.20}$$

And also this parameter is process and temperature dependent.

If the limiting amplifier is biased with a similar current reference to that in the current-mode filter design case depicted in Figure 6.24 so that the bias current is inversely proportional to β and to the square of R

$$I_{BB} \sim \frac{1}{\beta R^2}, \tag{7.21}$$

the temperature and process variation of the voltage gain can be cancelled out. However, the limiting level is now even more temperature and process variation dependent as before:

$$v_{lim} \sim \frac{1}{\beta R}. \tag{7.22}$$

Alternatively, in an n-well CMOS process, a PMOS differential pair with diode-connected PMOS load transistors can be used in order to achieve a process and temperature independent voltage gain [12]. In this case, the limiting voltage level can be stabilised with a bias current proportional to PMOS-transistor β. However, generating such a bias current generally leads to a rather complicated circuitry and, as a result, this process and temperature dependency is often left uncompensated [12].

7.2.2 Limiting CMOS current amplifiers

If a MOS current-mirror is used as a limiting amplifier, the gain of the amplifier is easy to set by scaling the aspect ratios of the transistor. Similarly the limiting level can be accurately set with the bias current but, however, this works only with a unipolar current signal [14]. By way of illustration, a current-mirror based pseudologarithmic

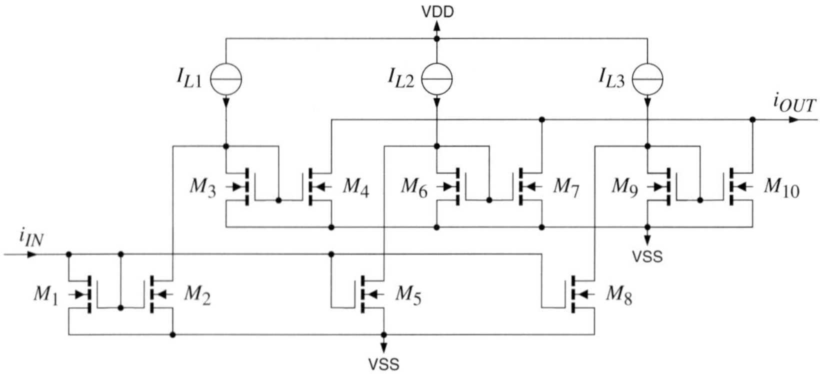

Figure 7.17 A three-stage current-mirror based pseudologarithmic amplifier.

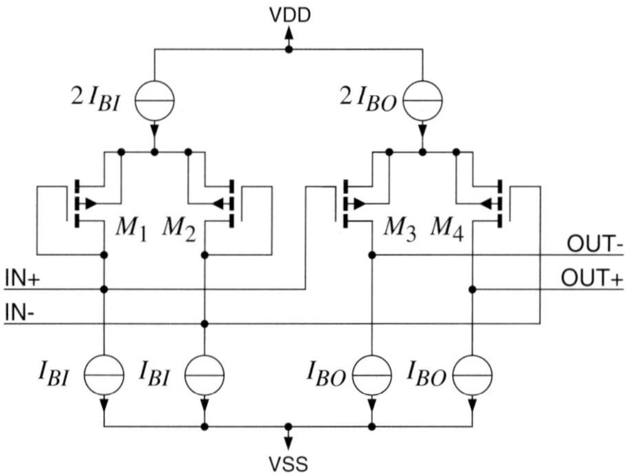

Figure 7.18 A differential limiting current amplifier.

amplifier is presented in Figure 7.17. For the sake of simplicity, this amplifier uses only three limiting levels. For a better accuracy, however, as many limiting stages as required can be added.

This simple circuit gives many levels of freedom to the designer. One design method is to scale the input current-mirror output transistors M_2, M_5 and M_8 and use identical aspect ratios for the three output current mirrors with identical limiting current levels. Alternatively, the output current mirrors and the limiting level current sources can be scaled while the input mirror transistor are implemented with equally sized transistors. However, for an optimal area and accuracy, it would be feasible to scale all transistors and current-sources.

It is also possible to implement limiting current amplifiers for bipolar current signals as the differential limiting current amplifier example of Figure 7.18. Once more,

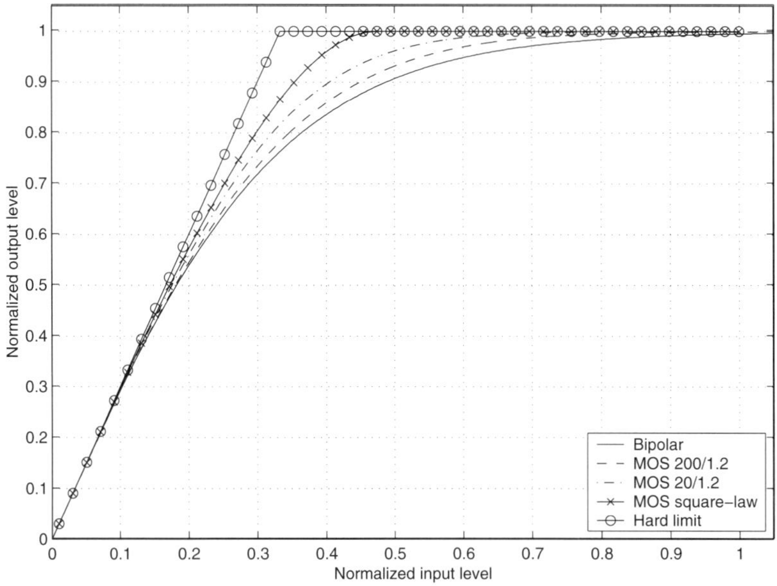

Figure 7.19 The theoretical and simulated limiting action of the MOS limiting voltage amplifier of Figure 7.16 compared to ideal hard clipping function and to the soft clipping behaviour of a bipolar differential pair. Both the input and output are scaled with the maximum output voltage.

the current gain A_i is set by scaling the aspect ratios so that

$$\frac{W_3}{L_3} = \frac{W_4}{L_4} = A_i\frac{W_1}{L_1} = A_i\frac{W_2}{L_2}, \tag{7.23}$$

$$I_{BO} = A_i I_{BI}. \tag{7.24}$$

Similarly, the limiting level is accurately set by the bias current, provided that these bias currents can be generated with low process and temperature dependencies.

7.2.3 Accuracy of the pseudologarithmic amplifier

The limiting current amplifier does not distort the signal before clipping, so hard clipping rather than soft-clipping occurs. Therefore, the piece-wise approximation of the logarithmic function is smoothed with voltage amplifiers and, therefore, with the same amount of limiting amplifier stages, the voltage-mode realisation results in a smaller maximum error than is the case in the current-mode realisation.

The smoothing effect of bipolar and MOS differential amplifier stages is demonstrated and compared with the hard clipping behaviour of limiting current amplifiers in Figure 7.19. In this case, the ideal clipping behaviour of bipolar and MOS limiting voltage amplifiers of Equations (7.16) and (7.17) is also compared with the simulated clipping behaviour of the MOS limiting voltage amplifier. For a more revealing comparison, all transfer curves are normalised by dividing both X- and Y-axis co-ordinates by the maximum output voltage level. Similarly, the load resistors are selected so that a gain of 3 is reached in all cases.

The simulations are repeated with two different aspect ratios using modern BSIM 3V3.1 simulation models so that the MOS transistor behaviour is also modelled adequately in the weak and moderate inversion. With an NMOS differential pair with an aspect ratio of $200\mu/1.2\mu$ and a bias current of 100 μA, the clipping action is almost as smooth as in the bipolar amplifier. Even in the case of MOS transistors, with ten times smaller aspect ratios and the same 100 μA bias current, the clipping action is smoother than the ideal square-law behaviour of Equation (7.17) predicts. Therefore, when even one of the input MOS transistors is operating in weak inversion, significantly smoother limiting behaviour occurs.

In Figure 7.20, the current-mode hard-clipping limiting amplifier based logarithmic amplifier is compared to the voltage-mode bipolar and MOS limiting functions of Equations (7.16) and (7.17). The pseudologarithmic amplifier is constructed of seven identical cascaded limiting amplifier stages with a gain of three with all outputs summed together to form the logarithmic output signal as in Figure 7.14a. Because of the tanh-type large signal behaviour of the bipolar differential pair, the piece-wise linear approximation is more accurate than any CMOS realisations. However, the simulations of the MOS limiting amplifier demonstrate that the clipping is smoother than predicted in Equation (7.17) and thus lower error results in the case of MOS limiting voltage amplifiers than Figure 7.20 shows.

In this example, identical liming amplifier stages are used and as the error will rise very rapidly above the piece-wise linear approximation range, the error can be further minimised by summing the input signal to the summed output. Alternatively, the output of the last output in the amplifier cascade can be rescaled to minimise this error.

7.2.4 Amplitude detection in pseudologarithmic amplifiers

CMOS rectifiers

When the CMOS pseudologarithmic amplifier is used to measure the amplitude of the received signal, additional non-linear signal processing is required. A simple way to produce an output signal proportional to its amplitude is to rectify the signal and then filter it. As an example, in Figure 7.21, the summing output of the CMOS limiting voltage amplifier of Figure 7.16 is modified to provide a half-wave rectified output [11]. This is achieved by converting the drain currents of transistors M_3 and M_4 into

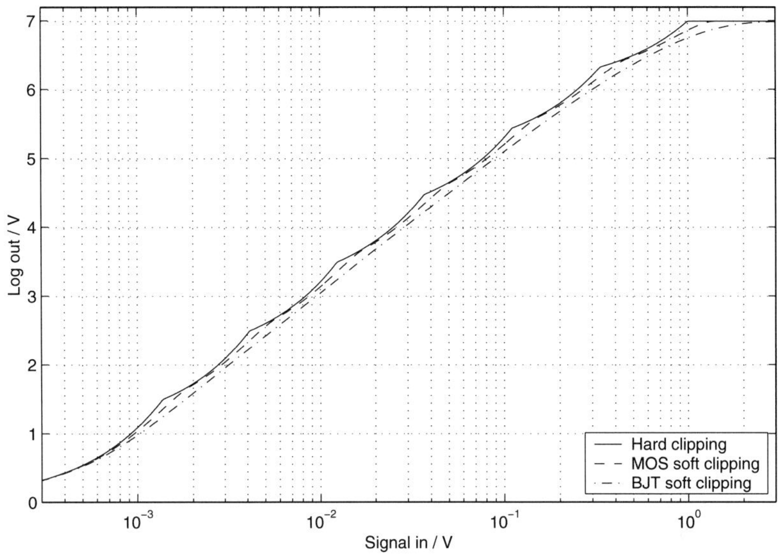

(a)

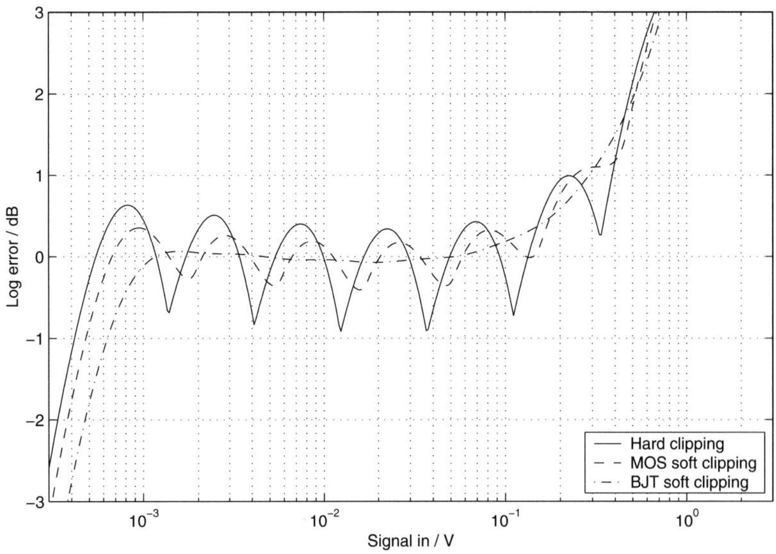

(b)

Figure 7.20 The operation of pseudologarithmic amplifiers with hard-clipping and MOS and bipolar soft-clipping. a) The output signal of the amplifier. b) The input referred error of the logarithmic output.

a single-ended bi-directional current with the current-mirror constructed of transistors M_5 and M_6. This signal is then fed into a second current-mirror (transistors M_7 and M_8) and as this current-mirror cannot sink current but only source it, the output signal is half-rectified.

Similarly, the current-mirror based pseudologarithmic amplifier of Figure 7.17 can be used to half-wave rectify the input signal. However, since in this case the input current-mirror has several scaled outputs, the total input capacitance of this amplifier will limit the frequency range, particularly in the case of low signal amplitudes. Therefore, a local rectifier mirror in individual limiting amplifiers provides more efficient high frequency operation when the signal amplitude variation is low and the parasitic capacitances are minimal.

CMOS squarers

As an alternative to rectifying the signal, it can be squared, thus proving a true RMS output signal after filtering. The square of the signal is easily achieved by using, for example, a four-quadrant multiplier such as a Gilbert-cell as the summing amplifier stage [12]. In this case, the input signal is simultaneously fed into both multiplier inputs to provide squared output.

However, there are also MOS circuits that provide the squaring function directly. By replacing the rectifier mirror transistor M_7 and M_8 in Figure 7.21 with the circuit in Figure 7.22 [15], a squared output signal is provided. If all transistors in this squarer are assumed identical, the output current is a non-linear function of the input current

$$i_{OUT} = 2I_Q + \frac{i_{IN}^2}{8I_Q}. \tag{7.25}$$

In order to prevent the transistors from turning off, the amplitude of input signal should be kept below $4I_Q$. When this squarer circuit is used instead of the current-mirror rectifier in the limiting amplifier of Figure 7.21, setting I_Q to one third of I_{BB} represents a convenient compromise between high frequency operation and the squarer output signal.

CMOS peak detectors

When both a logarithmically compressed AC-signal and a DC-signal proportional to the signal amplitude are required simultaneously, a conveniently modular approach is to use a peak rectifier to detect signal amplitude at the logarithmic output. The simplest way to implement a CMOS voltage peak detector is to use a voltage amplifier and a diode connected NMOS-transistor, as presented in Figure 7.23a.

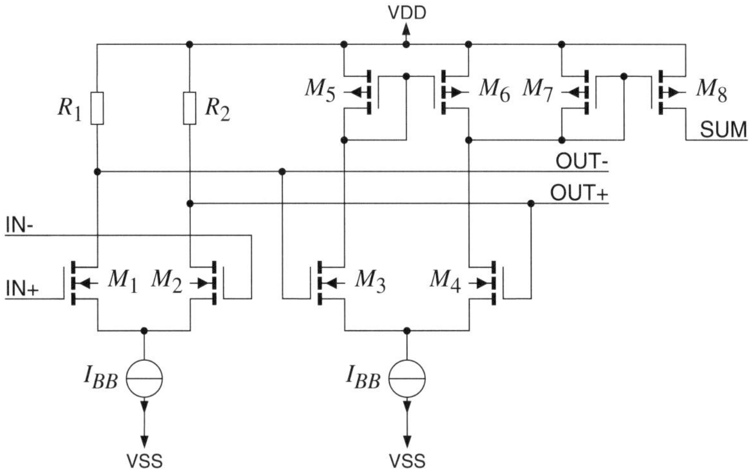

Figure 7.21 A CMOS limiting voltage amplifier stage with a half-wave rectified current output.

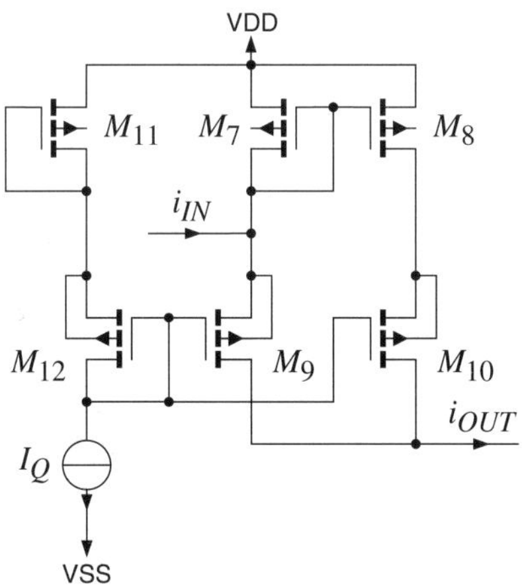

Figure 7.22 A simple squaring circuit optimised for an n-well CMOS-process [15].

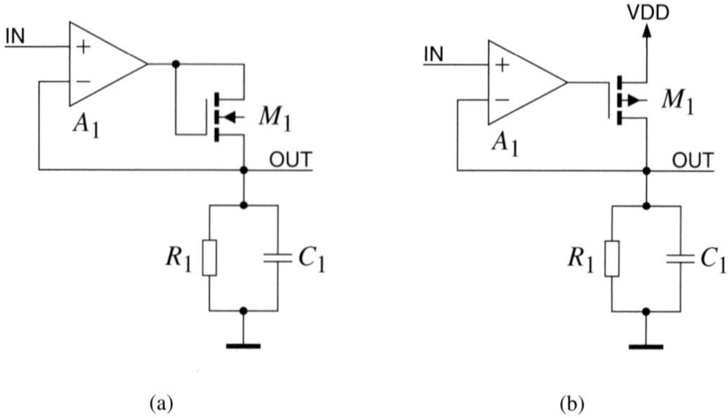

Figure 7.23 CMOS implementations of precision voltage peak detectors.

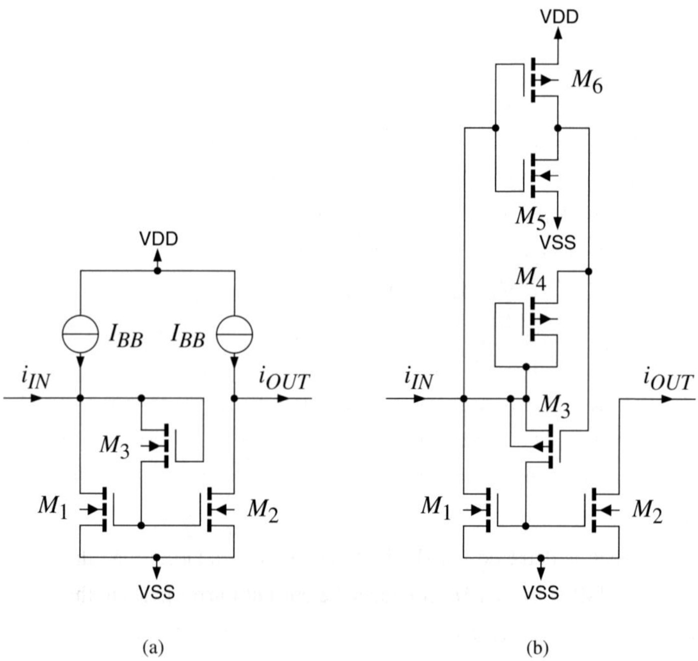

Figure 7.24 a) A simple CMOS current peak detector. b) A low voltage current peak detector.

However, in this approach the output voltage range is quite limited since the gate-source voltage increases rapidly with the output voltage deriving from the bulk effect. Furthermore, a floating-well MOS-device with source and bulk connected together cannot be used because the bulk-drain pn-junction begins to conduct when the MOS-diode is reverse biased. For these reasons, a circuit structure of Figure 7.23b is preferred [16]. In this case, the amplifier A_1 and PMOS-transistor M_1 together construct an amplifier that can source current to the load but not sink it.

The maximum voltage of the peak detector is limited by the common-mode range of the amplifier A_1. In high speed applications, the delay in peak detection must be minimised and consequently complex rail-to-rail common-mode range input stages cannot be used and the dynamic range will be relatively low with low supply voltages. The maximum speed is also limited by the hold capacitor C_1.

However, current peak detectors are also possible with a CMOS-technology. A simple CMOS current peak detector is presented in Figure 7.24a [17]. It is constructed of a current-mirror (transistors M_1 and M_2) and a diode-connected transistor M_3. The gate-source capacitances of M_1 and M_2 hold the gate-source voltage which corresponds to the peak value of the current.

There are several performance limitations in this circuit. The supply voltage cannot be very low because the diode-connected transistor must be biased on top of the mirror gate-source voltage. In addition, there are large voltage transients at the current-mirror input node which are not suppressed entirely even with additional cascode transistors. For a bi-directional input current, bias current sources are needed and thus mismatch of the bias currents degrade the detection accuracy at low signal levels. Additionally, the discharging of the gate-source capacitance of the current-mirror by a resistor results in a signal dependent discharge time constant.

The current peak detector in Figure 7.24b [18, 19] consists of an NMOS current mirror (M_1 and M_2), a CMOS inverter (M_5 and M_6), a PMOS switch (M_3), and a diode-connected PMOS transistor (M_4). The NMOS current mirror acts as a current-memory, which stores the peak current as in the first mirror. The CMOS inverter acts as a current comparator, which compares the incoming current i_{IN} with the drain current of the transistor M_1.

When a peak is detected i.e. the input current i_{IN} is larger than the drain current of the M_1, the PMOS switch M_3 connects the current mirror gates to the input and the current mirror follows the input current. When the input current begins to decrease, the inverter output rises and leaves the current mirror gate node floating and thus the current mirror holds the peak current. The diode-connected transistor M_4 then begins to supply the difference of the input current i_{IN} and the drain current of transistor M_1 to the input node.

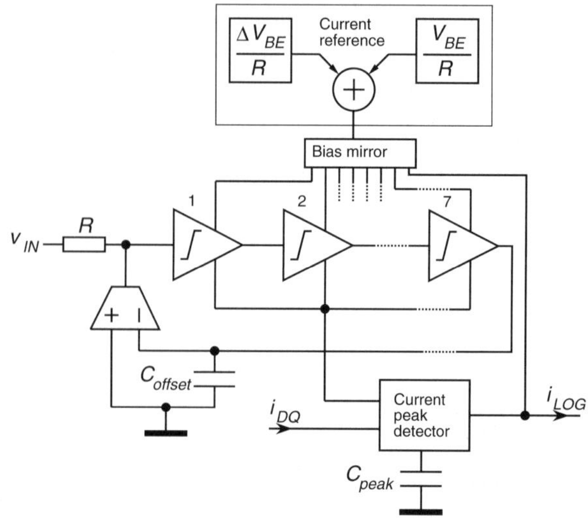

Figure 7.25 The block diagram of the designed pseudologarithmic amplifier.

At low current levels, the correct operation range is limited by the signal source impedance rather than the detector input impedance because the input impedance depends on the detected peak current and is thus very high at low current levels even without cascoding. A further limiting factor is the voltage gain of the CMOS inverter.

Because of the CMOS-inverter with a PMOS-diode feedback, this current peak detector can also operate with a bi-directional input current. In addition, the input voltage is held relatively constant. As a result, current mirror structures are not required because the gate-source capacitances of mirror transistors M_1 and M_2 should be moderately large in order to hold the current-level accurately, which results in a long channel and low channel length modulation for these transistors. At high current levels, the operation range is limited by the current sourcing capability of the diode-feedback inverter. The only drawback attached to this circuit is that discharging the gate-source capacitance of the current-mirror by a resistor similarly results in a signal dependent discharge time constant with this current peak detector.

7.2.5 Design case: A 2.5 V CMOS pseudologarithmic current amplifier

A pseudologarithmic amplifier with 60 dB of logarithmic range and a 2.5 V single supply is implemented with a 1.2 μm CMOS-process [18]. For this logarithmic amplifier, limiting current amplifiers are chosen for accurate control of current gain and limiting level. The block diagram of this circuit is presented in Figure 7.25. The logarithmic behaviour is approximated by seven cascaded limiting current amplifier stages with a current gain of three. The main output of the limiting amplifier chain is used

Transistors	Aspect ratio / μm/μm		
	1st	$2^{nd} \ldots 6^{th}$	7th
MB1-2, MS1		60/3	
MB3		120/6	
MB4		40/3	
MI1	180/3	60/3	60/3
MI2-3	180/6	60/6	60/6
MI4-5	60/3	20/3	20/3
MO1	180/3	180/3	60/3
MO2-3	180/6	180/6	60/6
MO4-5	60/3	60/3	20/3
MS2-3		60/6	
MI6-9	60/3	-	-
MI10-11, MZ	40/3	-	-

Table 7.2 Transistor dimension of the limiting amplifiers in Figure 7.26.

exclusively for offset compensation purposes. The summed currents are processed further with a current peak-detector because the logarithmic amplifier is used for signal strength measuring applications. All limiting amplifiers are biased with a current reference with a low temperature dependency. A constant current is subtracted from the detected peak current so that a control signal to drive an AGC-loop to the desired signal level is obtained.

Limiting amplifier

The limiting current amplifier designed is presented in Figure 7.26a. It uses a single-ended topology and the output stage transistor aspect ratios (MO1-5) are three times the input stage transistor aspect ratios (MI1-5) for a current gain of three. The bias current I_{BIAS} supplied from the current reference via additional current-mirrors is set nominally to 10 μA leading to a current limiting level of ± 10 μA. Three additional transistors (MS1-3) provide two currents I_{SP} and I_{SM}. One current (I_{SP}) from each limiting amplifier are combined to form the logarithmic output. The other current output (I_{SM}) is unused and connected to the negative supply.

In order to achieve proper input voltage-to-current conversion with a single resistor, the input impedance of the first limiting amplifier stage is realised as a high-gain current amplifier in a closed-loop configuration as depicted in Figure7.26b. In this case, a 3 pF compensation capacitor C_C in needed to stabilise the feedback loop.

The final limiting amplifier stage is used only in the offset compensation loop and, since the open-loop gain is already sufficiently high, the gain of this amplifier stage is set to one rather than three. The transconductance amplifier feeding the offset compensation feedback signal to the limiting amplifier cascade is implemented with a simple

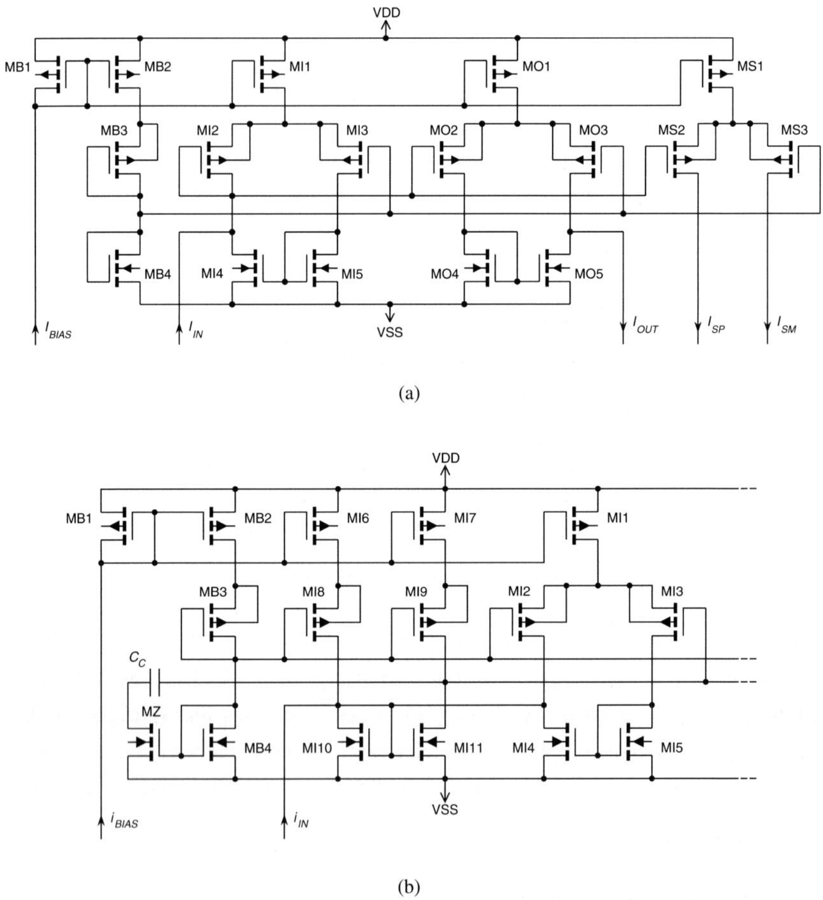

Figure 7.26 a) The designed CMOS current limiting amplifier. b) The circuit modifications in the first limiting amplifier stage in order to provide lower input impedance for input voltage-to-current conversion.

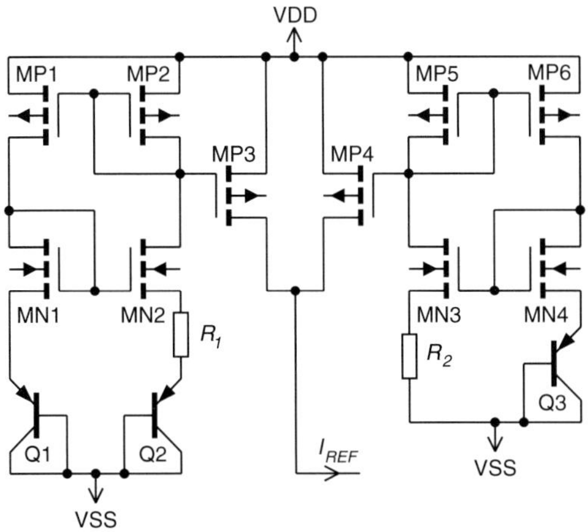

Figure 7.27 The temperature compensated current-reference.

Device	Measure	Value
MN1-4	Aspect ratio	$600\mu/6\mu$
MP1-2, MP5-6	Aspect ratio	$100\mu/6\mu$
MP3-4	Aspect ratio	$50\mu/6\mu$
Q1	Relative area	1
Q2	Relative area	10
Q3	Relative area	4
R1	Resistance	1.5 kΩ
R2	Resistance	15 kΩ

Table 7.3 Transistor dimensions and resistance values of the current-reference in Figure 7.27.

PMOS differential amplifier stage with identical device dimension to the limiting amplifier transistors MI1-5.

Current reference

Current references with low temperature dependencies are conventionally realised with a bandgap voltage reference and a V/I converter. With a low supply voltage converting accurately a 1.2 V reference voltage to current becomes increasingly difficult. Therefore, in the present approach the bandgap voltage is totally omitted by combining the output currents of separate PTAT and V_{BE} current references in order to realise a bandgap-like current reference as shown in Figure 7.27. The PMOS current-mirrors are implemented as cascode current-mirrors in order to minimise errors deriving from channel length modulation but they are omitted in the schematic for simplicity. For the same reason, the start-up and power-down circuitry is similarly omitted.

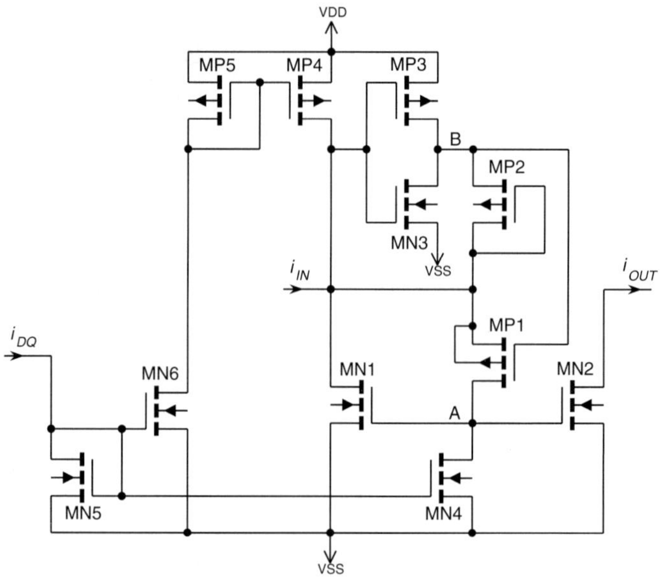

Figure 7.28 The current peak detector.

Transistors	Aspect ratio / μm/μm
MN1-2	500/10
MN3	20/1.2
MP1-3	60/1.2
MN5	300/5
MN4, MN6	30/20
MP4-5	30/6

Table 7.4 Transistor dimensions of the current peak detector in Figure 7.28.

The equation for the output current can be expressed as

$$I_{REF} = \frac{I_{PTAT} + I_{VBE}}{2} = \frac{kT}{2nqR_1} \ln \frac{A_{E2}}{A_{E1}} + \frac{V_{BE3}}{2R_2}. \tag{7.26}$$

In this case, the positive 3300 ppm/K temperature dependency of PTAT-reference current can be almost cancelled out by the negative approximately -2 mV/K temperature dependency of the base-emitter voltage of the transistor Q3 by adjusting the resistor ratio $\frac{R_2}{R_1}$ and the emitter area ratio $\frac{A_{E2}}{A_{E1}}$ [13]. Because transistor mismatches easily degrade the accuracy of the reference current, the emitter area ratio $\frac{A_{E2}}{A_{E1}}$ should be as large as possible. The pnp-type bipolar transistors used in this n-well CMOS-process are implemented by realising the emitter with p-diffusion and the base with n-well while the substrate acts as the collector. Typically, in CMOS-processes, two floating wells in different potential cannot be placed close to each other, which results in a very large die area for the bipolar transistor array.

In order to minimise the temperature dependencies in the logarithmic output, the resistors R_1 and R_2 are fabricated from identical resistive material (polysilicon) to the input voltage-to-current conversion resistor. Similarly, the output current-to-voltage conversion must be realised with a similar integrated resistor and therefore an additional 15 kΩ polysilicon resistor is added to the chip. This resistor can then be used as a feedback impedance for an off-chip current-to-voltage converter, resulting in almost temperature and process independent output voltage.

Current peak detector

The designed current peak detector in Figure 7.28 is based on the circuit principle of Figure 7.24b. In addition to the current memory (MN1-2), the current comparator (MN3, MP3), the PMOS-switch (MP1) and the PMOS-diode (MP2), an additional discharging time constant circuitry (MN4-6, MP4-5) is added.

The discharging time constant is realised by sinking a small amount of current from the current memory gates of transistors MN1 and MN2 by the current mirror MN4 and MN5. The same amount of current must be added to the peak detector input in order to prevent a peak detection error using transistors MN6, MP4 and MP5. The time constant can be further adjusted with an off-chip capacitor. However, the discharging time constant is signal dependent but in this application the detected peak varies from only 35 μA to 70 μA and thus the time constant signal dependency is small.

Experimental results

The microphotograph of the fabricated chip is presented in Figure 7.29, where the area of the chip is 3.2 mm^2. The integrated resistors and the bipolar transistor array consume almost one third of active circuit area. The circuit operates with as low as a 2.2 V single supply voltage and the power consumption is 3 mW with the nominal 2.5 V supply.

The measured logarithmic output voltage vs. input signal level in different temperatures is presented in Figure 7.30. The logarithmic output current is converted into voltage with an external operational amplifier and an on-chip polysilicon resistor. This resistor has the same 15 kΩ nominal resistance value as the input voltage-to-current conversion resistor. In the measurement set-up a sinusoidal 100 kHz signal, a discharge current of 1 μA, and a 150 pF external hold capacitor were used. Similarly, a 100 nF DC-decoupling capacitor was added at the input.

The logarithm accuracy is within ± 3 dB in a 60 dB dynamic range and within ± 1 dB in a 42 dB dynamic range. The offset compensation does not seem to work effectively enough to reduce errors at low signal levels. However, at these amplitude

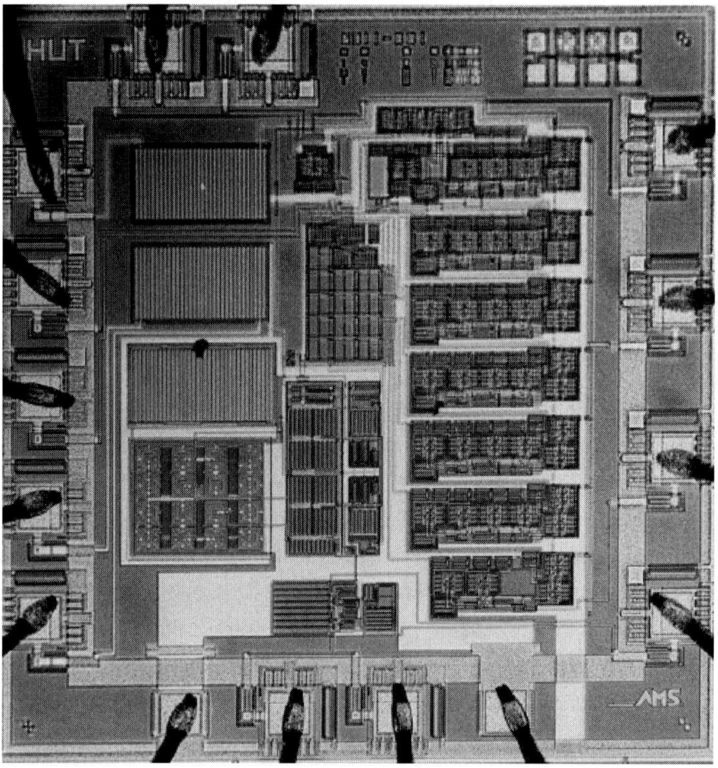

Figure 7.29 The microphotograph of the logarithmic amplifier.

levels many other effects such as crosstalk and limited power supply rejection begins to limit the operation. In this logarithmic amplifier, all limiting amplifiers are identical and thus the piece-wise linear approximation is not optimal for the full dynamic range. The error with a large input signal amplitude can be reduced by resizing the summing currents of the first two limiting amplifiers.

The temperature dependency of the detected signal amplitude is within ± 1 dB in a temperature range from -20 °C to +80 °C. As can be seen from the output voltage vs. the temperature plot in Figure 7.31, the operation of the current reference is not yet optimal. Temperature behaviour of the devices is not generally modelled accurately enough, so this is the accuracy expected in the first process run. By resizing MP3 and MP4 in the current reference, it would be possible to reduce the temperature dependency to 10-20 ppm/K rather than the -150 ppm/K in this test chip.

The temperature dependent error can be regarded as output offset and almost no gain error is present in this logarithmic amplifier. However, in the other published MOS pseudologarithmic amplifiers [11, 12], significant temperature dependent gain

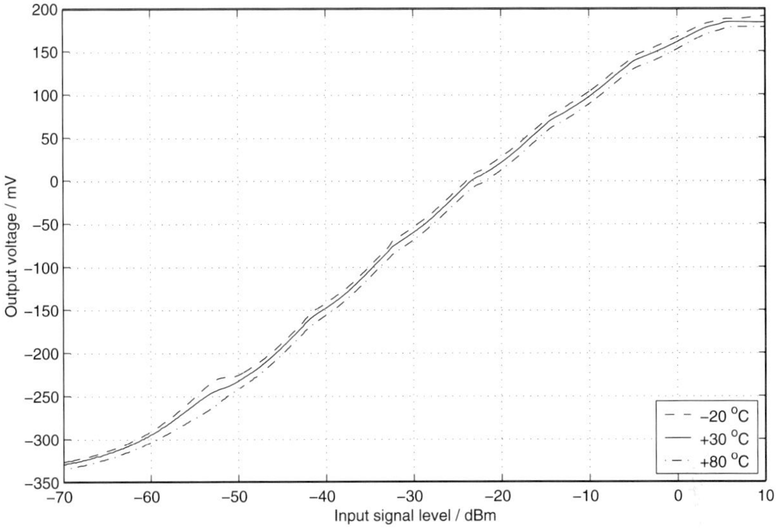

(a)

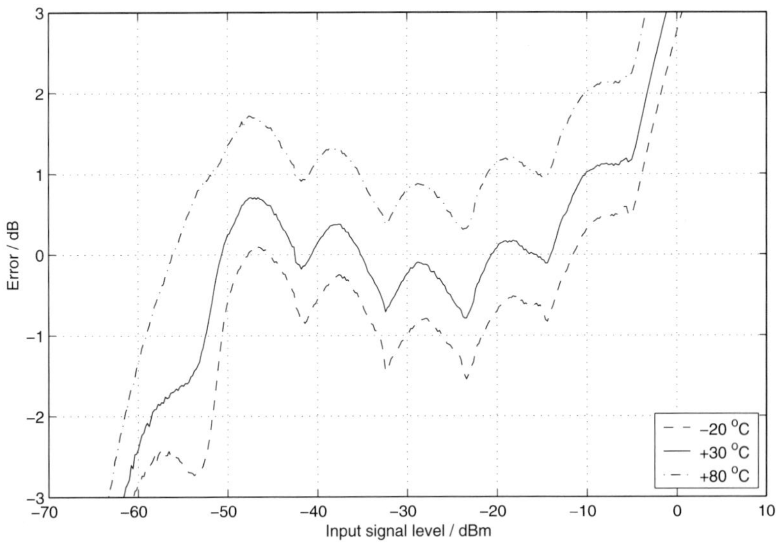

(b)

Figure 7.30 The measured logarithm accuracy in the temperature range of -20...+80 °C. a) The logarithmic output voltage. b) The input referred error in decibels.

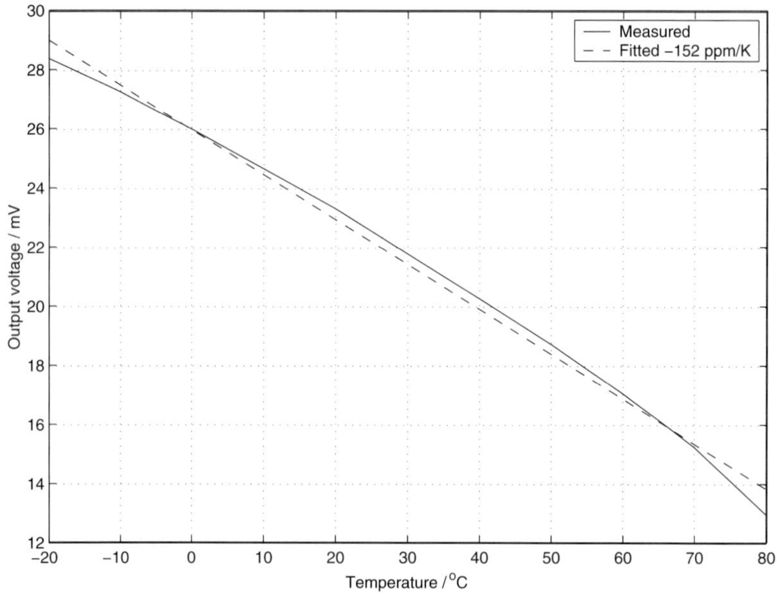

Figure 7.31 The output voltage vs. temperature with a 100 kHz -20 dBm input signal.

error is present caused by the temperature and process dependent liming voltage level. Furthermore, in [12] the measurements are obtained with an accurate external bias current rather than an on-chip reference bias circuit. Therefore, if the piece-wise linear approximation error is reduced by using more amplifier stages with a lower gain and rescaling the summing outputs, very accurate logarithmic amplifiers can be realised with this circuit topology.

If the supply voltage were further reduced, a differential limiting current amplifier topology of 7.18 could be used as in this case the NMOS current-mirrors in the input and output differential stages can be changed into NMOS current-sources. These NMOS current-mirrors also cause asymmetric settling behaviour when the signal amplitude exceeds the limiting level of the amplifier. This asymmetric operation causes errors in the offset compensation network, thus limiting accuracy at high frequencies. This error mechanism is not present in the differential limiting current amplifier and thus also more satisfactory high frequency operation is achieved.

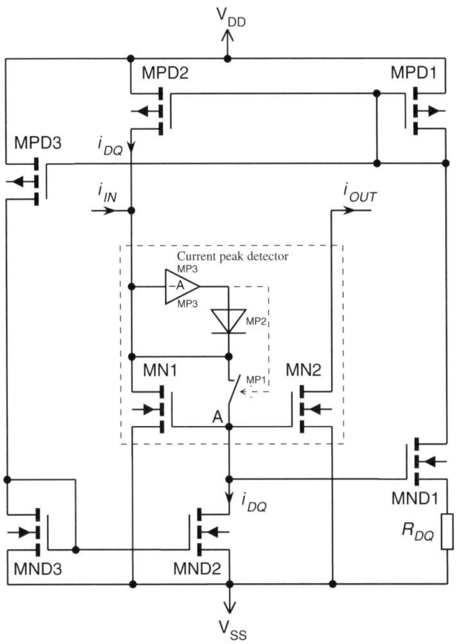

Figure 7.32 The peak detector discharge time constant adjustment principle.

Transistors	Aspect ratio / μm/μm
MN1-2	500/10
MP3	20/0.5
MP1	60/0.5
MP2-3	120/0.5
MND1	250/5
MND2-3	10/10
MPD1	20/10
MPD2-3	4/10

Table 7.5 The device dimension of the current peak detector in Figure 7.32.

7.3 Other approaches

Current peak detector with enhanced discharging time constant adjustment

A discharging time constant can be added in the traditional voltage-mode peak detector by adding a resistor R parallel to the peak voltage holding capacitor C. In this case, the output voltage can be expressed as a function of time as

$$v_{OUT} = \hat{v} e^{-\frac{t}{RC}}. \tag{7.27}$$

This means that the output voltage decreases with a rate dependent on the initial signal peak in addition to the time constant itself

$$-\frac{dv_{OUT}}{dt} = \frac{\hat{v}}{RC} e^{-\frac{t}{RC}}. \tag{7.28}$$

In the current peak detector a discharging resistor becomes too large to be integrated. Furthermore, because of the square-law voltage-to-current conversion characteristics of the MOS-transistor, the discharging operation is significantly different. In the current peak detector used in the designed pseudologarithmic amplifier (Figure 7.28 and Table 7.4), the current memory gate charge is discharged with a constant current I_{DQ} resulting in an output current equation

$$i_{OUT} = \frac{\beta}{2}\left(\sqrt{\frac{2\hat{i}}{\beta}} - \frac{I_{DQ}}{C}t\right)^2. \tag{7.29}$$

Then, the output current decreases with a rate

$$-\frac{di_{OUT}}{dt} = \frac{\beta I_{DQ}}{C}\left(\sqrt{\frac{2\hat{i}}{\beta}} - \frac{I_{DQ}}{C}t\right) \approx \frac{I_{DQ}}{C}\sqrt{2\beta\hat{i}}. \tag{7.30}$$

Thus, the output current decreases at an almost fixed rate. Therefore, at high input signal levels, the detected peaks are held for a very long time while at low signal levels, these detected peaks are discharged very rapidly. In the designed pseudologarithmic amplifier, the variation of current amplitude was relatively low so this nonideality did not cause any problems. However, if the dynamic range of the detected signal is large, this discharging behaviour is not acceptable.

This problem can be avoided by the enhanced current peak detector in Figure 7.32 [19]. In this case, the transistor MND1 and the resistor R_{DQ} convert the voltage at the peak current memory node A into a signal dependent discharge current. This current is further attenuated in the PMOS current-mirror (MPD1-3). The discharge current I_{DQ} is supplied both into the input and the current memory node in order to avoid offset current in the detected peak current.

The discharge time constant can be controlled by the resistor R_{DQ}, by the mirroring ratio of the PMOS current memory MPD1-3 or by adding external capacitance to the current memory node A. In order to maximise the speed without sacrificing adjustability internal capacitance and an external resistance are preferred.

Simulation results The current peak detector is designed and simulated with a modern 0.5 μm CMOS-process, using BSIM 3V3.1 simulation models for the transistors. An additional 1 pF capacitor is added to the input to simulate the effect of parasitic capacitances. The resistance R_{DQ} controlling the discharge time constant is set to 100 kΩ while other device dimensions are listed in Table 7.5.

The simulated waveforms of the peak detector with the enhanced discharge time constant circuitry are presented in Figure 7.33. The frequency of the sinusoidal input frequency is 500 kHz and the input peak current varies from 1 μA to 1 mA. Furthermore, the supply voltage is set to 3 V. The simulations show that the discharge time constant is almost constant as long as the detector itself operates correctly. Similarly, as the discharging ramp is almost linear in a logarithmic scale, the discharging behaviour is similar to the exponential discharging behaviour of the conventional voltage-mode peak detector.

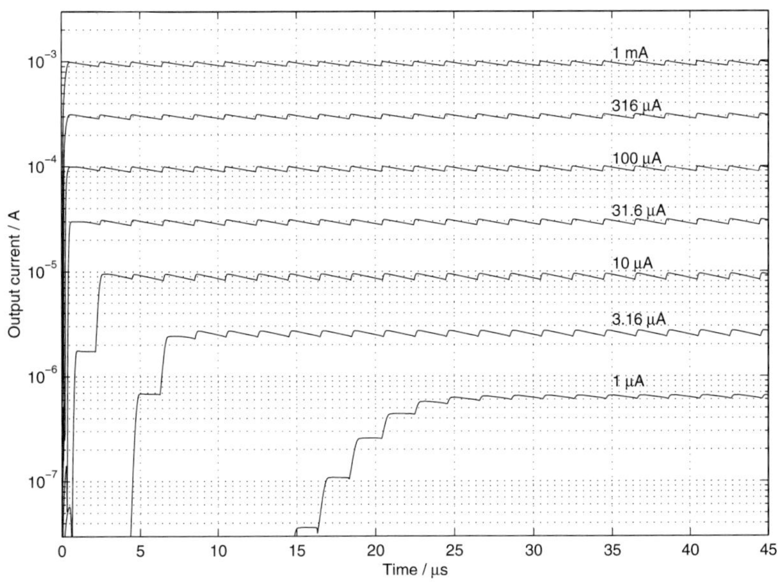

Figure 7.33 The simulated peak detector output current with the input current amplitude varied from 1 μA to 1 mA.

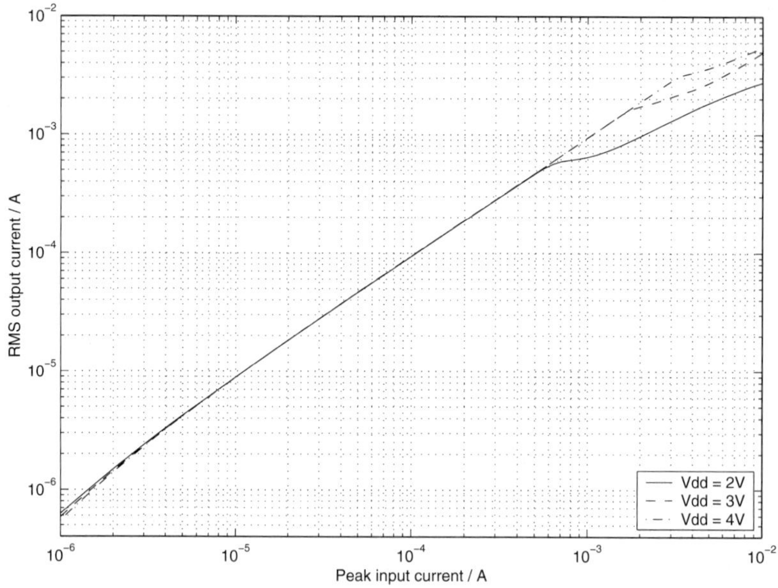

(a)

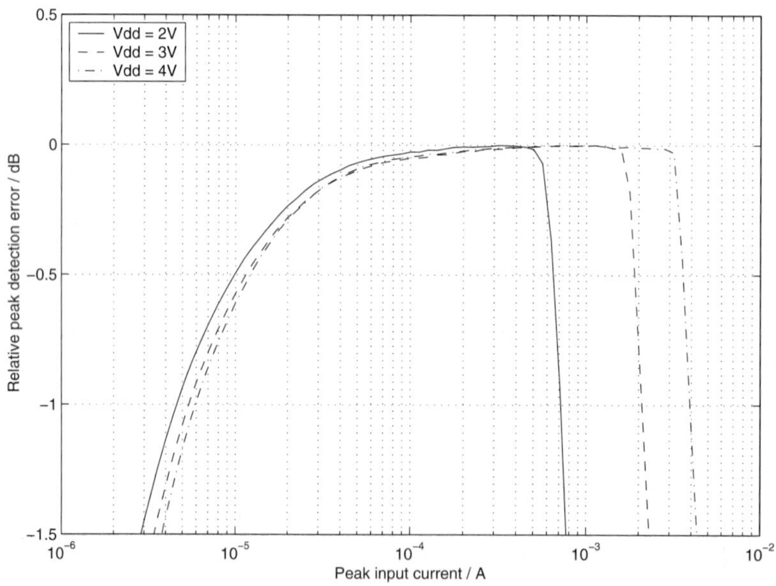

(b)

Figure 7.34 a) The simulated RMS output current vs. input signal amplitude. b) The simulated peak detection error vs. supply voltage.

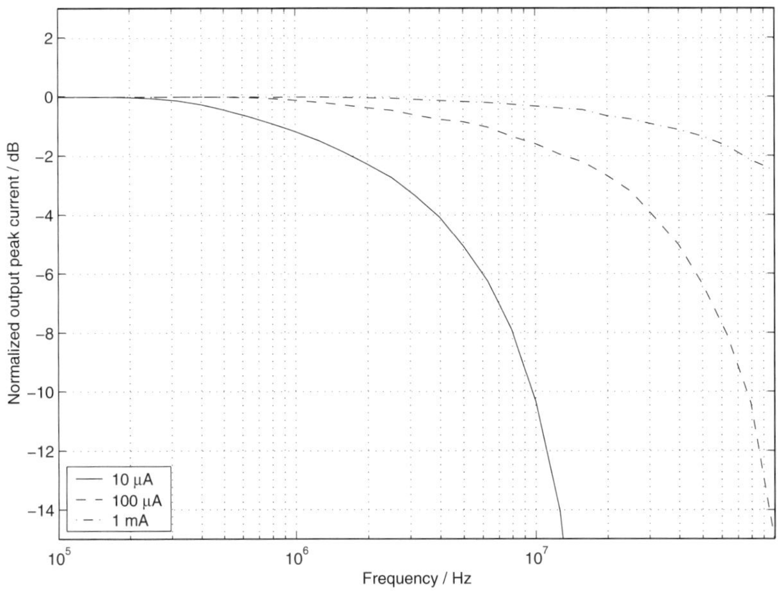

Figure 7.35 The frequency response of the peak detector with 10 μA, 100 μA and 1 mA input amplitudes.

A more efficient measure for the peak detection accuracy is the RMS value of the detector output current in Figure 7.34a. At high negative current peaks, the inverter with the diode-feedback (MP2-3 and MN3) fails to source enough current and in worst case situation the inverter output voltage drops to a level at which the PMOS-switch MP1 begins to conduct again and thus the current memory transistors MN1 and MN2 fail to hold the detected peak value.

The simulated RMS output current is further used to show the peak detection error in Figure 7.34b. The approximately 0.4 dB systematic error deriving from the output voltage ripple is subtracted from this error plot. Peak detection accuracy better than 1 dB is possible in a dynamic range of 44 dB with a 2 V supply voltage. With a 3 V supply voltage, the dynamic range increases to 51 dB and with larger supply voltages, even wider dynamic range is reached.

The frequency response of the peak detector is simulated by varying the input signal frequency in transient analysis with three different input signal amplitudes (10 μA, 100 μA, and 1 mA). The peak output current values are displayed in Figure 7.35 and the results are normalised with the input current amplitude and expressed in decibels. The simulations show that the -3 dB corner frequency is approximately 3 MHz with a 10 μA signal amplitude and this corner frequency exceeds 100 MHz at a 1 mA signal amplitude. The detection corner frequency depends strongly on the hold capacitance and thus by using smaller gate areas for the current memory transistors, faster operation is achieved. However, in this case, the tuning range of the discharge current is limited by transistor leakage currents and additionally errors caused by the switching transients will limit the detection accuracy.

Applications The dynamic range is limited by the current driving capabilities of the inverter when bi-directional input currents are used. Therefore, the dynamic range depends strongly on the supply voltage. However, with a unipolar input current, this peak detector operates well even with larger currents. In any event, this peak detector exhibits relatively large dynamic range and wide bandwidth with very simple circuitry. Therefore, in logarithmic signal strength measuring applications this peak detector could be placed before the logarithmic amplifier. In this case, the logarithmic amplifier must process only slowly varying unipolar currents resulting in much simpler logarithmic amplifier realisations such as the diode-feedback logarithmic amplifier of Figure 7.1 or the current-mirror based pseudologarithmic amplifier of Figure 7.17.

References

[1] B. Gilbert, "Translinear circuits: a proposed classification," *Electronics Letters*, vol. 11, pp. 14-16, 1975.

[2] B. Gilbert, "Current-mode circuits from a translinear viewpoint: a tutorial," In: C. Toumazou, F.J. Lidgey and D.G. Haigh (ed), *Analogue IC design: the current-mode approach*, Peter Peregnirus, London, 1990, pp. 11-91 (Chapter 2).

[3] B. Gilbert, "Translinear circuits: an historical review," *Analog Integrated Circuits and Signal Processing*, vol. 9, pp. 95-118, March 1996.

[4] E. Seevink, R. Wiegerink, "Generalized Translinear Circuit Principle," *IEEE J. Solid State Circuits*, vol. 26, pp. 1098-1102, 1991.

[5] K. Koli, K. Halonen, "BiCMOS current-feedback operational amplifier with 60 dB constant bandwidth range," *International Symposium on Circuits and Systems*, London, May 1994, pp. 525-528.

[6] K. Koli, K. Halonen, "Temperature compensated logarithmic peak detector", *Analog Integrated Circuits and Signal Processing*, vol. 9, pp. 131-140, March 1996.

[7] B. Gilbert, "Bipolar current mirrors," In: C. Toumazou, F.J. Lidgey and D.G. Haigh (ed), *Analogue IC design: the current-mode approach*, Peter Peregnirus, London, 1990, pp. 239-296 (Chapter 6).

[8] J. Greer, "Error analysis for pseudologarithmic amplification", *Measure and Science Technol.*, vol. 3, pp. 939-942, 1992.

[9] W. Barber, E. Brown, "A true logarithmic amplifier for radar IF applications," *IEEE J. Solid State Circuits*, vol. SC-15, June 1980, pp. 291-295.

[10] B. Gilbert and B. Clarke, "Monolithic DC-120 MHz log amp is stable and accurate (AD640)," *Analog Devices*, vol.23, no. 3, pp 3-6, 1989.

[11] K. Kimura, "A CMOS logarithmic IF amplifier with unbalanced source-coupled pairs," *IEEE Journal of Solid-State Circuits*, vol. SC-28. pp. 78-83, 1993

[12] S. Khorram, A. Rofougaran, A. Abidi, "A CMOS limiting amplifier and signal-strength indicator," *Symposium on VLSI Circuits*, 1995, pp. 95-96.

[13] P. Gray, R. Meyer, *Analysis and Design of Analog Integrated Circuits*, second edition, New York, John Wiley & Sons, 1984, 771 p.

[14] J. Ramírez-Angulo, E. Sánchez-Sinencio, A. Rodríguez-Vázquez, "A piecewise-linear function approximation using current mode circuits," Proc. IEEE Int. Symposium on Circuits and Systems (ISCAS-92), San Diego, May 1992, pp. 2025-2028.

[15] K. Bult, H. Wallinga, "A class of analog CMOS circuits based on the square-law characteristic of an MOS transistor in saturation," *IEEE J. Solid State Circuits*, vol. SC-22, pp. 357-365, June 1987.

[16] B. Linares-Barranco, A. Rodríguez-Vázquez, E. Sánchez-Sinencio, J. Ramírez-Angulo, "Generation, design and tuning of OTA-C high-frequency sinusoidal oscillators", IEE Proceedings-G, vol. 139, no. 5, pp.557-568, Oct. 1992.

[17] T. Fiez, R. Croman, E. Schneider, M. Goldenberg, "Current-mode signal processing," in M. Ismail, T. Fiez (ed.), *Analog VLSI Signal and Information Processing*, MacGraw Hill , 1994, pp. 248-311 (Chapter 6).

[18] Koli, K., Halonen, K., "A 2.5 V temperature compensated CMOS logarithmic amplifier," *Proc. IEEE Custom Integrated Circuits Conference* (CICC'97), Santa Clara, May 1997, pp.79-82.

[19] Koli, K., Halonen, K., "Low voltage MOS-transistor-only precision current peak detector with signal independent discharge time constant," *Proc. IEEE International Symposium on Circuits and Systems* (ISCAS'97), Hong Kong, June 1997, pp. 1992-1995.

Conclusions

The push-pull second-generation current-conveyor realised with a complementary bipolar integration technology is probably the most appropriate choice as a building block for low-distortion current-mode signal processing applications when supply voltages are relatively large (± 5 V or larger). Furthermore, this amplifier is already commercially available. There are current-feedback operational amplifiers such as AD844 and OPA 660 in the market that can additionally be used as a current-conveyor.

In modern low-voltage CMOS-processes push-pull second-generation current-conveyors have numerous shortcomings. First, it has a very limited input voltage range because of the bulk effect. Furthermore, the linearity of this amplifier depends considerably on the matching between PMOS- and NMOS-transistors. Finally, the X-terminal impedance of CMOS second-generation current-conveyors is usually too high for high-frequency low-distortion applications.

Although most low-gain CMOS current amplifiers operating in open-loop attain only moderate performance, there are high-gain CMOS current amplifier topologies that work well even with low supply voltages. Both current-mode operational amplifiers and high-gain current-conveyors provide a large open-loop current gain with a wide bandwidth with a relatively simple circuitry. These amplifiers exhibit distortion performance comparable to voltage-mode CMOS operational amplifiers. When low load and feedback impedance levels are required, these current amplifiers have better linearity than voltage-mode operational amplifiers since the linearity of current-mode operational amplifiers and high-gain current-conveyors is virtually independent of impedance level. Additionally, these current amplifiers can reach a higher full power bandwidth than voltage-mode operational amplifiers and while the CMOS-processes are further scaled down, this difference becomes even greater.

The current-mode operational amplifier and the high-gain current-conveyor are almost identical devices. They both have a second-generation current-conveyor as an input stage although in the current-mode operational amplifier, the input stage is a positive conveyor whereas in the high-gain conveyor, this input stage is a negative conveyor. Similarly, the output stage of a high-gain conveyor could also be a differential amplifier stage, in which case only the output ports must be interchanged. There are in addition many applications in which the noninverting output is grounded so that half of the output stage current is unused and thus a high-gain current-conveyor could be used in preference.

Voltage-mode operational amplifier based circuits can be converted by the adjoint principle to current-mode circuits using current-mode operational amplifiers as active elements. According to the adjoint principle, the single-ended voltage output of the voltage-mode operational amplifier is converted to a single-ended current input in the case of the current-mode operational amplifier. However, in many applications, a low impedance input without a well defined voltage level is difficult to use. Therefore, the hidden input conveyor Y-terminal should be available as an additional input terminal, rendering the input of the current-mode operational amplifier identical to the input of the high-gain current-conveyor.

Based on these facts, it would be convenient to include the current-mode operational amplifier in the category of high-gain current-conveyors. In this case, only the output stage topology can be selected according to the requirements of the application, which, however, is often also the case in the design of voltage-mode operational amplifier based circuits.

The design examples of current-mode filters implemented with a differential high-gain conveyor with a linearized output stage in Chapter 6.9 show promising performance with low supply voltages. Because the used output stage topology is easily scalable, programmable current-mode signal processing applications can also be realised with this differential high-gain conveyor. Similarly, this conveyor can be used as a highly linear amplifier in a closed-loop configuration.

Limiting current amplifiers is a simple and relatively process variation independent solution to piece-wise approximation of non-linear functions in CMOS-technology. Similarly, many other non-linear signal processing functions such as peak rectifiers can provide relatively wide bandwidth and dynamic range.

Although current-mode circuit techniques represent an efficient way to realise CMOS circuits there are however certain applications in which voltage-mode techniques are a more appropriate option. For example voltage followers are best realised with voltage-mode operational amplifiers with rail-to-rail input and output voltage swing. Similarly, in applications where low input offset currents are required voltage amplifiers perform more efficiently than current amplifiers. When designing integrated circuits, there are no limits in choosing the design techniques. It is hoped that this thesis may have provided sufficient information to help the circuit designer choose the most appropriate circuit technique for the application in hand. The correct current amplifier in the correct application may provide far better performance than the conventional "low risk" approach.

Appendix A

Basic distortion definitions

A.1 Harmonic distortion

The nonlinearity of weakly non-linear circuits can be modelled by means of a power series, in which case, the nonlinearity of a voltage amplifier can be modelled as

$$v_{OUT} = a_0 + a_1 v_{in} + a_2 v_{in}^2 + a_3 v_{in}^3 + a_4 v_{in}^4 + \ldots \tag{A.1}$$

and similarly, the nonlinearity of a current amplifier can be modeled as

$$i_{OUT} = a_0 + a_1 i_{in} + a_2 i_{in}^2 + a_3 i_{in}^3 + a_4 i_{in}^4 + \ldots \tag{A.2}$$

If one assumes that, in the case of the current amplifier, the input signal is

$$i_{in} = \hat{\imath} \sin \omega t, \tag{A.3}$$

the harmonic components can be collected with simple trigonometric equations as

$$\begin{aligned} i_{OUT} \approx \; & a_0 + \frac{1}{2} a_2 \hat{\imath}^2 + \left(a_1 \hat{\imath} + \frac{3}{4} a_3 \hat{\imath}^3 \right) \sin \omega t \\ & - \frac{1}{2} a_2 \hat{\imath}^2 \cos 2\omega t - \frac{1}{4} a_3 \hat{\imath}^3 \sin 3\omega t, \end{aligned} \tag{A.4}$$

where polynomial terms higher than third order are neglected as they seldom are significant anyway in weakly non-linear circuits. In this case, the second order harmonic distortion can be expressed as a ratio of spectral components at frequencies 2ω and ω as

$$HD2 \approx \frac{\hat{\imath}}{2} \frac{a_2}{a_1}, \tag{A.5}$$

if one assumes $a_1\hat{i} \gg \frac{3}{4}a_3\hat{i}^3$. Respectively, third order harmonic distortion is expressed as a ratio of spectral components at frequencies 3ω and ω, relying on the same assumptions as

$$HD3 \approx \frac{\hat{i}^2}{4}\frac{a_3}{a_1}. \tag{A.6}$$

A.2 Intermodulation distortion

Let us assume an input signal contains two frequencies ω_1 and ω_2 with the same amplitude $\hat{i}$:

$$i_{in} = \hat{i}\left(\sin\omega_1 t + \sin\omega_2 t\right). \tag{A.7}$$

By using similar trigonometric manipulations, one can collect the terms at frequencies $\omega_1 \pm \omega_2$ and divide them either with the term at ω_1 or ω_2 to form the second order intermodulation distortion as

$$HD2 \approx \frac{a_2}{a_1}\hat{i}. \tag{A.8}$$

Similarly, collecting the term at frequencies $2\omega_1 \pm \omega_2$ and $\omega_1 \pm 2\omega_2$, results in a third order intermodulation distortion as

$$HD2 \approx \frac{3}{4}\frac{a_3}{a_1}\hat{i}^2. \tag{A.9}$$

When the harmonic and intermodulation distortion equations are compared, it is clear that there is a simple ratio between them

$$IM2 = 2HD2, \tag{A.10}$$

$$IM3 = 3HD3. \tag{A.11}$$

Therefore, in weakly non-linear circuits there is a one-to-one correspondence between harmonic and intermodulation distortion, only one which needs to be calculated or measured. In many cases, it is easier to measure intermodulation distortion because the distortion of the signal source and the device under test can be shifted to different frequencies. Moreover, harmonic distortion measurements may be impossible in narrow-band applications such as bandpass filters.

A.3 Distortion in feedback amplifiers

A.3.1 Distortion in quasi-static feedback amplifiers

The block diagram of a voltage amplifier with feedback is presented in Figure A.1. The main amplifier is assumed to have a polynomial nonlinearity

$$v_o = a_1 v_i + a_2 v_i^2 + a_3 v_i^3 + ... \tag{A.12}$$

in open-loop configuration if the amplifier is assumed quasi-static so that there are no time constants affecting the circuit behaviour in the frequency range of interest. This assumption is valid considerably below the dominant pole of the amplifier. A part (or all) of the output signal $v_f = fv_o$ is subtracted from the signal v_s coming from the source so that the signal v_i at the amplifier input is

$$v_i = v_s - fv_o, \tag{A.13}$$

which leads to a recursive equation for the output signal

$$v_o = a_1 (v_s - fv_o) + a_2 (v_s - fv_o)^2 + a_3 (v_s - fv_o)^3 + ... \tag{A.14}$$

Once more, the feedback is assumed frequency independent.

An alternative power series expression for the closed-loop output voltage can be written as

$$v_o = a'_1 v_s + a'_2 v_s^2 + a'_3 v_s^3 + ... \tag{A.15}$$

where the coefficients are obtained by the Taylor series approximation by

$$a'_n = \left. \frac{1}{n!} \frac{\partial^n v_o}{\partial v_s^n} \right|_{v_s=0}. \tag{A.16}$$

Because $v_o = f(v_s)$ the coefficients a'_n can be obtained from Equation (A.14) by intrinsic derivation, relations between open-loop a_n and closed-loop a'_n nonlinearity coefficients can be derived as

$$a'_1 = \frac{a_1}{1 + fa_1}, \tag{A.17}$$

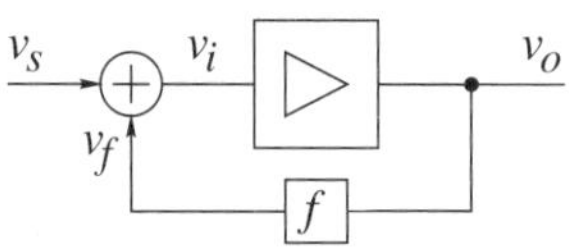

Figure A.1 The block diagram of an amplifier with feedback.

$$a_2' = \frac{a_2}{(1+fa_1)^3}, \tag{A.18}$$

$$a_3' = \frac{a_3(1+fa_1)-2fa_2^2}{(1+fa_1)^5}. \tag{A.19}$$

The closed-loop harmonic distortion can be expressed by using these coefficients in conjunction with Equations (A.5) and (A.6). Thus, the second-order harmonic distortion is

$$HD2 = \frac{\hat{v}_s}{2}\frac{a_2}{a_1(1+fa_1)^2}, \tag{A.20}$$

and the third-order harmonic distortion is

$$HD3 = \frac{\hat{v}_s^2}{4}\frac{\left|a_3(1+fa_1)-2fa_2^2\right|}{a_1(1+fa_1)^4}. \tag{A.21}$$

The equations show that, if the loop-gain $T = fa_1$ is sufficiently large, the feedback effectively reduces distortion. Moreover, there is third-order distortion present in the closed-loop amplifier, even if the open-loop equation has only second order distortion because a part of the distorted signal is fed back to the amplifier input, creating an infinite number of harmonics, even with an ideal square law nonlinearity. It is also possible that, with certain nonlinearity and feedback coefficient values, cancellation of third order distortion may occur. However, this minimum in distortion is a very narrow region and process variation renders it impossible to use this feature in distortion reduction.

A.3.2 Distortion in dynamic feedback amplifiers

In high gain feedback amplifiers, the open-loop gain decreases with frequency, starting from the dominant pole of the transfer function, which may be several decades below the closed-loop corner frequency. Problematically, the derived harmonic distortion equations assume that all polynomial coefficients are independent of frequency and therefore Equations (A.20) and (A.21) cannot accurately predict the distortion at frequencies above the dominant pole of the amplifier open-loop gain.

In high gain amplifiers, the signal amplitude constantly increases in the signal path from the input to output. The distortion is strongly dependent on the signal amplitude

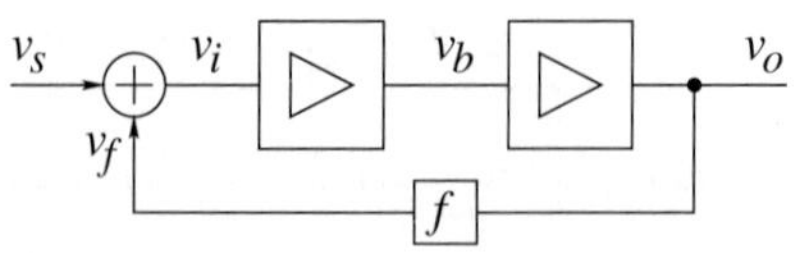

Figure A.2 The block diagram of a two stage amplifier with feedback.

and therefore, in most cases, it is safe to assume that the output stage of the amplifier dominates the nonlinearity of the amplifier. In this case, the amplifier can be divided to produce a cascade of two amplifiers whereby a linear frequency dependent amplifier drives a non-linear frequency independent amplifier, as depicted in Figure A.2.

The non-linear output amplifier is assumed to have a polynomial nonlinearity

$$v_o(t) = b_1 v_b + b_2 v_b^2 + b_3 v_b^3 + ... \tag{A.22}$$

The small-signal open loop gain of the whole amplifier between the dominant and the nondominant pole can be approximated as

$$A_{ol}(s) = \frac{v_o}{v_s} = \frac{\omega_0}{s}. \tag{A.23}$$

Then the output voltage of the linear input amplifier stage as a function of time is

$$v_b(t) = \frac{\omega_0}{b_1} \int v_s - f v_o dt. \tag{A.24}$$

If one assumes sinusoidal input voltage

$$v_s(t) = \hat{v}_s e^{j\omega t}, \tag{A.25}$$

it can be surmised that the voltage at the input of the output amplifier stage is

$$v_b(t) = V_1 e^{j\omega t} + V_2 e^{2j\omega t} + V_3 e^{3j\omega t}. \tag{A.26}$$

It is evident that higher order harmonic components are present in the feedback amplifier but they can be neglected if low distortion conditions are assumed. With this input voltage, the output voltage of the whole amplifier is

$$\begin{aligned} v_o(t) &= b_1 V_1 e^{j\omega t} + \left(b_1 V_2 + b_2 V_1^2\right) e^{2j\omega t} \\ &\quad + \left(b_1 V_3 + 2b_2 V_1 V_2 + b_3 V_1^3\right) e^{3j\omega t} + ... \end{aligned} \tag{A.27}$$

Similarly, v_i can be expressed as a function of v_b as

$$v_i(t) = \frac{b_1}{\omega_0} \frac{dv_b(t)}{dt} \tag{A.28}$$

$$= b_1 V_1 \frac{j\omega}{\omega_0} e^{j\omega t} + 2b_1 V_2 \frac{j\omega}{\omega_0} e^{2j\omega t} + 3b_1 V_3 \frac{j\omega}{\omega_0} e^{3j\omega t}. \tag{A.29}$$

Because $v_i = v_s - f v_o$, equations can be set up for the three frequencies

$$b_1 V_1 \frac{j\omega}{\omega_0} = \hat{v}_s - f b_1 V_1, \tag{A.30}$$

$$2b_1 V_2 \frac{j\omega}{\omega_0} = b_1 V_2 + b_2 V_1^2, \tag{A.31}$$

$$3b_1 V_3 \frac{j\omega}{\omega_0} = b_1 V_3 + 2b_2 V_1 V_2 + b_3 V_1^3, \tag{A.32}$$

and solve V_1, V_2 and V_3:

$$V_1 = \frac{\hat{v}_s}{b_1} \frac{\omega_0}{j\omega + f\omega_0}, \tag{A.33}$$

$$V_2 = -\frac{\hat{v}_s^2}{b_1^3} \frac{b_2 f \omega_0^3}{(j\omega + f\omega_0)^2 (2j\omega + f\omega_0)}, \tag{A.34}$$

$$V_3 = -\frac{f\omega_0^4 \hat{v}_s^3}{b_1^5} \frac{b_1 b_3 (2j\omega + f\omega_0) - 2fb_2^2 f\omega_0}{(j\omega + f\omega_0)^3 (2j\omega + f\omega_0)(3j\omega + f\omega_0)}. \tag{A.35}$$

As $v_b(t)$ is now solved, the coefficients V_1, V_2 and V_3 can be substituted to Equation (A.29) and the output voltage expressed as

$$v_o(t) = \frac{v_s - v_i}{f} \tag{A.36}$$

$$= \frac{\omega_0 \hat{v}_s}{j\omega + f\omega_0} e^{j\omega t} + \frac{2\hat{v}_s^2}{b_1^2} \frac{b_2 j\omega\omega_0^2}{(j\omega + f\omega_0)^2 (2j\omega + f\omega_0)} e^{2j\omega t}$$
$$+ \frac{3\omega_0^3 \hat{v}_s^3}{b_1^4} \frac{b_1 b_3 (2j\omega + f\omega_0) - 2fb_2^2 f\omega_0}{(j\omega + f\omega_0)^3 (2j\omega + f\omega_0)(3j\omega + f\omega_0)} e^{3j\omega t} \tag{A.37}$$

The magnitudes of these coefficients are needed to express the harmonic distortion of that amplifier. The second order harmonic distortion is therefore derived as before as

$$HD2 = \hat{v}_s \frac{b_2}{b_1^2} \frac{\omega\omega_0}{\sqrt{\omega^2 + f^2\omega_0^2}\sqrt{4\omega^2 + f^2\omega_0^2}}, \tag{A.38}$$

and respectively the third order harmonic distortion is

$$HD3 = \frac{3\hat{v}_s^2}{4} \frac{\omega\omega_0^2}{(\omega^2 + f^2\omega_0^2)\sqrt{9\omega^2 + f^2\omega_0^2}} \left| \frac{b_3}{b_1^3} - \frac{2b_2^2}{b_1^4} \frac{f\omega_0}{\sqrt{4\omega^2 + f^2\omega_0^2}} \right|. \tag{A.39}$$

Since the open-loop amplifier is assumed as an integrator, Equations (A.38) and (A.39) predict the harmonic distortion adequately only at a frequency range between the dominant and the nondominant pole of the amplifier open-loop transfer function. However, because of stability and settling time requirements, the nondominant pole of the amplifier is normally two or three times the unity gain frequency ω_0 and the harmonic distortion is predicted accurately enough at frequencies below ω_0. Similarly, at low frequencies Equations (A.20) and (A.21) can be used to predict the harmonic distortion of the amplifier.

Since the focus here is on the distortion within the closed-loop bandwidth of the amplifier, if $\omega < \frac{\omega_0}{2f}$ Equation (A.38) simplifies to

$$HD2 = \frac{\omega}{f\omega_0}\frac{b_2}{b_1^2}\hat{v}_s. \tag{A.40}$$

Letting $\hat{v}_s = f\hat{v}_o$ and $\hat{v}_o = b_1\hat{v}_b$, the distortion in respect to the open-loop distortion can be rewritten

$$HD2 = \frac{2\omega}{f\omega_0}HD2_b = \frac{HD2_b}{f\left|A_{ol}(2\omega)\right|}, \tag{A.41}$$

where

$$HD2_b = \frac{\hat{v}_b}{2}\frac{b_2}{b_1}, \tag{A.42}$$

is the open-loop second order distortion of the output amplifier. Therefore, the closed-loop second order distortion of the entire amplifier is equal to the open-loop second order distortion of the output amplifier divided by the loop gain at the frequency of the harmonic component. Similarly, the third order harmonic distortion can be rewritten as

$$HD3 = \frac{3}{4}\frac{\omega}{f^3\omega_0}\hat{v}_s^2\frac{\left|b_3b_1 - 2b_2^2\right|}{b_1^4} \tag{A.43}$$

$$= \frac{\left|HD3_b - 2HD2_b^2\right|}{f\left|A_{ol}(3\omega)\right|}, \tag{A.44}$$

if $\omega < \frac{\omega_0}{3f}$. Once more, in this case the distortion is attenuated by the loop gain at 3ω. However, the closed-loop third order distortion depends both on the second and third order distortion of the output amplifier.

References

[1] D. Pederson, K. Mayaram, *Analog Integrated Circuits for Communication (Principles, Simulation and Design),* Norwell, MA, Kluwer Academic Publishers, 1991, 568 p.

[2] W. Sansen, "Distortion in Elementary Transistor Circuits," *IEEE Trans. Circuits and Systems-II*, vol. CAS-46, pp. 315-325, March 1999.

[3] F. Op't Eynde, P. Ampe, L. Verdeyen, W. Sansen, "A CMOS Large-Swing Low-Distortion Three-Stage Class AB Power Amplifier," *IEEE J. of Solid-State Circuits*, vol. SC-25, pp. 265-273, Feb. 1990.

Appendix B

Distortion in push-pull current amplifiers

B.1 Class-A operation

At signal current amplitudes significantly lower than the quiescent current (class-A operation), the input signal is divided approximately equally between the upper and lower half-circuit:

$$i'_{IN} \approx I_Q + \frac{i_{IN}}{2}, \tag{B.1}$$

$$i''_{IN} \approx I_Q - \frac{i_{IN}}{2}. \tag{B.2}$$

The nonlinearity of the two separate signal paths can be modelled with power series $f(x)$ and $g(x)$:

$$f(x) = a_0 + a_1 x + a_2 x^2 + a_3 x^3 + a_4 x^4 + \ldots, \tag{B.3}$$

$$g(x) = b_0 + b_1 x + b_2 x^2 + b_3 x^3 + b_4 x^4 + \ldots, \tag{B.4}$$

so that the harmonic distortion of one of the half-circuits can be approximated by letting $x = \hat{i} \sin \omega t$ and collecting the harmonic components as

$$HD2 \approx \frac{\hat{i}}{2} \left| \frac{a_2}{a_1} \right| \approx \frac{\hat{i}}{2} |a_2|, \tag{B.5}$$

$$HD3 \approx \frac{\hat{i}^2}{4} \left| \frac{a_3}{a_1} \right| \approx \frac{\hat{i}^2}{4} |a_3|. \tag{B.6}$$

In this case, the output current can be approximated as

$$i_{OUT} = f(i'_{IN}) - g(i''_{IN})$$

$$\approx a_0 - b_0 + (a_1 - b_1) I_Q + \frac{a_1 + b_1}{2} i_{IN} + \frac{a_2 - b_2}{4} i_{IN}^2 + \frac{a_3 + b_3}{8} i_{IN}^3 + \frac{a_4 - b_4}{16} i_{IN}^4 + \dots \quad \text{(B.7)}$$

By letting $i_{IN} = \hat{i} \sin \omega t$ and collecting the harmonic components the second harmonic distortion in a class-A push-pull current amplifier is

$$HD2 \approx \frac{\hat{i}}{4} \left| \frac{a_2 - b_2}{a_1 + b_1} \right| \approx \frac{\hat{i}}{8} \left| a_2 - b_2 \right|, \quad \text{(B.8)}$$

and the third harmonic is

$$HD3 \approx \frac{\hat{i}^2}{16} \left| \frac{a_3 + b_3}{a_1 + b_1} \right| \approx \frac{\hat{i}^2}{32} \left| a_3 + b_3 \right|. \quad \text{(B.9)}$$

As a consequence of the signal division at the input, the signal amplitudes in the two half-circuits are half those of the original signal, which attenuates the second order harmonic distortion by 6 dB and the third harmonic distortion by 12 dB when compared to the distortion of the half-circuits. If the nonlinearities of the half-circuits are correlated, the even order harmonic components are efficiently attenuated. Problematically, most of the distortion generation mechanisms in the half-circuits are relatively random processes and therefore exact cancellation is not possible. Furthermore, since the upper and lower half-circuits are fabricated with opposite type of transistors, the matching of the nonlinearities between the two half-circuits is difficult to achieve. Furthermore, as a result, the input current signal division is not exactly symmetrical, which increases at least the second order distortion.

The operation of the push-pull amplifier is similar to differential amplifier structures, which similarly reject even nonlinearities. However, in differential structures, the two half-circuits have closer matching since both amplifiers use the same type of transistors and have equal device sizes. The decision whether to use push-pull or differential structures depends on the requirements of the application:

- Push-pull structures need twice the supply voltage of differential structures while differential structures need twice the supply current of push-pull structures.
- The even nonlinearities are more accurately cancelled in differential circuits.
- The differential structures are usually limited to class-A operation while most push-pull structures can operate with much larger signal currents.

B.2 Class-AB operation

In a push-pull connected class-AB current amplifier operating with signal amplitudes significantly larger than the quiescent current the class-A region rapidly becomes a small zero crossing region. In the time domain, therefore, the class-AB amplifier is in the class-A region for a very small fraction of the signal cycle time. Moreover, class-AB amplifiers are typically very linear at the class-A region, as explained in Appendix B.1. Because the operation in the class-B region is usually strongly non-linear, the distortion of a class-AB amplifier becomes almost equal to the distortion of the same amplifier biased as a class-B amplifier as signal amplitudes increase. There is additionally cross-over distortion present in a class-B amplifier, but it is not included in these calculations as they include transients, which depend on the signal amplitude and frequency and circuit topology and are therefore very difficult to predict.

Let us assume a current amplifier that amplifies positive halves of the input signal by a current gain of a_1 and the negative halves of the input signal by a current gain of b_1. Additionally, second order nonlinearity is assumed and thus the output current is

$$i_{OUT}(t) = \begin{cases} a_1 i_{IN} + a_2 i_{IN}^2 & \text{if } i_{IN} > 0, \\ b_1 i_{IN} - b_2 i_{IN}^2 & \text{if } i_{IN} < 0, \end{cases} \tag{B.10}$$

where $i_{IN} = \hat{i}\sin\omega t$. The positive and negative amplification coefficients differ from each other only slightly so that $a_n \approx b_n$. In this case, the harmonic content of the output current can be calculated by deriving the complex coefficients of the Fourier series:

$$\begin{aligned} C_n &= \frac{2}{T}\int_0^T i_{OUT}(t)e^{-jn\omega t}\,dt \\ &= \frac{2}{T}\left[\int_0^{T/2} \left(a_1\hat{i}\sin\omega t + a_2\hat{i}^2\sin^2\omega t\right)e^{-jn\omega t}\,dt\right. \\ &\quad \left. + \int_{T/2}^{T} \left(a_1\hat{i}\sin\omega t - b_2\hat{i}^2\sin^2\omega t\right)e^{-jn\omega t}\,dt\right], \end{aligned} \tag{B.11}$$

where $T = \frac{2\pi}{\omega}$ represents the period of the fundamental input frequency and n is a positive integer.

After a process of integrating and making simplifications, the equation reduces to

$$\begin{aligned} C_n &= -a_1\hat{i}\frac{e^{-j\pi n}+1}{\pi(n^2-1)} + b_1\hat{i}\frac{e^{-2j\pi n}+e^{-j\pi n}}{\pi(n^2-1)} \\ &\quad -2ja_2\hat{i}^2\frac{e^{-j\pi n}-1}{\pi n(n^2-4)} - 2jb_2\hat{i}^2\frac{-e^{-2j\pi n}+e^{-j\pi n}}{\pi n(n^2-4)} \\ &= \hat{i}C_{n1} + j\hat{i}^2C_{n2}, \end{aligned} \tag{B.12}$$

where C_{n1} combines the first two sum term due to the gain coefficients a_1 and b_1, and C_{n2} combines the last two sum terms due to the second order nonlinearity coefficients a_2 and b_2. The term C_{n1} is zero when n is odd. Furthermore, C_{n1} is undefined when $n = 1$. By substituting $n = 1 + \delta$, where $\delta \ll 1$, and approximating $e^{aj\pi\delta} \approx 1 + aj\pi\delta$, C_{n1} can be rewritten near $n = 1$:

$$\begin{aligned} C_{n1} &= -a_1 \frac{e^{-j\pi(1+\delta)} + 1}{\pi\delta(\delta+2)} + b_1 \frac{e^{-2j\pi(1+\delta)} + e^{-j\pi(1+\delta)}}{\pi\delta(\delta+2)} \\ &\approx \frac{-a_1 j\pi\delta - b_1 j\pi\delta}{\pi\delta(\delta+2)} = -j\frac{a_1+b_1}{\delta+2} \underset{\delta\to 0}{\longrightarrow} -j\frac{a_1+b_1}{2}. \end{aligned} \tag{B.13}$$

Accordingly, the term C_{n2} is zero when n is odd and undefined when $n = 2$. When $n = 2 + \delta$ is substituted and the same series approximation as with C_{n1} is used, C_{n2} can be rewritten near $n = 2$:

$$\begin{aligned} C_{n2} &= -2a_2 \frac{e^{-j\pi(2+\delta)} - 1}{\pi\delta(\delta+2)(\delta+4)} - 2b_1 \frac{e^{-2j\pi(2+\delta)} + e^{-j\pi(2+\delta)}}{\pi\delta(\delta+2)(\delta+4)} \\ &\approx -2\frac{-a_2 j\pi\delta + b_2 j\pi\delta}{\pi\delta(\delta+2)(\delta+4)} = 2j\frac{a_2-b_2}{(\delta+2)(\delta+4)} \underset{\delta\to 0}{\longrightarrow} j\frac{a_2-b_2}{4}. \end{aligned} \tag{B.14}$$

Consequently, these results can be collected for all positive values of n:

$$C_n = \begin{cases} -\frac{j\hat{i}}{2}(a_1+b_1) - \frac{4j\hat{i}^2}{3\pi}(a_2+b_2) & \text{if } n = 1, \\ -\frac{2\hat{i}}{3\pi}(a_1-b_1) - \frac{\hat{i}^2}{4}(a_2-b_2) & \text{if } n = 2, \\ -\frac{4j\hat{i}^2}{\pi n(n^2-4)}(a_2+b_2) & \text{if } n \text{ is odd and } n \geq 3, \\ -\frac{2\hat{i}}{\pi(n^2-1)}(a_1-b_1) & \text{if } n \text{ is even and } n \geq 4. \end{cases} \tag{B.15}$$

These results reveal that the second order distortion depends on the matching of the gain and nonlinearity of the positive and negative signal paths. Furthermore, push-pull operation converts even nonlinearities of the two signal paths to odd harmonics in the output current. At low signal amplitudes, the gain mismatch dominates the distortion and therefore it is not necessary to calculate the effects of any higher order nonlinearities.

The second order harmonic distortion of the push-pull amplifier is

$$HD2 \approx \frac{4}{3\pi}\left|\frac{a_1-b_1}{a_1+b_1}\right|, \tag{B.16}$$

because the terms due to the distortion coefficients a_2 and b_2 become significant compared to the terms constructed from the current gain coefficients a_1 and b_1 solely with very large signal amplitudes. Moreover, third or even higher order distortion were assumed for the push-pull amplifiers, the contribution of these higher order distortion coefficients to the total distortion would have been insignificant as these coefficients would be initially small and decrease very rapidly with the signal amplitude.

According to the same assumptions as with the second order harmonic distortion, the third order harmonic distortion is

$$HD3 \approx \frac{8\hat{i}}{15\pi}\left|\frac{a_2+b_2}{a_1+b_1}\right|. \tag{B.17}$$

If this is compared to the initial nonlinearity of the push-pull connected amplifiers the third order harmonic distortion is

$$HD3_{push-pull} \approx \frac{16}{15\pi}HD2_a \approx \frac{16}{15\pi}HD2_b. \tag{B.18}$$

Thus, the push-pull connection converts second order distortion to third order distortion and attenuates it to approximately 9 dB. However, this is somewhat difficult to verify because the harmonic distortion of a half-circuit cannot be simulated or otherwise calculated at higher signal amplitudes than the quiescent current.

Appendix C

Distortion in CMOS operational amplifiers

C.1 Miller-compensated unbuffered operational amplifier

In Figure C.1, a typical two-stage operational amplifier realised by using an n-well CMOS-process is presented. It uses Miller-capacitance for frequency compensation and has no output voltage buffer, so it is referred to as an unbuffered operational amplifier. Because of the lack of output voltage buffer, the voltage gain of the output stage is relatively small and therefore a large voltage gain in the input differential stage (transistors M_1, M_2, M_3, M_4 and M_5) is required. For the same reason, the only significant source of nonlinearity in the amplifier is the output stage (transistors M_6 and M_5). In this case, the non-linear output current can be expressed as

$$i_{OUT} = v_{GS6}\sqrt{2\beta_6 I_{BO}} + \frac{1}{2}\beta_6 v_{GS6}^2. \tag{C.1}$$

This output current is then converted into a non-linear output voltage

$$v_{OUT} = R'_L i_{OUT}, \tag{C.2}$$

where R'_L is the effective output load impedance, including the output impedance of the amplifier and the loading effect of the feedback network in addition to the actual load impedance R_L.

At low frequencies, the output voltage can be expressed as a function of the input voltage as

$$v_{OUT} = A_{vd} R'_L v_{in}\sqrt{2\beta_6 I_{BO}} + \frac{1}{2}\beta_6 A_{vd}^2 R'_L v_{in}^2, \tag{C.3}$$

where A_{vd} is the voltage gain of the input differential stage. If $v_{in} = v_s - f v_{OUT}$, the closed loop harmonic distortion of the amplifier can be expressed by using equations

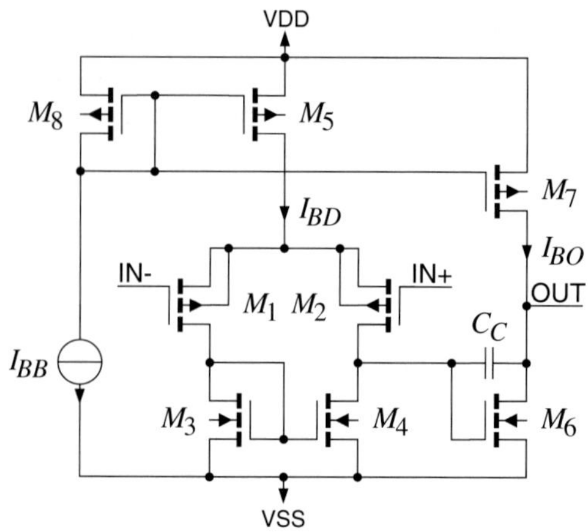

Figure C.1 An unbuffered Miller-compensated operational amplifier realised with a n-well CMOS-technology.

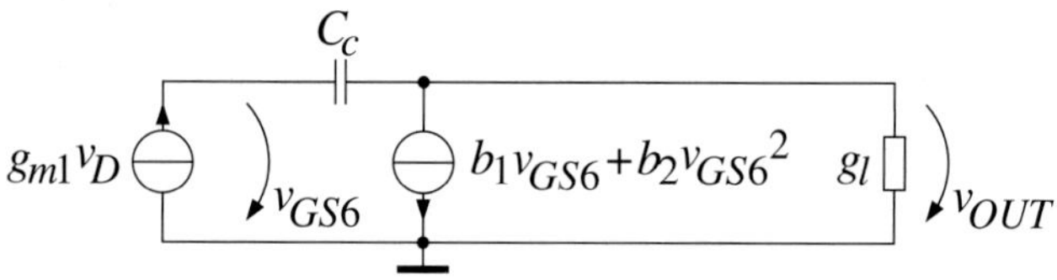

Figure C.2 Equivalent circuit for high frequency distortion calculations.

(A.20) and (A.21), derived in Appendix A

$$HD2(0) = \frac{\hat{v}_s}{8I_{BO}R'_L} \frac{A'_{ol}}{\left(1+fA'_{ol}\right)^2}, \tag{C.4}$$

where $A'_{ol} = A_{vd}R'_L\sqrt{2\beta_6 I_{BO}}$ is the loaded open-loop DC-gain of the operational amplifier. In this case, it is more convenient to express the harmonic distortion in terms of output current amplitude rather than the input voltage amplitude and thus letting $\hat{v}_s = fR'_L\hat{i}_{out}$ results in a second order harmonic distortion equation

$$HD2(0) = \frac{1}{8}\frac{\hat{i}_{out}}{I_{BO}} \frac{fA'_{ol}}{\left(1+fA'_{ol}\right)^2} \approx \frac{1}{8fA'_{ol}}\frac{\hat{i}_{out}}{I_{BO}}. \tag{C.5}$$

Similarly, the third order harmonic distortion can be expressed in terms of output current amplitude, by means of Equation (A.21), as

$$HD3(0) = \frac{1}{32}\frac{\hat{i}_{out}}{I_{BO}} \frac{f^3A'^3_{ol}}{\left(1+fA'_{ol}\right)^4} \approx \frac{1}{32fA'_{ol}}\left(\frac{\hat{i}_{out}}{I_{BO}}\right)^2. \tag{C.6}$$

Since the compensation capacitor acts as a feedback element for the output stage at high frequencies, Equations (A.38) and (A.39), derived in Appendix A cannot be used to depict the distortion performance of the amplifier. Therefore, the high frequency distortion based on the equivalent circuit of Figure C.2 is calculated. The effect of channel length modulation in this equivalent circuit is ignored because it is important exclusively at low frequencies and the low frequency distortion equations are already derived.

In this case, if one assumes that the voltages v_{GS6} and v_{GS6} contain two harmonics in addition to the fundamental frequency:

$$v_{GS6} = V_{g1}e^{j\omega t} + V_{g2}e^{2j\omega t} + V_{g3}e^{3\omega t}, \tag{C.7}$$

$$v_{OUT} = V_{o1}e^{j\omega t} + V_{o2}e^{2j\omega t} + V_{o3}e^{3\omega t}. \tag{C.8}$$

Thus, Kirchoff's law can be expressed in both nodes and the terms with the same frequency can be equated to solve coefficients $V_{g1}...V_{g3}$ and $V_{o1}...V_{o3}$. Additionally, the load conductance in the equivalent circuit is $g_l = \frac{1}{R'_L}$ and the global feedback is taken into account by letting

$$v_D = \hat{v}_s e^{j\omega t} + f v_{OUT}. \tag{C.9}$$

The output fundamental frequency component calculated is

$$V_{o1} = \frac{g_{m1} R'_L \hat{v}_s (b_1 - j\omega C_c)}{b_1 f g_{m1} R'_L + j\omega C_c (1 + b_1 R'_L - f g_{m1} R'_L)}. \tag{C.10}$$

Both $b_1 R'_L$ and $f g_{m1} R'_L$ are significantly greater than one and in most cases the input transconductance g_{m1} is significantly smaller than $b_1 = g_{m6}$. Therefore, one can approximate the equation as

$$V_{o1} \approx \hat{v}_s \frac{g_{m1}}{b_1} \frac{b_1 - j\omega C_c}{f g_{m1} + j\omega C_c} = \hat{v}_s \frac{\omega_0 - \frac{g_{m1}}{b_1} j\omega}{j\omega + f\omega_0}, \tag{C.11}$$

where $\omega_0 = \frac{g_{m1}}{C_c}$. Based on the same assumptions, the coefficients for the two harmonic frequencies are:

$$V_{o2} \approx -\frac{2\hat{v}_s^2 b_2 j\omega\omega_0^2 (1 + j\omega R'_L C_c)^2}{b_1^3 Z_L'^2 (j\omega + f\omega_0)^2 (2j\omega + f\omega_0)}, \tag{C.12}$$

$$V_{o3} \approx -\frac{6\hat{v}_s^3 b_2^2 j\omega\omega_0^3 (1 + j\omega R'_L C_c)^3}{b_1^5 R_L'^3 (j\omega + f\omega_0)^3 (3j\omega + f\omega_0)}. \tag{C.13}$$

Letting $b_1 = g_{m6} = \sqrt{2\beta_6 I_{BO}}$, $b_2 = \frac{1}{2}\beta_6$and $\hat{v}_s = f R'_L \hat{i}_{out}$ results in a second order harmonic distortion equation

$$HD2(\omega) = \frac{|V_{o2}|}{2|V_{o1}|} = \frac{1}{4}\frac{\hat{i}_{out}}{I_{BO}} \frac{f\omega\left(\omega_0^2 + \omega^2 g_{m1}^2 R_L'^2\right)}{R_L'\sqrt{\omega^2 + f^2\omega_0^2}\sqrt{4\omega^2 + f^2\omega_0^2}\sqrt{g_{m1}^2\omega^2 + g_{m6}^2\omega_0^2}}. \quad \text{(C.14)}$$

At frequencies below $\frac{f\omega_0}{2}$, this equation can be approximated by a simpler expression

$$HD2(\omega)\Big|_{\omega \ll \frac{f\omega_0}{2}} = \frac{1}{2}\frac{\hat{i}_{out}}{I_{BO}} \frac{1 + \omega^2 R_L'^2 C_c^2}{g_{m6} R_L'} \frac{1}{f\,|A_{ol}(2\omega)|}. \quad \text{(C.15)}$$

Therefore, at frequencies below $\omega_b = \frac{1}{R_L' C_c}$, the distortion is proportional to frequency but above ω_b the distortion is proportional to the third power of the frequency.

Similarly, the third order distortion is

$$HD3(\omega) = \frac{3}{32}\left(\frac{\hat{i}_{out}}{I_{BO}}\right)^2 \frac{\omega f^2\left(\omega_0^2 + \omega^2 g_{m1}^2 R_L'^2\right)^{\frac{3}{2}}}{R_L'\left(\omega^2 + f^2\omega_0^2\right)\sqrt{9\omega^2 + f^2\omega_0^2}\sqrt{g_{m1}^2\omega^2 + g_{m6}^2\omega_0^2}}. \quad \text{(C.16)}$$

Once more, at frequencies below $\frac{f\omega_0}{3}$, this equation can be approximated by a simpler expression

$$HD3(\omega)\Big|_{\omega \ll \frac{f\omega_0}{3}} = \frac{1}{32}\left(\frac{\hat{i}_{out}}{I_{BO}}\right)^2 \frac{\left(1 + \omega^2 R_L'^2 C_c^2\right)^{\frac{3}{2}}}{g_{m6} R_L'} \frac{1}{f\,|A_{ol}(3\omega)|}. \quad \text{(C.17)}$$

Since the distortion is dependent on the output current amplitude with very high impedance feedback and load, the distortion is very low. However, in order to maximise the power efficiency, the maximum current and voltage swing should be reached almost simultaneously. The local feedback in the output stage caused by the compensation capacitor is similarly efficient only in the case of high impedance loads.

At high frequencies, the gain at the output stage is low deriving from the compensation capacitor C_c, in which case the nonlinearity of the differential stage plays a larger role in the nonlinearity of the amplifier. Generally, in Miller-compensated operational amplifiers, a resistor is added in series with the compensation capacitor C_c to compensate the phase shift arising from the right half-plane zero. This resistor reduces the local feedback in the output stage at high frequencies. Similarly, the load can no longer be assumed to be resistive near ω_0. Moreover, capacitive load reduces the voltage gain of the output stage and thus the high frequency distortion performance is altered.

Taking these effect into account leads problematically to extremely complicated equations. In any event, several experimental and theoretical results show that the Miller-compensated operational amplifiers have a large distortion peak near the corner

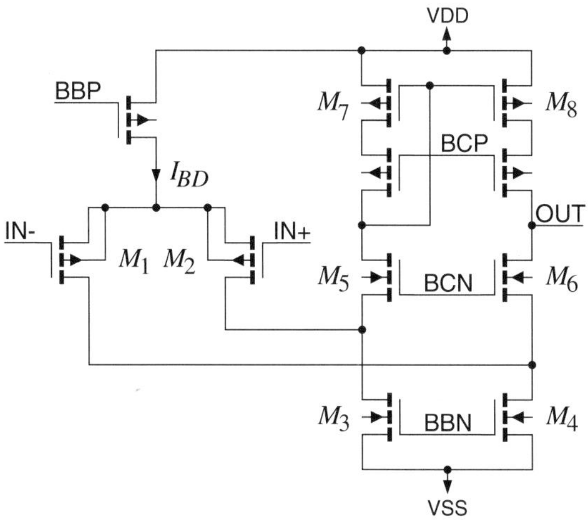

Figure C.3 A folded cascode operational transconductance amplifier realised with a n-well CMOS-process.

frequency [2,3], as the derived equations show. Moreover, below the corner frequency the output transistor M_6 remains the dominant source of distortion [3].

C.2 Folded cascode operational transconductance amplifier

A typical folded cascode operational transconductance amplifier (OTA) is presented in Figure C.3. In this amplifier, the nonlinearity of the output current comes from the differential pair since there is no subsequent current gain in the circuit. If one assumes the differential pair transistors to be ideally matched, the differential input voltage is

$$v_{in} = (v_{GS1} - V_t) - (v_{GS2} - V_t) = \sqrt{\frac{2i_{D1}}{\beta}} - \sqrt{\frac{2i_{D2}}{\beta}}. \tag{C.18}$$

Because it can be assumed that $I_{BD} = i_{D1} + i_{D2}$, both input transistor currents can be solved as

$$i_{D1} = \frac{I_{BD}}{2} + \frac{v_{in}}{4}\sqrt{\beta\left(4I_{BD} - \beta v_{in}^2\right)}, \tag{C.19}$$

$$i_{D2} = \frac{I_{BD}}{2} - \frac{v_{in}}{4}\sqrt{\beta\left(4I_{BD} - \beta v_{in}^2\right)}. \tag{C.20}$$

Since i_{D2} is mirrored to the output, the total output current results in

$$i_{OUT} = i_{D1} - i_{D2} = \frac{v_{in}}{2}\sqrt{\beta\left(4I_{BD} - \beta v_{in}^2\right)}, \tag{C.21}$$

if the errors due to the PMOS current-mirror are neglected. This equation holds true only if $|v_{in}| \leq \sqrt{\frac{2I_{BD}}{\beta}}$.

Using Taylor-series approximation, the non-linear output current can then be written as

$$i_{OUT} = v_{in}\sqrt{\beta I_{BD}} - \frac{\beta v_{in}^3}{8}\sqrt{\frac{\beta}{I_{BD}}} - \dots. \tag{C.22}$$

Since there is no quadratic term in the equation, in theory there is no second order distortion. At low frequencies, this output current is then converted into a non-linear output voltage

$$v_{OUT} = Z_L' i_{OUT} = Z_L' v_{in}\sqrt{\beta I_{BD}} - \frac{\beta Z_L' v_{in}^3}{8}\sqrt{\frac{\beta}{I_{BD}}} - \dots, \tag{C.23}$$

where Z_L' is the effective output load impedance including the output impedance of the amplifier and the loading effect of the feedback network in addition to the actual load impedance Z_L. In this case, letting $v_{in} = v_s - f v_{OUT}$ and using equation (A.21), the third order harmonic distortion at low frequencies expressed in terms of output current amplitude $\hat{i}_{out} \approx \frac{\hat{v}_s}{f}\sqrt{\beta I_{BD}}$ is

$$HD3(0) = \frac{1}{32}\left(\frac{\hat{i}_{out}}{I_{BD}}\right)^2 \frac{f^2}{(1+fA_{ol}')^3} \approx \frac{1}{32 f A_{ol}'^3}\left(\frac{\hat{i}_{out}}{I_{BD}}\right)^2, \tag{C.24}$$

where $A_{ol}' = Z_L'\sqrt{\beta I_{BD}}$ represents the loaded DC-gain of the amplifier. If this equation is compared to the third order distortion of the Miller-compensated operational amplifier described in equation (C.6), the closed-loop distortion of the OTA displays a much stronger dependency on the loaded open-loop voltage gain. Because the open-loop voltage gain of an OTA depends strongly on the output load impedance, it is to be expected that the distortion increases rapidly if A_{ol}' decreases.

For a wide frequency range a one-pole approximation $\frac{\omega_0}{s}$ with a unity gain frequency

$$\omega_0 = \frac{\sqrt{\beta I_{BD}}}{sC_L}, \tag{C.25}$$

represents an accurate model for the open-loop voltage gain of the OTA. Similarly, the non-linear model for the OTA can be thought of as a cascade of a non-linear amplifier stage and an integrator, as depicted in Appendix A. In this case, however, the non-linear amplifier precedes the integrator and thus Equations (A.38) and (A.39) cannot be used to depict the distortion performance of the OTA.

If it is assumed that the input voltage of the OTA contains three frequencies

$$v_{in}(t) = V_1 e^{j\omega t} + V_2 e^{2j\omega t} + V_3 e^{3j\omega t}, \tag{C.26}$$

the output voltage of the OTA can be expressed as

$$v_{OUT}(t) = \frac{\omega_0}{b_1}\int i_{OUT}\,dt \tag{C.27}$$

$$= \frac{\omega_0 V_1}{j\omega}e^{j\omega t} + \frac{\omega_0 V_2}{2j\omega}e^{2j\omega t} + \frac{\omega_0}{3j\omega}\left(V_3 + \frac{b_3}{b_1}V_1^3\right)e^{3j\omega t}\ldots, \tag{C.28}$$

where

$$b_1 = \sqrt{\beta I_{BD}}, \tag{C.29}$$

$$b_2 = -\frac{\beta}{8}\sqrt{\frac{\beta}{I_{BD}}}. \tag{C.30}$$

By substituting Equations (C.26) and (C.28) with (C.9) and equating at all three frequencies the coefficient V_1, V_2 and V_3 can be solved as

$$V_1 = \hat{v}_s\frac{j\omega}{j\omega+\omega_0}, \tag{C.31}$$

$$V_2 = 0, \tag{C.32}$$

$$V_3 = -\hat{v}_s^3\frac{b_3\,(j\omega)^3 f\omega_0}{b_1\,(j\omega+\omega_0)^3\,(3j\omega+\omega_0)}. \tag{C.33}$$

Finally, the frequency components of the output voltage can be expressed by means of Equation (C.9), resulting in a third order distortion as

$$HD3(\omega) = \left|\frac{\frac{V_3}{f}}{\frac{\hat{v}_s - V_1}{f}}\right| = \frac{1}{32}\left(\frac{\hat{i}_{out}}{I_{BD}}\right)^2\frac{\omega^3}{\left(\omega^2+f^2\omega_0^2\right)\sqrt{9\omega^2+f^2\omega_0^2}}, \tag{C.34}$$

where the harmonic distortion is once more expressed in terms of output current amplitude $\hat{i}_{out} \approx \frac{\hat{v}_s}{f}\sqrt{\beta I_{BD}}$. At frequencies significantly below $f\omega_0$, this equation can be approximated by using a simpler expression

$$HD3(\omega)\Big|_{\omega \ll \frac{f\omega_0}{3}} = \frac{1}{32}\left(\frac{\hat{i}_{out}}{I_{BO}}\right)^2\left(\frac{\omega}{f\omega_0}\right)^3. \tag{C.35}$$

Similarly, at frequencies above ω_0, the third order distortion can be approximated as

$$HD3(\omega)\Big|_{\omega \gg \omega_0} = \frac{1}{96}\left(\frac{\hat{i}_{out}}{I_{BO}}\right)^2. \tag{C.36}$$

At high frequencies, the cascode transistors M_5 and M_6 in addition to the PMOS cascode current-mirror, begin to generate distortion, leading to a non-zero second order distortion. However, in most cases, the closed-loop bandwidth is low compared to the non-dominant poles deriving from the cascodes and the current-mirror and thus the amount of added high frequency distortion is similarly low.

References

[1] F. Op't Eynde, P. Ampe, L. Verdeyen, W. Sansen, "A CMOS Large-Swing Low-Distortion Three-Stage Class AB Power Amplifier," *IEEE J. of Solid-State Circuits*, vol. SC-25, pp. 265-273, Feb. 1990.

[2] D. Webster, D. Haigh, G. Passiopoulos, A. Parker, Distortion in short channel FET circuits." In: G. Machado (ed), *Low-power HF microelectronic: a unified approach*, IEE, 1996, pp.929-958 (Chapter 24).

[3] P. Wambacq, G. Gielen, P. Kinget, W. Sansen, "High-Frequency Distortion Analysis of Analog Integrated Circuits," *IEEE Trans. Circuits and Systems-II*, vol. CAS-46, pp. 335-344, March 1999.

[4] A. Sedra, K. Smith, *Microelectronic Circuits*, New York, Holt Reinhart and Winston, 1987.

[5] P. Allen, D. Holberg, *CMOS Analog Circuits Design*, New York, Holt, Rinehart and Winston, Inc, 1987, 701 p.

Appendix D

Distortion in a dual current-mirror integrator

D.1 Single-ended integrator

A lossless inverting integrator can be constructed from a cascade of two current-mirrors, as depicted in Figure D.1 [1]. In order to simplify calculations, all transistors are assumed to be ideally matched. Similarly, a second-order nonlinearity in the form of

$$i_{Dn} = b_1 v_{GSn} + b_2 v_{GSn}^2; \; n = 1 \ldots 5, \tag{D.1}$$

is assumed for each transistor. As the gate-source capacitances are relatively linear when the transistors are operating in the saturation region, the total mirror input capacitances are represented as two linear capacitances, C_1 and C_2.

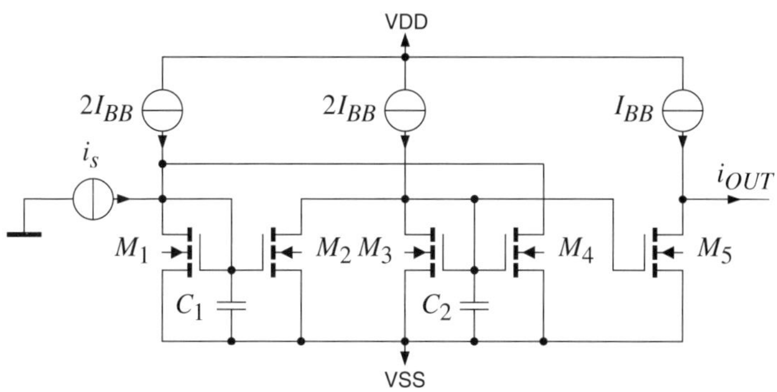

Figure D.1 The principle of a dual current-mirror lossless integrator [1].

If one assumes a sinusoidal input current

$$i_s(t) = \hat{i}_s e^{j\omega t}, \tag{D.2}$$

the voltages at the inputs of the two current-mirror inputs can be estimated to be in a form of

$$\begin{aligned} v_{GS1}(t) &= V_{11}e^{j\omega t} + V_{12}e^{2j\omega t} + V_{13}e^{3j\omega t}, && \text{(D.3)} \\ v_{GS3}(t) &= V_{31}e^{j\omega t} + V_{32}e^{2j\omega t} + V_{33}e^{3j\omega t}. && \text{(D.4)} \end{aligned}$$

Higher order harmonic components are present caused by the feedback in the mirror input transistors, but they can be neglected if low distortion conditions are assumed. In this case, the Kirchoff's current equation can be employed in the input node of the first current-mirror as

$$i_{D1}(t) + i_{D4}(t) + i_{C1}(t) = i_s(t). \tag{D.5}$$

By collecting signal components of the same frequency, one can set up equations for three frequencies

$$\begin{aligned} b_1\left(V_{11}+V_{31}\right) + j\omega C_1 V_{11} &= \hat{i}_s, && \text{(D.6)} \\ b_1\left(V_{12}+V_{32}\right) + b_2\left(V_{11}^2+V_{31}^2\right) + 2j\omega C_1 V_{12} &= 0, && \text{(D.7)} \\ b_1\left(V_{13}+V_{33}\right) + 2b_2\left(V_{11}V_{12}+V_{31}V_{32}\right) + 3j\omega C_1 V_{13} &= 0. && \text{(D.8)} \end{aligned}$$

Similarly, Kirchoff's current equation in the input node of the second current-mirror is

$$i_{D2}(t) + i_{D3}(t) + i_{C2}(t) = 0. \tag{D.9}$$

By collecting signal components of the same frequency, one can set up a further three equations as

$$\begin{aligned} b_1\left(V_{11}+V_{31}\right) + j\omega C_2 V_{31} &= 0, && \text{(D.10)} \\ b_1\left(V_{12}+V_{32}\right) + b_2\left(V_{11}^2+V_{31}^2\right) + 2j\omega C_2 V_{32} &= 0, && \text{(D.11)} \\ b_1\left(V_{13}+V_{33}\right) + 2b_2\left(V_{11}V_{12}+V_{31}V_{32}\right) + 3j\omega C_2 V_{33} &= 0. && \text{(D.12)} \end{aligned}$$

These six equations can be used to solve V_{11}, V_{12} , V_{13}, V_{31}, V_{32} and V_{33}. The resulting frequency components for the first mirror gate voltage are

$$\begin{aligned} V_{11} &= \frac{\hat{i}_s}{j\omega} \frac{b_1 + j\omega C_2}{b_1\left(C_1+C_2\right) + j\omega C_1 C_2}, && \text{(D.13)} \\ V_{12} &= \frac{b_2 C_2 \hat{i}_s^2}{\omega^2} \frac{2b_1^2 + 2j\omega b_1 C_2 - C_2^2\omega^2}{\left(b_1\left(C_1+C_2\right) + j\omega C_1 C_2\right)^2 \left(b_1\left(C_1+C_2\right) + 2j\omega C_1 C_2\right)}, && \text{(D.14)} \end{aligned}$$

$$V_{13} = \frac{2b_2^2 C_2 \hat{i}_s^3}{j\omega^3} \frac{2b_1^2 + 2j\omega b_1 C_2 - C_2^2\omega^2}{\left(b_1\left(C_1+C_2\right)+j\omega C_1 C_2\right)^3} \times \frac{b_1\left(C_1-C_2\right)+2j\omega C_2^2}{\left(b_1\left(C_1+C_2\right)+2j\omega C_1 C_2\right)\left(b_1\left(C_1+C_2\right)+3j\omega C_1 C_2\right)}. \quad \text{(D.15)}$$

Similarly, the resulting frequency components for the second mirror gate voltage are

$$V_{31} = -\frac{\hat{i}_s}{j\omega} \frac{b_1}{b_1\left(C_1+C_2\right)+j\omega C_1 C_2}, \quad \text{(D.16)}$$

$$V_{32} = \frac{b_2 C_1 \hat{i}_s^2}{\omega^2} \frac{2b_1^2 + 2j\omega b_1 C_2 - C_2^2\omega^2}{\left(b_1\left(C_1+C_2\right)+j\omega C_1 C_2\right)^2\left(b_1\left(C_1+C_2\right)+2j\omega C_1 C_2\right)}, \quad \text{(D.17)}$$

$$V_{33} = \frac{2b_2^2 C_1 \hat{i}_s^3}{j\omega^3} \frac{2b_1^2 + 2j\omega b_1 C_2 - C_2^2\omega^2}{\left(b_1\left(C_1+C_2\right)+j\omega C_1 C_2\right)^3} \times \frac{b_1\left(C_1-C_2\right)+2j\omega C_2^2}{\left(b_1\left(C_1+C_2\right)+2j\omega C_1 C_2\right)\left(b_1\left(C_1+C_2\right)+3j\omega C_1 C_2\right)}. \quad \text{(D.18)}$$

In this case, letting $C_1 = \frac{b_1}{\omega_1}$ and $C_2 = \frac{b_1}{\omega_2}$ and substituting V_{31}, V_{32}, and V_{33} to the drain current equation of the output transistor M_5, one can simplify and collect signal components of the same frequency, resulting in an output current

$$i_{OUT} = I_1 e^{j\omega t} + I_2 e^{2j\omega t} + I_3 e^{3j\omega t} + \ldots + I_6 e^{6j\omega t}, \quad \text{(D.19)}$$

where the fundamental frequency component

$$I_1 = \hat{i}_s \frac{\omega_1\omega_2}{j\omega(j\omega+\omega_1+\omega_2)}, \quad \text{(D.20)}$$

the second harmonic component

$$I_2 = \hat{i}_s^2 \frac{b_2\omega_1^2\omega_2\left(\omega^2+\omega_1\omega_2-\omega_2^2\right)}{b_1^2\omega^2(j\omega+\omega_1+\omega_2)^2(2j\omega+\omega_1+\omega_2)}, \quad \text{(D.21)}$$

and the third harmonic component

$$I_3 = \hat{i}_s^3 \frac{2b_2^2\omega_1^3\omega_2\left(\omega^2+2j\omega\omega_2-2\omega_2^2\right)\left(j\omega\left(\omega_1+3\omega_2\right)+2\omega_1\omega_2\right)}{b_1^4 j\omega^3(j\omega+\omega_1+\omega_2)^3(2j\omega+\omega_1+\omega_2)(3j\omega+\omega_1+\omega_2)}. \quad \text{(D.22)}$$

In order to realise an accurate integrator with this circuit, one of the mirror input capacitances must be significantly larger than the other. Normally, this is realised by placing the integrating capacitance at the input of the second current-mirror and thus the capacitance C_2 is much larger than the capacitance C_1. Therefore, the second harmonic component can be simplified by assuming $\omega_2 \ll \omega_1$, resulting in

$$I_2 \approx -\hat{i}_s^2 \frac{b_2}{b_1^2} \frac{\omega_2^2}{\omega^2}. \tag{D.23}$$

By means of simple calculations, it is clear that this result is identical to the second order harmonic component of the lossless integrator realised by a single MOS-transistor shown in Figure 6.9b. However, if the integrating capacitor is placed at the input of the first current-mirror, one can assume that $\omega_2 \gg \omega_1$ and Equation D.21 simplifies to

$$I_2 \approx \hat{i}_s^2 \frac{b_2}{b_1^2} \frac{\omega_1^2}{\omega^2}. \tag{D.24}$$

Therefore, this harmonic component would cancel the second harmonic component of the single MOS-transistor integrator. It would, however, be easier to use differential integrators to get the same effect.

When the same assumptions are used to simplify the third harmonic component of Equation D.22, it is clear that

$$I_3\Big|_{\omega_2 \ll \omega_1} \approx I_3\Big|_{\omega_2 \gg \omega_1} \approx 0. \tag{D.25}$$

Therefore, the nonlinearity of the dual current-mirror lossless integrator is almost identical to a single MOS-transistor lossless integrator when one of the mirror capacitances is significantly larger than the other.

Additionally, there is a notch in the second harmonic component at

$$\omega = \sqrt{\omega_2^2 - \omega_1 \omega_2}, \tag{D.26}$$

if $\omega_2 \geq \omega_1$. Problematically, the notch is very near ω_2 and is thus at too a high frequency to be useful in any application. Furthermore, in practical applications, an integrator capacitance ten times greater than the plain current-mirror input capacitance readily results in a capacitor of at least 10 pF thus the two mirror poles cannot be moved far away from each other. Thus, a significant third harmonic component remains present and thus the integrator generates more distortion than a single MOS-transistor integrator.

D.2 Differential integrator

The two current-mirror loop can be rearranged to a differential lossless integrator, as seen in Figure D.2 [2, 3]. In this case, the Kirchoff's current equation can be used in both input nodes as

$$i_{D1}(t)+i_{D4}(t)+i_{C1}(t) = i_s(t). \tag{D.27}$$

$$i_{D2}(t)+i_{D3}(t)+i_{C2}(t) = -i_s(t). \tag{D.28}$$

By comparing these equations to the current equations of the single-ended circuit, it is clear that Equations (D.6), (D.7), (D.8), (D.11) and (D.12) apply similarly to this differential circuit; Equation (D.10) is only slightly modified, resulting in

$$b_1\left(V_{11}+V_{31}\right)+j\omega C_2 V_{31} = -\hat{i}_s. \tag{D.29}$$

In an ideally matched case, one can assume $C_1 = C_2 = C$ and thus solve V_{11}, V_{12}, V_{13}, V_{31}, V_{32} and V_{33} from these following equations

$$V_{11} = -V_{31} = \frac{\hat{i}_s}{j\omega C}, \tag{D.30}$$

$$V_{12} = V_{32} = \frac{b_2\hat{i}_s}{\omega^2 C^2\left(b_1+j\omega C\right)}. \tag{D.31}$$

$$V_{13} = V_{33} = 0. \tag{D.32}$$

Similarly, one can let $C = \frac{b_1}{\omega_0}$ and substitute V_{11}, V_{12}, and V_{13} to the drain current equation of the output transistor M_5 and V_{31}, V_{32}, and V_{33} to the drain current equation of the output transistor M_6, leading to a differential output current with merely a third order harmonic component

$$i_{OUT} = i_{OUT+} - i_{OUT-} = I_1 e^{j\omega t} + I_2 e^{3j\omega t}, \tag{D.33}$$

where the fundamental frequency component

$$I_1 = -\hat{i}_s\frac{2\omega_0}{j\omega}, \tag{D.34}$$

and the third harmonic component

$$I_3 = \hat{i}_s^3\frac{4b_2^2\omega_0^2}{b_1^4 j\omega^3\left(j\omega+\omega_0\right)}. \tag{D.35}$$

In a simple OTA based current integrator (Figure D.3a), where the non-linear output current of the OTA is assumed as

$$i_{OUT} = a_1 v_{in} + a_2 v_{in}^2 + a_3 v_{in}^3, \tag{D.36}$$

the integrator output current with a sinusoidal current excitation $i_s(t) = \hat{i}_s e^{j\omega t}$ results

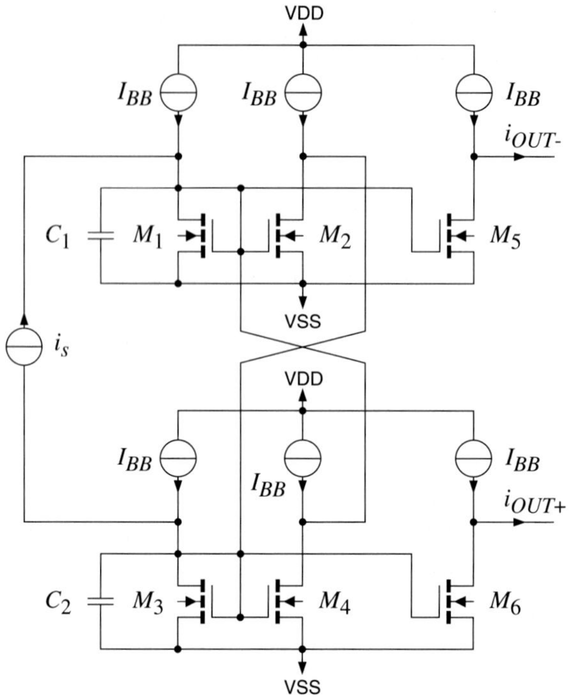

Figure D.2 Current-mirror based differential lossless integrator [2, 3].

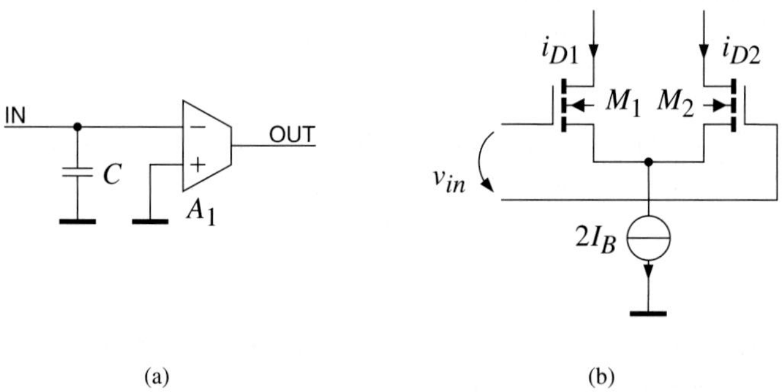

Figure D.3 a) Simple transconductance amplifier based current integrator. b) A simple NMOS differential pair.

in

$$i_{OUT} = \frac{a_1 \hat{i}_s}{j\omega C} e^{j\omega t} - \frac{a_2 \hat{i}_s^3}{\omega^2 C^2} e^{j2\omega t} - \frac{a_3 \hat{i}_s^3}{j\omega^3 C^3} e^{j3\omega t}. \tag{D.37}$$

Therefore, the linear integrator and non-linear transconductance can similarly be separated in the case of the current-mirror based differential lossless integrator. Thus, the non-linear output current without the integrating function is

$$i_{OUT}(v_{in}) = i_{OUT+}\left(\frac{v_{in}}{2}\right) - i_{OUT-}\left(-\frac{v_{in}}{2}\right) = b_1 v_{in} - \frac{1}{2}\frac{b_2^2}{b_1 + j\omega C} v_{in}^3. \tag{D.38}$$

This result can be compared to the nonlinear output current of a simple NMOS differential pair of Figure D.3b. This equation was previously derived in Appendix C as Equation (C.22) and can be rewritten for more convenient comparison as

$$i_{OUT}(v_{in}) = i_{D1}\left(\frac{v_{in}}{2}\right) - i_{D2}\left(-\frac{v_{in}}{2}\right) \approx b_1 v_{in} - \frac{1}{2}\frac{b_2^2}{b_1} v_{in}^3 + \ldots \tag{D.39}$$

It can be seen that Equations (D.38) and (D.39) are almost equal. There remains, however, an interesting difference in that the nonlinearity of the current-mirror based differential lossless integrator begins to decrease above the integrator unity-gain frequency $\omega_0 = \frac{b_1}{C}$.

References

[1] S. Lee, R. Zele, "CMOS continuous-time current-mode filter for high-frequency applications," *IEEE J. of Solid-State Circuits*, vol. SC-28, pp. 323-329, March 1993.

[2] S. Smith, E. Sánchez-Sinencio, "Low voltage integrators for high-frequency CMOS filter using current mode techniques," *IEEE Trans. Circuits and Systems-II*, CAS-43, Jan. 1996, pp. 39-48.

[3] R. Zele, D. Allstot, "Low-power CMOS continuous-time filters," *IEEE J. of Solid-State Circuits*, vol. SC-31, pp. 157-168, Feb. 1996.